HISTORY, PROPHECY, AND THE STARS

T0256780

Theologian and astrologer in harmonious discussion. From Pierre d'Ailly, *Concordantia astronomie cum theologia. Concordantia astronomie cum hystorica narratione. Et elucidarium duorum precedentium.* (Augsburg: Erhard Ratdolt, 1490), verso of the title page.

HISTORY, PROPHECY, AND THE STARS

THE CHRISTIAN ASTROLOGY

OF PIERRE D'AILLY,

1350–1420

Laura Ackerman Smoller

PRINCETON UNIVERSITY PRESS   PRINCETON, NEW JERSEY

Library of Congress Cataloging-in-Publication Data
Smoller, Laura Ackerman, 1960–
History, prophecy, and the stars : the Christian astrology
of Pierre d'Ailly, 1350–1420 / Laura Ackerman Smoller.
p.   cm
Includes bibliographical references and index.
ISBN 0-691-08788-1 (alk. paper)
1. Ailly, Pierre d', 1350–1420?—Contributions in
astrology.   2. Astrology—History.   3. Occultism—Religious
aspects—Christianity—History.   I. Title.
BF1679.8.A453S56   1994
133.5'092—dc20   93-44907

For Bruce

Contents

List of Illustrations   ix

Acknowledgments   xi

Chapter One
Introduction   3

Chapter Two
The Medieval Debate about Astrology   25

Chapter Three
The Making of an Astrologer: The Development of Pierre d'Ailly's
Thought on the Stars   43

Chapter Four
Astrology and the Narration of History   61

Chapter Five
The Great Schism and the Coming of the Apocalypse   85

Chapter Six
Astrology and the Postponement of the End   102

Chapter Seven
The Concordance of Astrology and Theology   122

Appendix 1
A Note on the Availability of d'Ailly's Writings on Astrology   131

Appendix 2
A Chronology of d'Ailly's Works Dealing with Astrology   136

Abbreviations   138

Notes   139

Select Bibliography   209

Index   227

Illustrations

Theologian and astrologer in harmonious discussion. From Pierre d'Ailly, *Concordantia astronomie cum theologia. Concordantia astronomie cum hystorica narratione. Et elucidarium duorum precedentium* (Augsburg: Erhard Ratdolt, 1490), verso of the title page. (By permission of the Rare Books and Manuscripts Division, The New York Public Library, Astor, Lenox and Tilden Foundations)   *Frontispiece*

Figure 1. A schematic representation of the epicycle-deferent system. From a fifteenth-century German translation of Sacrobosco's *De sphera*. (The Pierpont Morgan Library, New York. M 722, fol. 18r)   14

Figure 2. Christ's nativity chart. From Pierre d'Ailly, *Tractatus de imagine mundi et varia ejusdem auctoris et Joannis Gersonis opuscula* (Louvain: Johann de Westphalia, ca. 1483), fol. ee2v. (By permission of the Houghton Library, Harvard University)   18

Figure 3. Saturn-Jupiter conjunctions. After John D. North, in *Centaurus* 24 (1980): 187   21

Figure 4. The horoscope of Creation from the *Vigintiloquium*. From Pierre d'Ailly, *Tractatus de imagine mundi et varia ejusdem auctoris et Joannis Gersonis opuscula* (Louvain: Johann de Westphalia, ca. 1483), fol. [bb5r]. (By permission of the Houghton Library, Harvard University)   67

Acknowledgments

I HAVE incurred many debts to individuals and institutions in the course of writing this book, and it gives me pleasure to be able to acknowledge them here. I have had the opportunity to work at three excellent libraries. To the staff of the Houghton Library at Harvard, I owe a special thanks for their kind and attentive responses to innumerable requests for information and materials. Similarly, the librarians from the Rare Books and Manuscripts Division of the New York Public Library often went beyond the call of duty in retrieving lost or miscatalogued books for me. Finally, I must express my gratitude to Sonia Moss of the Interlibrary Loan Office at Stanford University's Green Library for her courteous attention to my numerous queries.

To Jean-Pierre Codaccioni of the Bibliothèque Municipale of Marseilles, and to Marian Zwiercan of the Biblioteka Jagiellonska in Cracow, I am deeply grateful for their generosity in providing me with copies of manuscripts of d'Ailly's works. I also owe my thanks to Dr. Julian Plante of the Hill Monastic Manuscript Library for his assistance in procuring for me microfilm copies of two manuscripts from the Österreichische Nationalbibliothek in the library's film collection, as well as to the photographic division of the Bibliothèque Nationale in Paris.

A number of scholars have been kind enough to discuss this work with me, and I am greatly indebted to them for their advice and comments. Professors Anthony Grafton of Princeton and James Hankins of Harvard read portions of my manuscript in its earliest version. They are both due special thanks for their critical suggestions. So, too, am I grateful to the anonymous reviewers at Princeton University Press for their careful readings of my manuscript and for their perceptive and extremely helpful criticisms. Brian MacDonald edited the manuscript with an eye far more perceptive than my own, and I thank him for his intelligent precision. I owe an enormous debt to Professor Noel Swerdlow of the University of Chicago. He has generously shared his time and his expertise with me and saved me from a number of technical errors. Finally, I can express only inadequately my gratitude to Professors Steven Ozment and John Murdoch of Harvard University, who saw this project through from its beginnings as my doctoral dissertation. They have been exemplary mentors and editors, as well as kind and caring advisors. Without these two men's attentive supervision, the writing of my thesis would not have been the richly rewarding experience that it was.

To my long-suffering family, I offer both thanks and apologies for the many hours in which my mind was absorbed in medieval astrology. I am indebted in many ways to my parents, who have always encouraged and supported my academic pursuits. My two sons, Jason and Gabriel, have patiently put up with my typing at the computer and with distracted responses to their questions. To my husband, Bruce, I owe the most gratitude, for his support and encouragement throughout the years, for his enthusiasm for my work, and for reading my manuscript with his incomparable penetrating insight.

Stanford, California
September 27, 1993

HISTORY, PROPHECY, AND THE STARS

CHAPTER ONE

Introduction

FIVE HUNDRED YEARS ago Christopher Columbus set sail across the Atlantic Ocean with the hope of reaching the Indies. Scholars have traditionally depicted Columbus as a discoverer, an enlightened adventurer whose "new" geography and navigational skills helped usher in the modern age.[1] More recent authors have argued that the desire for plunder and glory was a major motivation behind his voyages.[2] Yet Columbus was also stirred by a curious blend of astrological prognostications and apocalyptic fervor. As he explained in a letter to his patrons Ferdinand and Isabella of Spain, astrology dictated that the world would endure only some 155 years to come. Preceding its destruction, however, Columbus told the monarchs, all of the races of the world would be converted to Christianity. He saw his own voyages as part of the universal missionizing of the last days.[3]

Columbus's ideas were neither the rantings of a wild genius nor the flattering propaganda of a skilled courtier. His vision of the end of the world and its astrologically determined time came from the same volume in which he found the excessively small estimate of the earth's circumference that emboldened him to undertake his journey. Columbus's source was a collection of treatises by the French scholar and churchman Pierre d'Ailly (1350–1420), printed in Louvain in 1483 under the title *Tractatus de imagine mundi et varia ejusdem auctoris et Joannis Gersonis opuscula* (Treatise on the image of the world, and various works by the same author and by Jean Gerson). In that volume Columbus found both geographical works and treatises dealing with astrology and the end of the world. He read and annotated them all and obviously absorbed their meaning, enough so to believe that he was living near the end of time and to see his voyages in that context.[4]

To find Columbus dabbling in astrology may be surprising to many; it may be even more shocking, however, to learn that the source for his star lore was a noted cardinal and theologian. To speak of astrology today generates mainly negative images of banal advice columns in newspapers, charlatans pretending to foresee the future, or New Age witches absorbed in an engaging world of make-believe.[5] And yet there was a time when astrology was considered a science, even the highest of sciences, capable of assisting the noble study of theology. Such was the view of the stars presented by Columbus's intellectual progenitor Pierre d'Ailly, and he was not alone in his beliefs. D'Ailly was a careful man, one who prided himself on the moderation of his opinions, typical of his age. Far from being an embarrassment or an anomaly in his use of astrology, d'Ailly stood squarely within a traditional and rational view of the world.

What makes d'Ailly both so unusual and so interesting is the fact that he left an abundant record of his opinions about astrology. Whereas most astrologers composed only prognostications and textbooks, d'Ailly's works are as much justifications of the science as they are examples of its use. In sermons, lectures, and treatises, d'Ailly detailed and defended his evolving views on the stars. D'Ailly lived during a time in which belief in the power of the stars flourished despite religious and scientific objections to astrology. He understood well the paradoxes involved in the acceptance of astrology, for he was familiar with both theological and rational arguments against the science. Nevertheless, d'Ailly came to embrace the science of the stars. Through his example, we may understand a time when even the best minds turned to astrology for advice.

D'Ailly's ideas about the stars were related to his changing interpretations of the Great Schism (1378–1414), that horrendous division of western Christendom that pitted first two, then three, rival popes and their followers against each other. In an age filled with troubles, it was the salient crisis. In the early years of the Schism, d'Ailly viewed the situation strictly in accordance with Scripture. He deplored the lack of peace in the church and assumed, following the standard interpretation of 2 Thessalonians, that the Schism signaled the approach of Antichrist. Even as d'Ailly preached the likely imminence of the End, he stressed that this event could not be predicted by any human means, and certainly not by astrology.

With the passage of time, however, and the long duration of the Schism without Antichrist's having appeared, d'Ailly began to revise his interpretation of the situation. Increasingly he became hopeful that human efforts (with divine assistance) would bring the Schism to an end. With peace restored and the church reformed, he believed, the apocalypse would be forestalled. In the years between 1400 and 1414—the latter marking the opening of the Council of Constance, which healed the Schism—d'Ailly formulated and refined his new view of the crisis. He read deeply in apocalyptic literature, seeking to discover if his own times were indeed the final days. And he turned to astrology. In 1414, d'Ailly produced the culmination of his astrological studies: the prediction that Antichrist would arrive in the year 1789. The End was not at hand, and astrology had confirmed d'Ailly's new view of events. It was a message of comfort for those who feared the apocalypse and hope for those who sought to heal the divided church.

D'Ailly's example adds an interesting chapter to the history of astrology in the Middle Ages, a subject that only recently has begun to receive serious scholarly attention. Despite such notable exceptions as Lynn Thorndike's monumental *A History of Magic and Experimental Science*, previous generations of historians of science had preferred to concentrate on developments in medieval *astronomy* rather than on what George Sarton termed "superstitious flotsam." After Otto Neugebauer's 1951 plea for "the study of wretched subjects," however, both

Neugebauer and a number of followers began to incorporate astrological texts into the investigation of astronomy's history.[6] These scholars have produced highly technical works ably demonstrating the complexities of medieval astronomy and firmly establishing astrology's place as an important aspect of that science. Through their extensive study of the history of astrology, they have pointed out the lines of transmission whereby Indian and Sassanian star lore (itself influenced by Greek astrology) penetrated into the medieval West.

A number of notable authors have combined the study of medieval astronomy with that of astrology. E. S. Kennedy and David Pingree, for example, have elucidated both the sources of and the mathematics undergirding some of the more important medieval Arabic astrological texts, works that in their Latin translations had a great influence in Europe. Together they have reconstructed the lost astrological history of the eighth-century Jewish astrologer Masha'allah (Messahalla to the Latin West). Similarly, Pingree has pieced together the traces of a missing work by the important ninth-century astrologer Abu Ma'shar (Albumasar).[7] John D. North has authored fascinating studies of the fourteenth-century astrologer Richard of Wallingford and of the varied uses of astrology by Chaucer, as well as descriptions of various aspects of medieval astrology.[8] No one can read these works without an appreciation of what a sophisticated and demanding science was practiced by the medieval astrologers.

Just as historians of science had tended to view astrology as a superstition unworthy of their notice, so too, medieval historians often seemed to regard astrology as an aberrant set of beliefs whose persistence had to be explained. Hence, many authors who considered medieval star lore showed less interest in the practice of astrology than in the prevailing attitudes toward the science, particularly those of medieval theologians. While their studies nicely illuminate the Catholic church's response to astrology, they say little about the opinions of persons who actually consulted the stars.[9] Other scholars were interested in the views of those few thinkers who totally repudiated astrology, presumably finding in their skepticism a progenitor of modern rationalism. Herbert Pruckner in 1933 and G. W. Coopland in 1952 produced annotated editions of anti-astrological treatises from the fourteenth century.[10] Although extremely valuable, these editions do not shed light on the many people who believed in astrology. In fact, they reveal less of the true place of astrology than do the surveys of the church's attitudes about the stars. And a catalogue of famous astrologers compiled in the late fifteenth century by the astrologer Symon de Phares, although published in 1929 by Ernest Wickersheimer, has until recently received little attention.[11]

Thus, while historians of science have illuminated the details of medieval astrological calculations and historians have explored the debates about the stars, many questions about the appeal of astrology in the Middle Ages and the extent of its practice have remained unanswered. Recently, a handful of authors have begun to address such issues. S. J. Tester, in *A History of Western Astrology*, attempts to discern broad patterns in the use of the stars from ancient Greece through modern times.[12] He gives lucid descriptions of the major texts in the history of astrology, yet the broad scope of his work and his own training as a classicist preclude a thorough analysis of astrology's role in the Middle Ages. Valerie Flint has examined clerical attitudes about astrology and magic in the early Middle Ages, arguing that churchmen in fact bolstered the study of astrology in order to combat more "dangerous" magical practices.[13] Maxime Préaud has looked at the lives of late medieval astrologers, examining both their practices and the circumstances of their careers.[14] Hilary Carey has investigated the use of astrology in English court society in the later Middle Ages. She suggests, however, that astrology may have had a much greater hold in Italy and France, so her data may have no real significance there.[15]

The case of Pierre d'Ailly, then, stands to reveal to us some genuine new insights into the appeal of astrology in the later Middle Ages. Because d'Ailly wrote so extensively on the topic, we have an in-depth record of one man's reactions to the study of the stars. Because he was "converted" to astrology late in his life, his writings give us an idea of why a person would turn to the stars for knowledge. For the same reason, d'Ailly's works illustrate a practical program of study in astrology. Just as his example demonstrates *why* a person would become involved with astrology, it also shows *how* one would go about gaining the necessary knowledge. Finally, because d'Ailly was a man of great prominence in ecclesiastical and intellectual circles, we can know that his opinions about the stars were safe. D'Ailly embraced and publicly advocated the use of astrology at the same time that he played a leading role at the Council of Constance, his greatest political triumph. If astrology were not a part of mainstream thought, d'Ailly would not have preached an astrological sermon at Constance.[16]

Although many aspects of his career have received abundant scholarly attention, d'Ailly's astrological pursuits have largely been ignored or misunderstood. Of d'Ailly's three modern biographers, two, Paul Tschackert and Louis Salembier, seem embarrassed by the cardinal's interest in the stars and discuss his astrological treatises only to condemn their "superstitious" contents.[17] The third, Bernard Guenée, correctly links d'Ailly's pursuit of astrology to his anxiety about the future, but he fails to see the changes in d'Ailly's thinking about both the stars and the apocalypse.[18] Some scholars have misread d'Ailly entirely and have placed him in the camp of astrology's opponents, confused, understandably, by the title of d'Ailly's most important defense of astrology, *De legibus et sectis contra superstitiosos astronomos* (On the laws and the sects, against the superstitious astrologers).[19] Most, however, have passed over d'Ailly's astrological interests entirely to focus on such topics as his philosophy, his theology, his political thought and activities, and his geographical writings.[20] Admittedly, d'Ailly is best known for his role at the Council of Constance and for his articulation of the conciliar theories that made the gathering possible. Few persons, however, have recognized the importance of astrological and apocalyptic notions in d'Ailly's view of the Schism. The Schism undoubtedly provided the occasion for the development of conciliar theories that foreshadowed later constitutional monarchies. But by focusing merely on the birth of these notions, scholars may fail to see that these ideas were driven by an urgent concern that the End might be at hand.

From d'Ailly's example, then, astrology emerges as an integral part of the rational view of the world in the fourteenth and fifteenth centuries. The belief that the heavenly bodies had some sort of influence on the earth below was just as pervasive as the notion that God had a plan for the world's destiny. At the time that d'Ailly plunged into the study of astrology, he was deeply interested in the world's fate. Quite naturally he turned to Scripture and prophecy for information, but also equally naturally he looked to astrology. D'Ailly saw astrology not as a magical art by which he could manipulate the future course of the world but rather as a rational science by which he could discern the broad patterns of earthly events. The great numbers of people who used astrology in medicine, in making their business decisions, and for political advice must have believed, also, that they were turning to science for knowledge.

Pierre d'Ailly was born in Compiègne in 1350 or 1351, the child of prosperous bourgeois parents.[21] A child of obvious intellectual talent, he was sent to Paris to study in 1363 or 1364, where he enrolled in the College of Navarre of the University of Paris. D'Ailly received the licentiate in arts in 1367 and became a student in the faculty of theology in September 1368, during which time he also taught in the faculty of arts. After six years of "audition" (attending lectures) in theology, d'Ailly lectured on the Bible in the years 1374–76 and on Peter Lombard's *Sentences* (the standard textbook of theology) in 1376–77. Then, after four years of required sermons, lectures, and disputations, d'Ailly received the *licentia docendi* (license to teach) and became a doctor of theology in April 1381.

D'Ailly distinguished himself not merely academically in those years. In 1372 he was elected proctor of the French "nation" (the faculty of arts was divided into four "nations" according to the student's origin). And in 1379 he was entrusted with the job of carrying the *rotulus* of the French "nation" to Pope Clement VII in Avignon. D'Ailly earned the friendship of two powerful men close to the French crown, Jean de la Grange and Philippe de Mézières. A suite of honors ensued, as d'Ailly became canon of the cathedral chapter of Noyon in 1381, rector of the College of Navarre in 1384, chaplain to the king in March 1389, and chancellor of the University of Paris in October 1389. D'Ailly obtained prominence in the years 1385–87 as he successfully pleaded the university's case in the affairs of Jean Blanchard, the corrupt chancellor of the university, and Juan de Monzon, whose denial of the Immaculate Conception the university sought to condemn.[22]

Two factors complicated d'Ailly's career in the years after he completed his doctorate, Schism in the church and infighting in the French royal house. From

1309 to 1378, the years of the so-called Babylonian Captivity, the papacy was centered not in Rome, but in Avignon, where it was perceived to be a tool of the French crown. Pope Gregory XI returned the Holy See to Rome in 1377, but died the following March. Under intense pressure from the Roman populace, the college of cardinals elected the Italian archbishop of Bari to succeed Gregory as Urban VI. Four months later twelve of the sixteen cardinals repudiated Urban's election and chose in his stead the Frenchman Robert of Geneva, who took residence in Avignon as Clement VII. Neither election was clearly valid, and both popes steadfastly refused to step down. The Schism continued with Urban's and Clement's successors; the abortive Council of Pisa of 1409 added yet a third line of popes. Christendom was divided for nearly forty years until the Council of Constance at last deposed all the rival claimants and elected one universally recognized pope.

Reactions to the Schism fell predictably along political lines. France and its allies supported the Avignonese popes, while England and the empire gave their allegiance to Urban and his successors. In Paris, the university was under intense pressure to follow the crown's lead, and it reluctantly gave its approval to Clement. Meanwhile, in 1380, the French king Charles V died, leaving a minor son, Charles VI, to succeed him. France was ruled by Charles's three powerful uncles, all supporters of Clement VII. Charles took his own power in 1388, yet ruled only four years before suffering the first of a series of bouts of insanity. Once again Charles's uncles, the dukes of Berry and Burgundy, took control. In 1394, the Aragonese cardinal Pedro de la Luna succeeded Clement VII as Benedict XIII; the French princes were divided in their allegiance to the Spaniard. By 1403 there was outright civil war in France between Louis, duke of Orleans (the king's brother), and Jean sans Peur, son of the duke of Berry. For better or worse, d'Ailly had aligned himself with the party of Orleans and Benedict; when their fortunes waned, he was forced to quit Paris, as in 1395 and in 1407.

The king sent d'Ailly as an emissary to greet Benedict upon his election in 1394. Benedict openly courted Charles's representative and granted d'Ailly the bishoprics of Le Puy in 1395 and Noyon in 1396, before naming him to head the enormous diocese of Cambrai in 1397. In 1398, in an attempt to force Benedict to resign the papacy, France withdrew obedience from the Avignonese pontiff. D'Ailly, now perceived as Benedict's tool, prudently stayed away from Paris until French obedience was restored in 1403. After repeated efforts to secure the resignations of the two rival popes had failed, d'Ailly broke with Benedict in 1408. In 1409, he threw in his lot with the dissident cardinals who met at the Council of Pisa, who declared the two rival popes deposed and elected a third claimant, Alexander V. It was Alexander's successor, John XXIII, who named d'Ailly cardinal in 1411; in 1413 he sent d'Ailly to Germany as papal legate. In this capacity d'Ailly arrived at the Council of Constance in 1414, where he was instrumental in the council's successful resolution of the Schism. After his final triumph at Constance, d'Ailly retired to Avignon as legate for the new pope Martin V. He died there on August 9, 1420.

If d'Ailly's public career was marked by political success and a prudent sense of timing, his intellectual achievements are characterized by an abundant output and a careful moderation in opinions. In his philosophy and theology, as expressed mainly in the commentary on Peter Lombard's *Sentences* of 1376–77, d'Ailly followed his professors at the University of Paris in his heavy dependence on William of Ockham (d. 1349). Like Ockham, d'Ailly stressed the omnipotence of God and denied that universal concepts had any real existence in the form of divine ideas. To say that Socrates was a man because he participated in a divine idea called "man"-ness was, to d'Ailly's way of thinking, unduly restrictive on the Creator. If such an idea really existed, then God would be obliged to create all subsequent men to fit that mold.[23] Following Ockham's reasoning, d'Ailly argued that knowledge must come from sensory experience via a process of abstraction. In holding such ideas d'Ailly supported the general teaching of the late scholastics.[24]

In his theology, again like most of his contemporaries, d'Ailly stressed the omnipotence and freedom of God. To explain many theological points, he distinguished between God's absolute and ordained power, drawing a line between God's absolute freedom of action (limited only by his inability to produce a contradiction) and the voluntary boundaries God had placed on his power in selecting the present order of the world. God's ordained power represented an implied covenant with man, by which the Creator agreed to act within certain limitations. God's covenant guaranteed consistent laws of salvation, morality, and grace. In this system there are no natural laws of morality or ethics, only the chosen, but binding, law laid down by God. Had he chosen, God could have enacted a system whereby murder and blasphemy were meritorious acts; God's choice, and not some intrinsic value in the deed, determines its worth.[25]

D'Ailly's political thought was closely linked to his philosophy and theology, although in the development of his conciliar views he relied more upon John of Paris (d. 1306) than he did upon William of Ockham. God has chosen to operate through secondary causes in the world, d'Ailly argued, and to associate legitimate authority with certain signs. The consent of the community, he maintained, was the true source of authority, and the common good was best expressed by a delegation of the wisest members of the community. In both church and state d'Ailly felt that the community had a right to resist its own division by judging and deposing a tyrant. For the church, the proper body for such an action was the General Council. Although d'Ailly's name is associated with the birth of this conciliar theory, his ideas were hardly new. D'Ailly borrowed heavily from John of Paris, who, in turn, relied upon the canon law commentators of the thirteenth century.[26]

All in all, d'Ailly was intelligent, prolific, and highly influential, without, however, being particularly original in his thought. His great indebtedness to earlier authors has, in fact, led some modern scholars to accuse him of plagiarism. This charge would be true were it not the case that d'Ailly constantly acknowledged the sources of his words, frequently by name, sometimes by the type of oblique reference common at the time ("Some say . . ."; "A *certain doctor says* . . .").[27] Many of his works were little more than collages, composed of bits and pieces of other writers' prose. Through all his borrowings, however, d'Ailly generally managed to convey his own opinion, which was sometimes quite different from that of his source.

There are two explanations for d'Ailly's generous liftings from earlier writers. First, such borrowing was common in scholarly works and in literature; if d'Ailly took large portions of his prose from other authors, there were later writers who copied extensively from d'Ailly.[28] In a consideration of German theological writers at the end of the fifteenth century, the historian Bernd Moeller has remarked on the intentional avoidance of originality and the pervasive respect for tradition.[29] Second, d'Ailly wrote prolificly on a wide variety of topics, all the while pursuing an active career in the church and the university. It is difficult to imagine how he could have composed so many treatises by any other method. On the whole, d'Ailly was a compiler and digester of others' thought. His later readership suggests that there was a vast need for this type of writing.[30]

For the historian of philosophy who expects to find some new development in the works of such an important figure, this method of composition can be maddening. More than one author has deplored d'Ailly's startling "lack of originality."[31] For the purpose of this present investigation into the general appeal of astrology in the later Middle Ages, however, d'Ailly's borrowings are an unexpected boon. In the first place, the fact that d'Ailly relied on other authors so heavily gives further proof of the representative quality of his thought. He clearly based his thinking on the same books that his contemporaries were reading, and he remained close to his sources. Second, d'Ailly's frequent use of others' prose makes it easy to follow his program of reading. To track down the borrowings in his astrological treatises is to observe at close range how a scholar went about his research in this period. We can follow as d'Ailly, drawing on what he must have learned while a student in Paris, set out to learn about the stars, moving beyond the basic textbooks to become an expert in astrology.

As a student in the faculty of arts at the University of Paris, d'Ailly was expected to complete courses in the seven liberal arts: grammar, rhetoric, dialectic, arithmetic, geometry, astronomy, and music. Both astronomy and astrology, as we know them today, made up the medieval "science of the stars," which was divided into the "science of movements" (our astronomy) and the "science of judgments" (our astrology). The science of movements taught how to describe and predict the motions of the heavens; knowledge of such movements was presupposed for the student of the science of judgments, which charted the heavens' influence. A student in the faculty of arts would acquire an elementary knowledge of the heavenly movements, in all likelihood using the textbook *De sphera* (On the sphere) of Johannes de Sacrobosco (John of Holywood, who

taught in Paris in the thirteenth century). D'Ailly authored a commentary on the
De sphera, which perhaps dates from his period of lecturing in the faculty of arts.
Sacrobosco's textbook taught, as well, a rudimentary vocabulary of astrology,
the names of the heavenly bodies, and their general characteristics. More specific
knowledge about the stars' motions and influences came in supplementary lectures, given mainly to students in the faculty of medicine, where astrology was
a required subject.[32]

Astrology, of course, required a more exact knowledge of heavenly motions
and their effects than the general description of spherical astronomy put forth by
Sacrobosco. To understand the planets' behavior, d'Ailly or any other would-be
astrologer most likely would have consulted the so-called *Theorica planetarum*,
attributed to Gerard of Cremona. Other treatises also circulated under this title,
such as the *Theorica planetarum* penned by Campanus of Novara, whom d'Ailly
cites. These treatises presented a detailed account of Ptolemaic planetary theory.
Campanus also included instructions for constructing an instrument (the *equatorium*) by which to determine the positions of the planets. Another important
conduit for Ptolemy's astronomy, and one of Campanus's sources, was the
ninth-century Arabic author al-Fargani (Alfraganus).[33] To calculate planetary
positions on his own, d'Ailly would have had to master the use of astronomical
tables as well. By d'Ailly's time the favored tables were those purportedly compiled for Alfonso X ("the Wise") of Leon and Castille in the thirteenth century.
They were accompanied by "canons" explaining their use. D'Ailly undoubtedly
would have consulted the canons composed at his own University of Paris in
1327 by John of Saxony. His canons were the basic university text for the higher
levels of astronomy.[34] Finally, he would have needed a more thorough introduction to astrological theory. One of the most popular of such texts was that by
the tenth-century Arabic author al-Qabisi, known to the west as Alchabitius
(among other spellings). It was translated into Latin in 1144 by John of Seville
and was also the subject of a commentary by John of Saxony in 1331.[35]

Late in his life, d'Ailly penned a series of treatises on cosmology, offering to
anyone who could read Latin the sort of basic information about the heavens
and earth that he had learned at the university. In the short treatise *Epilogus
mappe mundi* (Epilogue to the map of the world), he offered a simple outline of
medieval astronomy. "In the first place," he began, "one should suppose that
heaven is of a spherical or rotund shape."[36] Philosophers, theologians, and astronomers all held that the heavens in fact consisted of a number of nested
spheres, although the exact number was disputed. All agreed that there were at
least nine: one for each of the seven planets (among which the moon and sun
were numbered), an eighth sphere containing the so-called fixed stars (because
they did not "wander" like the planets), and a ninth sphere called the "prime
mover," which carried all the lower spheres around with it in a daily motion
from east to west. The earth sat, immobile, at the center of all these concentric
spheres. A person traveling, like Dante, from the earth upward would pass

consecutively through the spheres of the moon, Mercury, Venus, the sun, Mars, Jupiter, Saturn, and the fixed stars before reaching the ninth sphere of the prime mover.[37]

The ancient Greek astronomical theories that dominated the medieval world view attempted to describe the apparent motions of the heavens by means of a system of uniform circular movements.[38] For the viewer on earth, the sun, moon, stars, and planets all demonstrate a daily motion of rising and setting, that is, a circular rotation from east to west. The fixed stars complete this circuit in a length of time just short of the twenty-four hours of the solar day. As a result, each day the sun appears to lag farther behind the fixed stars' movement. This apparent contrary path of the sun, from west to east, describes a great circle on the heavenly sphere known as the ecliptic, which the sun traverses in a tropical year of just under 365¼ days. The daily rotation of the heavens takes place around an axis coinciding with a line through the earth's north and south poles. The sun's apparent path, however, is around a different axis, that is, the ecliptic is inclined to the earth's equator—or, to speak more properly, to the celestial equator—by about 23½ degrees. The sun spends about half of the year north of the celestial equator and about half to the south, thereby creating the seasons on earth.

Like the sun, the planets also appear to have a basic eastward movement in addition to their daily rising and setting. Their slow eastward paths, however, do not coincide exactly with the sun's course along the ecliptic. Rather the planets travel along paths within a band of the heavens 6 degrees to either side of the ecliptic known as the zodiac. Astronomers divided the 360 degrees of the zodiac into twelve signs (only roughly corresponding to twelve constellations) of 30 degrees each. Beginning at the point where the ecliptic intersects the equator on the sun's northerly journey and continuing along its apparent yearly path, the sun traverses the signs of Aries, Taurus, Gemini, Cancer, Leo, Virgo, Libra, Scorpio, Sagittarius, Capricorn, Aquarius, and Pisces. To describe the position of any of the seven planets, astronomers would speak in terms of signs and degrees. Thus, for example, the sun's entry into Aries 0° (the initial point in the sign of Aries) defines the vernal equinox, the beginning of spring. At Cancer 0°, the sun reaches the summer solstice, the northernmost point of the ecliptic. The autumnal equinox occurs at Libra 0°, and Capricorn 0° marks the winter solstice.

Were the motions of the sun, moon, and planets regular, the astronomers' goal of describing them with uniform circular movements would have been quite easy. The prime mover's daily rotation would explain the daily rising and setting of all of the heavenly bodies. And each planet's sphere would provide its own contrary motion through the zodiac. The planets, however, demonstrate a number of irregularities in their motions. First, the five planets do not travel uniformly through the zodiac. Rather, they move at times more rapidly, at times more slowly, and at times they seem to stop and then to move in the opposite direction. (This motion, in a sense contrary to that of the sun's path, is called retrograde; the planet's usual motion is called direct.) Even the sun and the moon, which display no retrograde movement, appear irregular in their motions. For example, the sun spends more than half of the year traversing the half of the ecliptic from the vernal equinox to the autumnal equinox. Finally, even the fixed stars do not appear to stay put, but show a gradual eastward drift with respect to the vernal equinox (known as precession of the equinoxes or, to medieval astronomers, the motion of the eighth sphere).

The most enduring solution to the problem of explaining these irregularities yet keeping uniform circular motions was that outlined in the *Almagest* of Ptolemy (Claudius Ptolemaeus) in the second century A.D. Medieval astronomers understood the heavens largely by means of Ptolemy's system. How did Ptolemy "save the phenomena"? His system is far too complex and mathematically sophisticated to do it justice here. In the most general terms, however, Ptolemy's theories maintained uniform circular motions, but the circles around which the heavenly bodies traveled were not always centered on the earth.

For the case of the sun, it was a relatively simple matter of making the sun's path eccentric. In other words, Ptolemy still had the sun describe a circle about the earth, moving at a uniform speed. But the center of the circle of the sun's course no longer coincided with the earth's center. Figure 1 displays a medieval artist's attempt to represent the rudiments of planetary motion in one diagram. The artist has placed the sun ("Sol," the large luminary at the top of the diagram) on the circumference of the circle labeled "deferens." (The label is more appropriate to planetary motion than to the sun's, but the artist is trying to represent both in one illustration.) The center of the sun's orbit, labeled "centrum deferentis," is the uppermost of the three points at the center of the diagram; it is clearly not the same as the earth's center, labeled "centrum mundi."[39] One can easily see that more than half of the sun's circle falls within the six signs from Aries to Virgo, producing the irregularity described earlier.

Ptolemy explained the retrogradations and stations (periods of no motion) of the planets by means of epicycles (see fig. 1). As with the sun, there is again a circle whose center is displaced from the center of the earth, properly termed the deferent (labeled "deferens"). The planet, however, travels on the circumference of yet another circle, the epicycle ("epiciclus"), whose center lies on the circumference of the deferent. Both deferent and epicycle rotate in the direction of the sun's journey through the zodiac, or, as shown in figure 1, in a counterclockwise sense. When the planet is on that portion of the epicycle nearest the earth, its motion will appear to be retrograde. At some distance on either side of the perigee of the epicycle, the planet will appear to stand still prior to changing direction (its first and second "stations").

Ptolemy added one further refinement, the equant point. The equant point lay in a line with the center of the earth and the center of the eccentric deferent, at a distance beyond the deferent's center equal to that point's distance from the

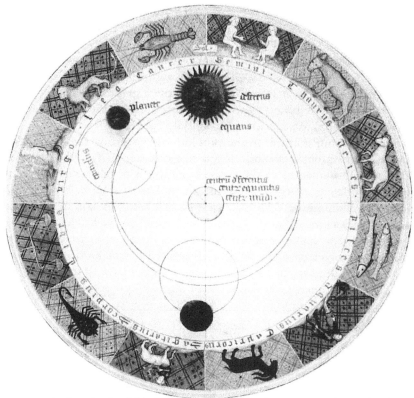

Figure 1. A schematic representation of the epicycle–deferent system. From a fifteenth-century German translation of Sacrobosco's *De sphera*.

earth's center.[40] The equant defined the center of uniform motion such that the epicycle center's motion was uniform with respect to the equant point and not with respect to the center of the deferent.[41] The planet's apparent motion was not uniform, while the center of its epicycle traveled along the deferent uniformly with regard to the equant point.

Because the axis of the earth moves slowly about a line parallel to the axis of the ecliptic, the fixed stars show the apparent slow drift to the east noted previously. Hipparchus, in the second century B.C., estimated the rate of motion as not less than one degree in one hundred years, which Ptolemy confirmed. Sacrobosco and many other medieval authors cited Ptolemy's rate of precession; another typical medieval figure was one degree in sixty-six years, found by al-Battani. Astronomers also held that the apogees (*auges*) of the sun and the planets, the points of their orbits most distant from earth, shared in the slow precessional movement of the eighth sphere. In d'Ailly's time astronomers placed the sun's apogee, for example, around the beginning of Cancer (as opposed to Ptolemy's figure of Gemini 5°30′).[42] The fifteenth-century illuminator of figure 1 displays the sun at its apogee, more or less correctly near the beginning of Cancer.

By d'Ailly's time the favored explanation for the motion of the eighth sphere was that attributed (falsely) to the ninth-century Arab astronomer and mathematician Thabit ibn Qurra and circulating under the Latin title *De motu octave spere*. According to this theory, the vernal and autumnal equinoctial points did not remain fixed but rather moved in small circles, thereby creating a motion referred to as trepidation or *accessus* and *recessus*. The model yielded a movable ecliptic with which moved the fixed stars and the apogees of the planets and the sun. One of the most important innovations of the Alfonsine tables' Parisian compilers was to incorporate both explanations of the eighth sphere's movement. The apogees and fixed stars were said to have a double motion—both of precession and of access and recess.[43] D'Ailly made reference to both "Thabit" 's theory and the traditional figures for precession, noting the association of the latter with the concept of the Great Year, a period after which all the stars would return to their original positions. And he incorporated the eighth sphere's *accessus* and *recessus* into his prediction of Antichrist's advent.

Such, then, were the basics of astronomy as it was understood in the Middle Ages. If astronomy represented theoretical knowledge about the heavenly motions, astrology was the "applied" version of that science. Scholars universally assumed that the stars and planets exerted an influence on earth below, an influence that varied with the stars' positions and mutual relations. To chart the stars' effects on earth, one needed to know the general characteristics of the heavenly bodies, the positions of the heavenly bodies at one or more specific points in time, and a complicated set of rules of interpretation.[44] The characters of the various heavenly bodies were familiar to most. Literature and popular prophecy often presupposed such knowledge.[45] Planetary positions could be derived from tables or by using a labor-saving device known as an equatory (*equatorium*).[46] Knowledge of the rules of astrology came from any of a number of handbooks, such as that of Alchabitius mentioned earlier.

The astrological theories that Alchabitius and other medieval texts passed on were of Hellenistic origin, although they contained elements of Babylonian, Egyptian, and Indian thought. A large omen literature had developed in Mesopotamia in the second millennium B.C. Lists of omens suggested correspondences between celestial phenomena and events on earth, without, however, drawing a causal link. The Babylonians gave names to constellations, which later would designate the zodiacal signs, and attributed characteristics to the planets, which would also become part of Hellenistic astrology. Omen lists spread to Greece and Egypt, where in the second century B.C. Greek mathematical astronomy, Egyptian calendrical theories, and Babylonian omen literature were combined to produce genethlialogy, the art of predictions based on the horoscope at birth. A number of authors contributed to the refinement of the science. Ptolemy, in the second century A.D., systematized the various elements and developments in his *Tetrabiblos (Quadripartitum)*. Just as his *Almagest* came to represent ancient astronomy for the Middle Ages, so too did Ptolemy become the great authority in astrology as well. Medieval European astrologers were also influenced by Near Eastern authors. Starting in the third century A.D., Sassanian scholars in particular incorporated elements of both Greek and Indian astrology; an important innovation was the development of theories allowing for more general predictions of the fate of a nation or the entire world.[47]

In this amalgamation of traditions, then, lay the science of astrology as d'Ailly and his contemporaries knew it. Fundamental to their predictions was the belief that each planet had particular characteristics. Saturn, for example, was a malevolent planet, cold and dry by nature, melancholic in temperament.[48] Jupiter, warm and moist by nature, was, by contrast, a fortunate planet, sanguine in temperament and a signifier of wealth. Mars's hot and dry nature was seen as having an evil influence; in particular astrologers believed that the planet could incite men to anger and war. The sun was also hot and dry by nature, yet its influence was held to be more benign than Mars's. Venus, too, was a fortunate planet, cold and moist by nature, phlegmatic in temperament. Mercury, astrologers taught, was of a changeable nature, taking on the properties of whatever planet had the strongest influence on it. Finally, the moon, like Venus, was cold and moist and fortunate in its influence.

The seven planets all traveled along the path of the zodiac, and the twelve signs which made up that band also were deemed to have their own characteristics. In one division of the zodiac, astrologers distributed the signs among four triplicities (*triplicitates*, also sometimes translated as trigons). The signs of each triplicity all shared the characteristics of one of the four elements (fire, earth, air, and water). The signs were assigned successively to one of the four triplicities, so that a planet in its path through the zodiac would pass first through a fiery sign, then through an earthy sign, then through an airy sign, and finally through a watery sign. There were three such series in any trip around the zodiac. The fiery triplicity consisted of the signs Aries, Leo, and Sagittarius. It was under the rule of the sun by day and Jupiter by night. The earthy triplicity contained Taurus, Virgo, and Capricorn, under the rulership of Venus and the moon. Gemini, Libra, and Aquarius made up the airy triplicity, under Saturn and Mercury. Finally, the watery triplicity comprised Cancer, Scorpio, and Pisces, with Mars ruling both day and night.

Further divisions of the zodiac defined some locations as masculine and others as feminine, the usefulness being, as d'Ailly explained in his commentary on Sacrobosco, that astrologers were sometimes called upon to determine the sex of a baby in the womb. In such cases, a signifier in a masculine place would indicate the birth of a boy, whereas a signifier in a feminine one would point to a baby girl's arrival.[49] Another classification denoted some signs to be mobile, some fixed, and some common. If the moon is in a fixed sign at the outset of any task, d'Ailly notes, that task will be of long duration; with the moon in a mobile sign, it will be accomplished quickly.[50] Further, each sign had significance over particular parts of the body and over particular geographical areas. According to Alchabitius, who reports the common traditions, Aries, for example, signified the head and the face and ruled over such areas as Babylonia, Palestine, and Arabia. At the same time, each planet had influence over a different part of the body depending upon its location. In Taurus, for example, the sun governed the knees, whereas in Capricorn it represented the back.[51]

When astrologers sought to measure the influence of all these various heavenly bodies, their work fell into one of four branches of the science: general predictions, nativities (genethlialogy), elections, and interrogations. *General predictions* looked at the stars' effects on society as a whole and forecast such events as the weather, war, famine, and plague. In making these predictions, astrologers considered such factors as the appearance of a comet or the impending conjunction of two or more planets. Also important for any general prediction was the position of the heavenly bodies at the beginning of the year, that is, at the moment of the sun's entry into the sign of Aries. An astrologer drawing up a *nativity* chart sought to predict the general characteristics and fortune of a person based upon the configuration of the heavens at the moment of the subject's birth. The horoscope so erected took into account both the exact hour and the precise location of a person's birth. Of utmost importance in the interpretation of the chart was the *horoscopus* or ascendant, the degree of the zodiac rising over the horizon at the instant of birth.

In the case of *elections* (or inceptions), the astrologer was asked to select the most propitious time at which his client might undertake some given action, such as embarking on a journey, having himself bled, or marrying. Naturally the astrologer would need to consider both the future state of the heavens and the positions of the stars at his client's birth in making such a recommendation. Finally, under the category of *interrogations*, the astrologer was asked to use his knowledge of the stars to answer some question put to him. He might, for example, be asked about the probable outcome of an undertaking begun at a certain time or be asked to predict the course of a given illness. In any case he would need to examine the present or future course of the heavens.

The fundamental tool of the astrologer was the astrological chart or horoscope (*figura celi*), essentially a diagram of the heavens at a given point in time in relation to a fixed point on earth (see fig. 2). In a horoscope, the astrologer plotted the positions of the heavenly bodies with respect to two scales. On the one hand, notations on the chart specified the locations of the seven planets within the zodiac. In figure 2, for example, Saturn is at Gemini 15°. On the other hand, the horoscope also shows the relation of the heavens to a particular location on earth, such as the place of birth in a nativity chart. To do so the astrologer superimposes on the heavens yet another division, this time into

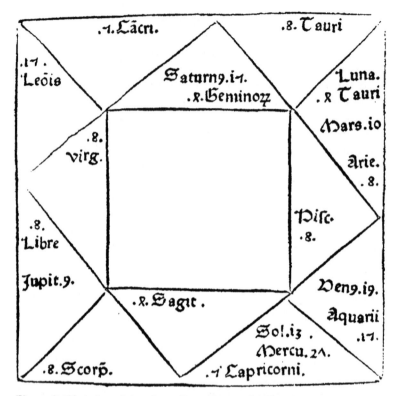

Figure 2. Christ's nativity chart. From Pierre d'Ailly, *Tractatus de imagine mundi et varia ejusdem auctoris et Joannis Gersonis opuscula* (Louvain: Johann de Westphalia, ca. 1483), fol. ee2v.

twelve houses (*domus*, sometimes referred to as mundane houses or places). Astrologers differed on the ways in which to erect the houses in an astrological figure, but almost all their systems began with four cardinal points: the ascendant (sometimes called the horoscope), the lower midheaven (*imum medium celum*), the descendant, and the midheaven (*medium celum*). These four points mark the cusps (initial boundaries) of the first, fourth, seventh, and tenth houses.[52]

To compute a horoscope, then, the astrologer would have to begin by locating these points. First would be the ascendant, the point of intersection of the ecliptic and the horizon. (The horizon is the great circle dividing the celestial sphere into two equal halves, the visible and the invisible.) The descendant lies 180 degrees distant from the ascendant. The *medium celum* and the *imum medium celum* are also points on the ecliptic, defined by the intersection of the ecliptic with the meridian and its continuation. (The meridian is a line running through the two celestial poles and the observer's zenith, or overhead point.) Astrological charts cast for different locations on earth, then, would not have the same configuration because their cardinal points would differ even if they were computed for the same time.

The most common procedure in the Middle Ages was then to divide each of the four quadrants marked by the cardinal points into three arcs of equal rising time. The three resulting houses were of an equal number of degrees when measured along the equator, but not along the ecliptic. Once the astrologer had determined the cardinal points, he could derive the cusps of the remaining houses from various tables or, most conveniently, using an astrolabe.[53] He would then enter the cusp of each house and the location of each of the seven planets in a chart like that shown in figure 2.[54] In the Middle Ages, the most common form for such charts was a square around whose edges ran a series of triangular subdivisions. Each of these triangles marked one of the twelve houses, beginning with the first, in the center of the left side of the square, and running counterclockwise around the square's perimeter. In figure 2, for example, the first house begins with Virgo 8°, the second with Libra 8°, and so on.

Houses were important to the astrologer because each house had significance for a different aspect of the subject's life. According to Alchabitius, for example, the first house, or horoscope, gave indications about the subject's life in general. The second house signified *substantia*, one's wealth, business, or property; the third, brothers and sisters. The fourth related to one's parents, the fifth to children, the sixth to sickness, the seventh to marriage, the eighth to death, the ninth to voyages and religion, the tenth to honors, the eleventh to friends, and, finally, the twelfth to enemies.[55] The planet that had the greatest influence in a particular house indicated how one might reasonably expect to fare in that area of one's life. Alchabitius gives an example in which he determines the sun to dominate in the second house in the case in which that house's cusp is Aries 5°.[56]

How did Alchabitius arrive at this conclusion? It was simply a matter of counting the dignities of the various planets for the sign of the zodiac beginning the house in question. Each planet, for example, had a particular degree in the ecliptic that was termed its "exaltation." A planet would score four points (or "dignities") if its exaltation fell within the sign being considered. Similarly, a planet would gain a score of three points if the sign was part of the triplicity over which it ruled. Each sign of the zodiac was divided into five "terms" and three "faces"; a different planet ruled each term and face. Astrologers allocated a planet two points if the sign contained one of its terms and one for one of its faces. In Alchabitius's example, the sun wins out with a score of seven points because Aries contains its exaltation (four dignities) and is in the sun's triplicity (three dignities).

Astrologers also made much of the planets' locations in reference to one another, particularly if they were in one of the five so-called aspects. A planet was said to be in aspect with (or to look at, *aspicere*) another planet if it was in one of the following angular relations with it: two planets separated by 180 degrees were said to be in opposition; by 0 degrees, in conjunction; by 120 degrees, they were trine to one another; by 60 degrees, sextile; and by 90 degrees, quartile. If two planets were in opposition or quartile, the aspect was held to be malign; if they were trine or sextile, it was benign. Conjunctions had their own set of rules for interpretation, and this was the area of astrology in which Pierre d'Ailly gained the most expertise. In general, the effects of a conjunction varied with the characters of the planets involved, the sign of the zodiac in which the conjunction occurred, and the time of the conjunction. In all aspects, Saturn and Mars radiated a malign influence, whereas Jupiter and Venus put forth fortunate effects.

Aside from these very rudimentary considerations, astrologers had a variety of complex calculations in their repertoire. They could project a chart into the future to predict the subsequent fate of their clients, for example, or offer an estimation of a person's hour of death.[57] By another calculation they named a "lord of the year" for individuals, cities, and countries, as well as for the population at large. In other cases, amulets, crafted of the proper material and engraved with suitable astrological images, aimed to capture the stars' influence. Even at the basic level of interpreting a single astrological chart, however, there was such a great amount of data to be considered and so many rules by which to analyze it that astrologers frequently disagreed in their predictions. The impossibility of obtaining consistent and accurate interpretations was a common argument put forth by astrology's critics. Yet this very complexity, the fact that a chart was open to seemingly endless analysis, must also have bolstered the science. A failed prediction indicated only a fault in the interpreter and not a flaw in the science itself.

Pierre d'Ailly was well aware of the difficulties involved in astrological forecasting. For that reason he counseled against making specific predictions concerning individuals and urged astrologers to limit themselves to general predictions. Accordingly, by all appearances d'Ailly never cast his own horoscope or consulted the stars in regard to his career. Rather, he concentrated on analyzing the large-scale effects of planetary conjunctions. As did his astrological sources, d'Ailly gave the greatest consideration to those conjunctions of the two superior planets, Saturn and Jupiter. Their exalted positions and slow motions meant that their conjunctions were of more universal and enduring significance than those of the other planets.[58]

Astrologers classified these conjunctions according to the signs and triplicities in which they occurred. Saturn completes its course through the zodiac in roughly thirty years, and Jupiter takes around twelve years to make the same circuit. Hence, the time between any two conjunctions of Saturn and Jupiter will be approximately twenty years, during which time Saturn will have traveled a little more than two-thirds of the way through the zodiac. Thus, in the astrologers' customary example, if the first conjunction of Saturn and Jupiter occurs in Aries, the second will be in Sagittarius, the third in Leo, and the fourth in Aries again. But, because the two planets do not complete their course through

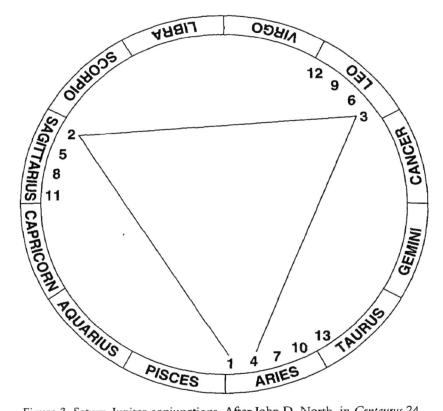

Figure 3. Saturn–Jupiter conjunctions. After John D. North, in *Centaurus* 24 (1980): 187.

the zodiac in exactly thirty or twelve years, they do not return to the same precise point in Aries for their fourth conjunction. Rather, they are joined some 2°25′ from the point of the initial conjunction, to take Albumasar's figure.[59] Hence a series of conjunctions of Saturn and Jupiter will show a gradual progression like that in figure 3. Eventually, a conjunction will happen in Taurus, the neighboring sign to Aries. Then the succession will begin again in another set of three signs.

Astrologers seized upon this pattern of conjunctions, for successive conjunctions all tended to be within the same triplicity. After approximately twelve conjunctions in one triplicity, the pattern will repeat itself in a sign adjacent to the first sign, and, hence, in a new triplicity. A conjunction appearing in, say, a fiery sign would have different effects than one in the earthy triplicity. Hence, astrologers believed the first conjunction of the new triplicity to be of special significance.

In d'Ailly's terminology, a change of triplicity is the greater (*maior*) conjunction, which happens every 240 years, and it signifies a change of sect in some region of the earth. After conjunctions have been formed in each of the four triplicities, that is, after approximately 960 years, Saturn and Jupiter are again joined in the original location, invariably defined as the first point of Aries. D'Ailly calls this conjunction the greatest (*maxima*) conjunction. It is said to signify "changes in empires and kingdoms, fiery impressions in the air, floods, earthquakes, and the supply of crops."[60] In all, d'Ailly delineated four types of Saturn-Jupiter conjunctions: the *coniunctio maxima* (every 960 years), the *coniunctio maior* (every 240 years), the *coniunctio magna* (every 60 years), and the *coniunctio minor* (every 20 years).[61] D'Ailly located such conjunctions throughout history and related them to the growth of new kingdoms and the rise of new religions. He used astrology, then, as a coherent principle by which to explain and to observe the course of the world's fate.

To sum up, astrology in the later Middle Ages was a science whose applications ran the gamut from the most practical to the most theoretical. On the one end, astrology was a fundamental part of the physician's training. Each sign of the zodiac ruled a different part of the human body, and it was the physician's job to know how to read the heavens so as to administer his medicine at the most efficacious moment.[62] On the most theoretical side, astrology demonstrated the framework of human history, as in d'Ailly's catalogue of Saturn-Jupiter conjunctions, or in the case of Robert Holkot (d. 1349), who related the history of the world to the signs of the zodiac.[63] In between, astrology offered information about weather, crop size, and business and personal matters.[64] All of these applications were guided by one fundamental principle, drawn from the reigning Aristotelian science: that the stars had an influence here below which men could measure and predict by a sufficient knowledge of the heavenly motions.

Three basic themes underlay the following examination of d'Ailly's study of the stars' effects. First, his writings demonstrate a fundamental interplay and essential harmony between his astrological and his theological beliefs. D'Ailly was deeply concerned with establishing a workable boundary between matters of astrological speculation and matters of faith, one that neither contradicted Christian teachings nor detracted from the power of astrology. His defense of the science would reflect the central importance of God's omnipotence in his theology. And his excursions into astrology attempted to define the role of the stars in the development of religions. With no mere rhetorical flourish, d'Ailly termed astrology "natural theology."

Second, his story underscores the importance of apocalyptic notions in Europe's reaction to the Great Schism. D'Ailly was not alone among conciliarists in maintaining and then abandoning a belief in the imminence of the End. Apocalyptic notions apparently had to be discarded before one could bring human means to bear upon ending the Schism. It may well be that one factor in the division's long duration was the slowness with which some churchmen gave up their belief that the Schism marked the approach of Antichrist. It was certainly a long process for d'Ailly, and his ultimate hopeful attitude was the fruit of many hours of reading and astrological calculation.

Finally, it is evident that the appeal of astrology lay in the fact that it offered useful information, sometimes unknowable by any other means, while it looked and operated like a science. Even astrology's critics agreed that the heavens exerted a real influence that produced various effects on earth. It was obvious to all that the sun caused warmth here below; that the other planets and stars would similarly have an effect on earth and that those effects could be determined and predicted by a person of knowledge seemed logically to follow. Astrologers of the later Middle Ages had strong scientific theory, as well as the evidence of any number of successful predictions, to back them up.

The complexity of astrology's rules and the disagreements among its practitioners served mainly to increase the respect accorded the science, just as failures and disagreements among modern physicians rarely cause the public to lose faith in medical science, only to seek a more experienced practitioner. This was certainly the case with Pierre d'Ailly. When faced with astrology's inconsistencies, he always assumed that more reading, more research, and greater knowledge of the heavens would provide the answer. D'Ailly believed that life on earth had a meaning, which was provided by God's plan for history, but he also believed that terrestrial events had an order. Astrology provided that order.

Scholars have sometimes viewed the history of astrology within the context of the conflict between magic and religion (as in Keith Thomas's *Religion and the Decline of Magic*) or the development of a split between science and the occult sciences (Wayne Schumaker's *The Occult Sciences in the Renaissance*). D'Ailly's case indicates that these categories are not valid for discussing the astrology of the later Middle Ages. Astrology was a science, and, while one could argue about which components of that science were adequately demonstrated, any attack on astrology as a whole belonged to a much larger conflict, that between the roles of reason and faith. In his defense of astrology and his promotion of the science as a "natural theology," d'Ailly attempted to come down on both sides of the debate. He believed that there were religious truths inaccessible to human reason and that faith had primacy in such matters. His vigorous use of the stars, however, stands as an eloquent testimony to his trust in the power of the human mind.

Such, then, were the concerns of a fifteenth-century churchman, a man who deplored the Schism in the church and whose greatest moment of triumph came at the council that would heal the division. Reputation and interest kept d'Ailly's works in circulation, and the printers of the later fifteenth century must have been confident of a demand for published editions of his writings, both scientific and theological—hence the Louvain edition that found its way into Columbus's hands. When Columbus read that volume of d'Ailly's writings, he was witness to a fair cross-section of the cardinal's interests. Geography, astronomy, astrology, the Schism, prophecy, the world's history, and its end all make their appearance in the treatises printed in 1483. Furthermore, these disparate subjects do not emerge as the mysterious manifestations of a mind whose possessor is paradoxically both scientist and mystic. Rather d'Ailly's treatises represent an ongoing effort to understand his own world and times. History, prophecy, and the stars were all routes to that end.

In the mind of Columbus, d'Ailly's reader, these seemingly contrasting elements of his thought may have merged as well. Following his source, Columbus concluded that the world was nearing its end, but that prior to its consummation all of its inhabitants would be converted to Christ. Dubbing himself "Christoferens" (Christ-bearer), he believed that God had sent him on his voyages to seek out and convert the world's pagan peoples. Was Columbus, discoverer and seeker of gold, stirred to an even greater extent by a mystical, fanatical Christianity, as Pauline Moffit Watts has suggested?[65] Certainly that element was present in his thought. But if he remained true to the intentions of his source, Columbus was also motivated by the most rational of science.

CHAPTER TWO

The Medieval Debate about Astrology

LIKE ANY THINKER of the later Middle Ages, Pierre d'Ailly had to confront two opposing sets of traditions about astrology. On the one hand were the attacks by the church fathers on a remnant of pagan superstition that brought into question man's freedom of will. The most systematic expression of this opinion, and the *locus classicus* for any attack on astrology throughout the Middle Ages, was that of Saint Augustine in *The City of God*. On the other hand was the widescale acceptance of astrology as a science from at least the twelfth century on. Curiosity about the stars' effects was fed by a renewed interest in nature and, in particular, in Aristotelian physics. In the twelfth century a huge body of ancient writings that had been preserved in Arabic entered the Latin West. Foremost among the new translations from the Arabic were the works of Aristotle, which found an eager and receptive audience in western Europe. Aristotelian cosmology brought with it the notion that the heavens affected events on earth. Behind astrological theories now stood the authority of "the Philosopher," Aristotle.

Heirs to these two conflicting traditions about the stars, the scholastic thinkers of the thirteenth century had worked out a compromise, a way to preserve both celestial influences and human free will. Many offered a solution along the lines of that put forth by Saint Thomas Aquinas. But even after Aquinas's careful exposition of the matter, numerous astrological propositions were among those condemned by the bishop of Paris in 1277. Further, the third quarter of the fourteenth century in Paris witnessed scathing attacks on judicial astrology by both Nicole Oresme and Henry of Langenstein. Pierre d'Ailly had all of this long history in mind when he wrote his series of treatises defending astrology. He addressed both the rational and the religious objections to astrology familiar from centuries of debate on the stars. But he also demonstrated a desire to harmonize astrology with the major theological concerns of the later fourteenth century. Occupied not merely by the science's potential conflict with human free will, d'Ailly also took pains to erect an astrological system that safeguarded the omnipotence of God.

THE EARLY MEDIEVAL VIEW OF ASTROLOGY

Church fathers attacked judicial astrology for two basic reasons. First, its practice bore traces of pagan superstition and star worship. Second, it denied the fundamental Christian principle of man's freedom of will. Quite simply, if the stars controlled all human actions, men were not responsible for their good or evil deeds, and the economy of salvation was meaningless. Consequently, the early Christians prohibited astrology, terming it a species of divination done with the aid of demons. Astrology was not merely wrongheaded, they argued; it could also expose one to the operation of hostile spirits.[1]

Christian objections to the science found a vigorous and systematic voice in the writings of Saint Augustine of Hippo (354–430). In *The City of God* Augustine brought together a number of ancient arguments against judicial astrology and underscored the science's incompatibility with Christian beliefs.[2] His sentiments would have been familiar to any medieval author discussing the stars. Augustine attacked astrology with three basic arguments. First, astrology impinged upon man's free will. Second, stellar control of earthly events raised questions about God's role in such happenings. Finally, the issue of twins born to different fates brought into doubt the accuracy and the rules of the science.[3]

Augustine decried the lack of human free will and, thus, accountability in a world where God rules the stars and the stars rule man.[4] "How is any room left for God to pass judgment on the deeds of men," he wrote, "if they are subject to astrological forces, and God is Lord both of stars and men?"[5] Further, Augustine found it injurious to God to suppose that either the Creator or the stars, as his agents, would decree wicked deeds to be done. The responsibility for human evil rested upon man and not upon God.[6] Finally, Augustine addressed the standard classical solution to the problem of astrology and free will, that is, the argument that the stars signify rather than cause future events. "This view has been held by men of no ordinary learning," he noted.[7] Augustine met this argument with a question: Even if the philosophers speak of the stars as signs and not causes, how do they explain the differences in the lives of twins, born at the same hour under the same constellation?[8]

The problem of twins led Augustine to question the possibility of making any accurate astrological predictions. First, he offered an example supposedly drawn from the writings of Hippocrates. The great physician had suspected two brothers to have been twins because they fell ill, convalesced, and recovered at the same time as one another.[9] Astrologers would attribute their simultaneous illness to the fact that they were born under the same constellation. This similarity, said Augustine, is rather to be explained as the physician would. The fetuses were nourished by the same mother at the same time and the children were raised together in the same environment. Many different kinds of beings with different fates may be born under the same sky, he added. Further, we know that twins may act differently, travel to different places, and suffer different illnesses.[10]

Augustine ridiculed the astrologers' attempts to explain why twins could have different fates.[11] Astrologers had argued that the small interval of time elapsing between the births of twins causes a great enough change in the heavens to account for a vast difference in their lives. If these *imperceptible* changes in the celestial sphere could make twins' lives so different, objected Augustine, how dared astrologers claim accuracy in *any* prediction based on a natal constellation?[12] If astrologers ascribed twins' differences to the time separating their births, they related their similarities to the position of the heavens at their simultaneous conception. Augustine weakened this proffered explanation by adducing the example of twins of different sexes. It was not improbable, Augustine granted, that there could be some stellar influence on bodies alone, though certainly not on men's wills (the soul being a higher order of being than the body). But when he looked at the case of twins of different sexes, he found even this position less than believable, for certainly nothing pertains more intimately to the body than its sex. "What could be more stupid to say or believe," asked Augustine, than that the position of the stars at their simultaneous conception could not prevent the difference in twins' sexes, yet the different constellations at their births could cause other great differences in their lives?[13]

Augustine finished his discussion of astrology with reference to the doctrine of elections, which he found entirely absurd. According to this practice, by selecting an astrologically propitious time to perform a certain act, a man could influence the consequences of that act. Thereby, Augustine objected, he might overcome the fate contained in the heavens at his birth, invalidating the birth horoscope. This possibility demonstrated the inconsistencies of the astrologers' rules. "Considering all this," he concluded, "the belief is justified that when astrologers miraculously give true replies, . . . this is due to the furtive prompting of evil spirits."[14] Having entirely rejected the principles of astrological prediction, Augustine reaffirmed the early Christian view: when men claimed to divine by aid of the stars, they did so with the intervention of demons.

Another important early source for medieval attitudes about astrology appears in the writings of Isidore of Seville (ca. 560–636), in particular in his encyclopedic *Etymologiae*. Although Isidore was firmly opposed to using astrology, his writings foreshadow the later ambivalence about the stars. The very position of astrology in the arrangement of Isidore's work points to its ambiguous status as both science and magic. Isidore discussed the art of prognosticating by the stars both in chapters devoted to the mathematical sciences and in the section of the *Etymologies* dealing with magic.[15] Further evidence of Isidore's dual view of astrology comes from his discussion of the terms *astrologia* and *astronomia*. Isidore's predilection for words and their origins led him to attempt to distinguish the two synonymous terms. By Isidore's definitions, much like today, *astronomia* dealt with the motions of the heavens and *astrologia* with their effects.[16] This neat distinction between the two words did not persist, however, and the terms were blurred, jumbled, and sometimes reversed throughout the Middle Ages. Pierre d'Ailly, for example, fairly consistently used *astronomia* for "astrology" and *astrologia* for "astronomy."

Isidore's writings did provide the source, however, for a distinction of great importance for medieval discussions about astrology. "Astrology," he wrote, "is part natural and part superstitious." Only the forecasting of the positions of the heavenly bodies themselves was acceptable as "natural" astrology.[17] Isidore's descriptions of the two types of astrology were sufficiently brief to allow interpretation, however, and later supporters of the art could draw the line between superstitious and natural astrology as they saw fit. Thus, d'Ailly could defend astrology in a treatise bearing the paradoxical subtitle *contra superstitiosos astronomos* ("against the superstitious astrologers").

Isidore treated divination by the stars as a species of magic. He referred to practitioners of the art variously as *genethliaci* (after genethlialogy, or the study of nativities), *mathematici*, and *magi*, but not *astrologi* or *astronomi*. During the course of his discussion, Isidore mentioned the Magi of the Gospels. Since the Magi became aware of Christ's birth by observing the heavens, this fact might be taken to validate astrology. Indeed, knowledge of astrology had been allowed, but only until the time of Christ, said Isidore. "From then on," he insisted, "no one would interpret any person's nativity from the heavens."[18]

Given the condemnations of astrology by the Christian fathers, many scholars have assumed that interest in and practice of astrology waned during the early Middle Ages. M.L.W. Laistner, however, has argued that church authorities were far more concerned with other superstitions than with the more scientific astrology. He points out that the major condemnations of astrology by the councils of Toledo (400) and Graga (560–65) were in fact directed against the heresy of the Priscillianists, who also believed in astrology. Indeed, Laistner maintains that only a lack of proper textbooks, and not the censure of the church fathers, led to astrology's decline in the early Middle Ages.[19] Until recently, scholars have concurred with Laistner's judgment and have seen little practice of astrology in the early medieval period.[20]

In a fascinating reappraisal of the place of magic in early medieval Europe, Valerie I. J. Flint has argued that there was indeed considerable practice of astrology in the early Christian centuries. Further, she concludes that church authorities eventually bolstered astrology (and other forms of magic) as an alternative to more alarming and dangerous forms of divination.[21] For Flint, Isidore's distinction between "natural" and "superstitious" astrology was a key step paving the way for a rehabilitation of magic in the early Middle Ages. In her view, Isidore was able to salvage some part of the old, prohibited magic that Augustine and others had censured in order to offer a "Christian, 'scientific' counter to non-Christian magical divination."[22] The success of the project Flint sees in an increase in manuscripts of texts dealing with topics close to astrology (such as treatises on the astrolabe, star charts, and tables of times for bloodletting) from the ninth century on; in a reappraisal of the Magi in biblical commentaries from the same period; and, finally, in the eleventh-century rediscovery of ancient astrological texts such as that of Firmicus Maternus. Thus, for Flint, the acceptance of astrology in the High Middle Ages owes less to the twelfth-century translations of Arabic sources than it does to the conscious manipulations of the churchmen of the early Middle Ages.[23] She counters Laistner's picture of astrol-

ogy's demise in these centuries by suggesting that sufficient knowledge to practice astrology could have circulated by word of mouth and via short-lived charts
and tables. Yet only a very rudimentary form of astrology could have survived
in this manner.[24]

Flint illuminates nicely the ways in which the writings of Christian authorities
contained the seeds for a rehabilitation of astrology, whether or not this was
their authors' express intention, as she so intriguingly suggests. Thus, Augustine's admission that the stars may have some effect on bodies and Isidore's
definition of a "natural" component of astrology pointed to ways in which
future students of the stars might present their activities as scientific and in harmony with Christian doctrine. It may well be, as Flint has argued, that such
interest in the stars began as early as the reign of the Carolingians.[25] Certainly
fascination with astrology rose greatly in the twelfth century with increased
speculation about the natural world and in the wake of translations from the
Arabic. Yet scholars would also have to reconcile their interest in the stars with
the condemnations of the early church fathers, canonized in such works as Gratian's *Decretum* in the 1140s.[26] In Isidore's distinction of natural and supernatural
lay the ammunition for their defense of astrology.

THE ACCEPTANCE OF ASTROLOGY IN
THE HIGH MIDDLE AGES

The early twelfth century, then, saw a revival of interest in the natural world, in
cosmology, and also in astrology. Adelard of Bath, for example, traveled extensively in pursuit of learning in the field of natural philosophy. He made a number of translations from Arabic sources, including the pseudo-Ptolemaic *Centiloquium* and the astronomical tables of al-Khwarizmi in the 1120s.[27] Bernard
Silvestris, in the 1140s, gave an account of the creation in which all events had
prior existence in the stars.[28] Contemporary scholars embraced the notions that
there were definite laws of nature, that the universe was animated by a "world
soul," and that man was a microcosm of the larger universe.[29] All of these theories supported the belief that the stars influenced happenings on earth.

Equally important for the rise of astrology was the influx of translations from
the Arabic beginning in the early twelfth century. This new corpus gave Western scholars the practical textbooks in astronomy and astrology they had previously lacked, as well as a new theoretical basis for a justification of astrology. By
the mid-twelfth century, the basic textbooks of astronomy (Ptolemy's *Almagest*)
and astrology (Albumasar's *Introductorium maius* and Ptolemy's *Quadripartitum*)
had been translated, along with a host of other astronomical and astrological
works.[30] Further, the weight of Aristotle now supported the study of astrology.
Statements in the *De generatione et corruptione* and the *Meteorologica* linked changes
on earth to motions of the sun and moon and, by interpretation, the planets.[31]

And Albumasar's *Introductorium maius* provided a model defense of astrology.[32] Even with this wealth of new materials, real expertise in astrology was likely rare in the twelfth century. Adelard of Bath was perhaps one of the only competent astrologers in early twelfth-century England. Yet surviving horoscopes now attributed to his hand are full of inaccuracies and deficiencies.[33]

By the thirteenth century, astrological knowledge was far more common, and a ruler such as the emperor Frederick II could depend upon the services of two excellent astrologers.[34] Writers of the twelfth and thirteenth centuries were forced to wrestle with the problem of reconciling the theory of celestial influences with the early Christian rejection of astrology. In general, scholars accepted astrology's place in medicine, meteorology, and alchemy, while wavering somewhat on the effects of the stars on human behavior. Not wishing to deny either celestial influences or human free will, most authors affirmed astrology while taking refuge behind the oft-quoted maxim "Sapiens dominabitur astris" ("The wise one will rule the stars").[35] Although nearly universal, it was an unsteady compromise, leaving it unclear whether free will or astral causality lay behind men's deeds.

Of the numerous thirteenth-century authors discussing astrology, among the most important for subsequent generations were Albertus Magnus, Roger Bacon, and Albert's pupil Thomas Aquinas. Albertus Magnus (?1193–1280) was one of the greatest and most influential minds of the thirteenth century. Known to contemporaries as an expert in the occult sciences, Albertus described a magic based entirely on natural causes and "occult virtues" as well as a demonic magic.[36] Albert followed Aristotle in positing a fifth element making up the sky and the heavenly bodies. This element served as a medium between the first cause and matter. Through its mediation, the motions of the heavenly bodies caused whatever existed on earth.[37] Thus, astrology was a valid and useful science. To preserve human free will, Albert drew a clear demarcation between man's body and his soul. The soul received its essence wholly from the first cause and hence was not like other inferior substances. Human reason and will were thus free from stellar control, although a man who did not resist the flesh would be swept along by the stars.[38] Having dealt with the objection that astrology denies man's freedom of will, Albert endorsed not just the divinatory practice of casting nativities but some more operative aspects of astrology as well. He approved of electing propitious hours to begin an undertaking and even of engraving astrological images on gems and minerals to work marvels.[39] In the *Speculum astronomiae* he went so far as to say that it was rash and working against the freedom of the will *not* to elect astrologically propitious times for one's actions.[40]

Roger Bacon's *Opus maius* (ca. 1266) and other works also made great claims for astrology.[41] His writings were to make an important impression on Pierre d'Ailly. As did Albertus Magnus, Bacon accepted the stars' influence on the body and denied them direct power over the soul. Although the individual might resist the heavens' influence, Bacon held that the masses rarely did and, thus, that astrological predictions regarding large groups were generally valid. In particular, Bacon was taken with charting the astrological causes of religions, following the doctrine of the great conjunctions introduced to the West by the writings of Albumasar.[42] Bacon described six religions based on conjunctions of Jupiter with each of the six other planets (including the sun and moon as planets). The conjunction with Mercury was associated with Christianity because of numerous favorable and mystical qualities of that planet. Bacon also described how the stars foretold Christ's birth and the advent of the Christian religion. He believed that such astrological descriptions of Christianity would strengthen the Christian's faith.[43] He also urged the church to avail itself of astrological elections and images to resist the infidel.[44]

Although Thomas Aquinas (1225–74) was less enthusiastic about astrology than were Albert and Bacon, his solution to the problem of astrology and free will became the classic one for later writers on the stars.[45] Like Albertus Magnus, Thomas insisted upon the difference between the soul and the body. While the heavenly bodies have no direct power over human reason and will, based upon the premise that "no corporeal being can make an impression on an incorporeal being,"[46] he granted that the stars did influence material things, and hence human bodies, directly.[47] They might thus incline men to action indirectly through the lower sensory powers, which in turn touched upon the intellect.[48] The will, however, was free to resist such impulses. "Thus these very astrologers," Aquinas explained, "say that the wise man is master of the stars in that he is the master of his own passions."[49] But most men followed their passions, according to Thomas. Hence a great many of the predictions of the astrologers were true.[50]

When Thomas asked whether divination by the stars was licit, he took a stand similar to that of the early church fathers.[51] He granted that many astrological forecasts were true because so few men resisted their bodily passions. But Aquinas worried that attempting to foretell some types of future events might lead men to mingle with demons. One such instance was the attempt to predict with certainty (*per certitudinem*) the future activities of men. This attempt was, in Isidore's old term, "superstitious" and wrong. But one might licitly use astrology to predict events determined by the heavenly bodies, such as rains and droughts. Did such licit use include nonspecific predictions about men's actions? Aquinas was silent, but left open the possibility of an affirmative answer.[52]

Thomas's solution, or some variation thereof, was the standard defense of astrology in the later Middle Ages. Granting the stars power over bodies saved the now universally accepted field of astrological medicine. That the stars had only an indirect influence on human behavior answered charges of astral determinism as well as the criticism that astrological predictions were often inaccurate. The inescapable fact that most men followed their bodily inclinations left room for astrological practice. If the wise men ruled the stars, then it followed that the great majority of men, who were not wise, were for the most part the heavens' subjects. This solution offered justification for the production of the astrological almanacs and annual predictions that would become so popular in later years, particularly after the invention of printing.[53] But the individual horoscope, in Aquinas's view, remained problematic.

It was an unsteady resolution of a conflict that was, in the final analysis, insoluble. If most men could not resist their bodily passions and if these inclinations were dictated by the stars, there was, in effect, no free will. If man's soul was immune to heavenly influences and he acted according to the dictates of his intellect and his will, then astrological prediction was impossible, beyond forecasting the weather, natural disasters, and major epidemics. Too great an insistence on either astrological or theological principles would undermine the compromise. The ambiguity of Aquinas's teaching on astrology is underscored by the treatment it received in the hands of Pierre d'Ailly. D'Ailly used the same passages from the *Summa theologiae* to attack astrology in the 1380s and to defend its use in 1419. The interesting fact about the position worked out by Aquinas and the other thirteenth-century scholastics is that, for most of their contemporaries and successors, it was good enough. No one wanted to deny either Christian theology or Aristotelian physics, and so they accepted a compromise, which, at least on the surface, preserved them both.

CRITIQUES OF ASTROLOGY AT PARIS IN THE LATER MIDDLE AGES

By the later thirteenth century, church authorities had begun to see the dangerous implications contained in much of the "new learning." In 1277, only three years after Thomas Aquinas's death, the bishop of Paris, Etienne Tempier, issued a list of condemned propositions, including a number of astrological precepts. Whatever the effect of these condemnations on later thought, they reiterated the old patristic objections to astrology and underlined the connection between astrology and its non-Christian sources.[54] On the other hand, the actual practice of astrology in high places was on the rise. When asked for an explanation of the plague of 1348–49, the faculty of medicine at Paris responded in astrological terms.[55] In the 1360s, with an endowment from the king, Charles V, Maître Chrétien Gervais founded a college of astrology and medicine at Paris.[56] The library of Charles V at his death in 1380 was full of volumes on astronomy and astrology.[57] Due in part to the condemnation of astrology in 1277 and to its increased use by the aristocracy, the latter half of the fourteenth century witnessed the most thoroughgoing attack on judicial astrology in the West since Augustine's. The assault came from the pens of Nicole Oresme and his younger contemporary Henry of Langenstein (or Hesse). Both men sought not simply to discredit astrology, but also to erect new systems by which to explain natural events without recourse to occult virtues or the influence of the stars.

Bishop Tempier's condemnations in 1277 roundly denounced the fatalistic aspects of astrology, although they left open a small amount of room for astral medicine.[58] Particularly repugnant was the notion that the heavenly bodies had any influence on the soul or the will, condemned, for example, in the propositions "That in the various signs of the heavens are signified various conditions in men, both of spiritual gifts and of temporal goods,"[59] and "That our will is subject to the power of the heavenly bodies."[60] Tempier also denounced the fatalism implied in the doctrine of the Great Year, the notion that the stars would all return to their original configuration in some specified number of years and cause the same series of effects to recur on earth.[61] Finally, the bishop attacked the theory behind the casting of nativities, declaring it an error to believe that there was "a disposition in man from the hour of his birth inclining him to certain actions or events." But here Tempier left a loophole; such a teaching was acceptable if "it was understood as concerning natural events and by way of disposition."[62] While the condemnations thus preserved the possibility of some astral influences on the body, they remained largely hostile to judicial astrology, behind which lurked a dangerous and condemned fatalism.

The philosopher and theologian Nicole Oresme (ca. 1320–82) translated a number of Latin works into French at the behest of Charles V.[63] Presumably with his royal friend and patron in mind, he also wrote treatises in Latin (*Tractatus contra iudiciarios astronomos*) and French (*Livre de divinacions*) arguing against the use of astrology by princes and rulers.[64] Further, he had attacked astrology in a number of his scientific works, in particular in the treatise *De commensurabilitate vel incommensurabilitate motuum coeli* (On the commensurability or incommensurability of the heavenly motions). Oresme's invective against astrology also appeared in his *Quodlibeta* (Quodlibetal questions) and in a *Quaestio contra divinatores horoscopis* (Question argued against those divining by horoscopes) argued (*determinata*) in Paris in 1370. Oresme offered all the traditional arguments against astrology as well as new reasons designed to show the impossibility of having accurate enough knowledge of the heavenly movements for astrological prediction. In addition, he developed an alternative theory to explain natural events usually attributed to occult virtue, demons, or astral influence.

Oresme offered largely standard arguments in the *Tractatus* and the *Livre* and in parts of the *Quodlibeta* and *Quaestio*.[65] The aim in the first two works was not so much to refute judicial astrology as to show that it was largely unprofitable for a king to use it.[66] Astrology denied human free will, Oresme argued, in practice if not in fact.[67] It had earned the censure of the Bible, the church fathers and doctors, and such philosophers of antiquity as Statius, Cato, and Cicero.[68] The rules by which the astrologers prognosticated had their origins more in fiction than in any rational basis.[69] Astrological predictions, even about the weather, were generally wrong,[70] and astrologers often gave deliberately ambiguous answers or resorted to fraud in order to make their forecasts come true.[71] Further, astrologers generally disagreed with one another.[72] Oresme cited the old objection about twins from Augustine and other classical authors.[73] In keeping with his theme, in both the *Tractatus* and the *Livre*, Oresme gave examples of kings who had come to ruin using astrology.[74]

Oresme's writings also included a number of novel attacks on astrology. In the *Quodlibeta*, for example, he attempted to demonstrate the incompatibility of astrology with Aristotelian physics. Hitherto, Aristotle had generally been taken as a proponent of astrology.[75] In the *Quaestio*, Oresme attacked Aquinas's notion of the heavenly bodies' influence on the sensitive soul and his contention that men follow their bodily inclinations.[76] Oresme even questioned Thomas's authority to speak on this matter because he was not an astrologer.[77] Oresme's most valuable new weapon against astrology, however, was the argument from incommensurability.

In a series of mathematical treatises, Oresme developed the hypothesis that the proportion of the velocities of the heavenly bodies would most likely be an irrational number and that their motions would therefore be incommensurable.[78] Since an irrational number cannot be known exactly, the incommensurability of the heavenly motions would mean that all of the tables and calculations upon which astrology was based were inaccurate. Astrologers taught that even a small change in the position of a planet could radically alter an astrological chart. Therefore, any uncertainty in our knowledge of the heavenly motions rendered accurate astrological judgments well-nigh impossible.[79] Incommensurability would also mean that no configuration of the heavens would ever repeat itself exactly. Hence the astrologers could not base their predictions on past experience of the heaven's effects.[80]

Oresme also developed a system by which to explain natural phenomena without recourse to the stars, his doctrine of "configurations." Oresme postulated that different geometrical figures could represent the various "qualities" making up an individual. The relationship (*convenientia* or *disconvenientia*) between the "configurations" of two objects, he argued, could explain effects not reducible to the actions of the four elements. "Configurations" could thus account for phenomena like the magnet's attraction for iron, normally referred to occult virtue or heavenly influence.[81] It is important to note that Oresme presented configuration as a theory only (*ymaginatio*) and that he believed these configurations would be largely unknowable by humans.[82] The theory does indicate, however, Oresme's desire to explain natural phenomena by means other than astrology.

Even so, Oresme himself did not consider his attack on astrology to be conclusive. In the work most hostile to judicial astrology, the *Quaestio* of 1370, he allowed the heavens to affect the earth by "light, motion, and influence" and posited a general influence from the heavens as a whole.[83] His arguments sought mainly to discredit reliance on astrological judgments by indicating the hopelessness of attaining sufficient knowledge for accurate predictions. As Oresme said in the *Tractatus*:

Although there may be in some way a diversity of complexions [dependent on] the heavens, which inclines the souls of men to various characters, although without the necessity of fate; nevertheless the astrologers are unable to foreknow it because the proportions of the heavenly motions are unknowable (as I have shown else-where), because the forces of the stars are unknown, and because their rules have been strengthened with frivolous persuasions.[84]

Like Oresme, Henry of Langenstein (1325–97) also sought to explain natural phenomena without recourse to heavenly influences.[85] He too posited a theory of "configurations" for demonstrating effects not attributable to the qualities of the four elements.[86] Henry did grant the stars some causative power, higher and nobler than that of the four elements, but entirely dependent upon the first cause, God.[87] In practice, however, the heavens' influence was limited to mak-ing objects hot or cold and moist or dry, and the stars had no occult virtues.[88] Further, Henry insisted that one must know both the disposition of the receiver and the means of irradiation of the quality in order to foresee the effect of any given astral influence. Since this knowledge was practically impossible, astrolog-ical divination was rendered extremely difficult.[89]

Henry addressed astrological questions specifically in two treatises, the *Quaes-tio de cometa* (after the comet of 1368) and the *Tractatus contra coniunctionistas* (Treatise against the conjunctionists, provoked by predictions based on a con-junction in 1373).[90] Here he attacked as irrational several key doctrines of astrol-ogy. In particular, he found it difficult to accept that the effects of a planet would vary with its position in the heavens or with its proximity to another planet.[91] Further, Henry rejected the doctrine that granted a greater influence to some conjunctions than to others. Why should the rarity of a phenomenon, he ques-tioned, have any impact on its effects?[92] Henry especially reproved the practice by which astrologers would attribute events on earth to the appearance of a comet (or to a conjunction) occurring months or even years previously.[93] Fi-nally, he argued that the human mind was incapable of knowing the qualitative structure of natural bodies or the laws of celestial motions with enough accuracy to make any kind of predictions based on the stars.[94] Although he granted a limited influence to the stars, Henry's rejection of all rules based on the positions of the planets rendered astrology as it was practiced essentially powerless.

For so forceful an attack, the writings of Nicole Oresme and Henry of Langenstein had remarkably little impact.[95] Oresme's anti-astrological works had a small circulation among his contemporaries, although borrowings from his writings showed up in Philippe de Mézières's *Songe du vieil pelerin*, Eustache Deschamps's *Demoustracions contre Sortileges*, and the *Somnium Viridarii*, as well as in the works on astrology of Henry of Langenstein, Pierre d'Ailly, Jean Gerson, Pico della Mirandola, and a handful of others.[96] Judging from the attacks on astrology by Gerson and Pico, Oresme's writings certainly seem to have had little effect on astrological practice. Nor did Henry of Langenstein manage to crush the use of astrology. Indeed, in the fifteenth century, Henry received the dubious and ironic distinction of having astrological apocalyptic prophecies attributed to him.[97]

The effect of the attacks on astrology in Paris in 1277 and the later fourteenth century, then, was not totally to discourage the use of astrological judgments. Rather the attacks served to reinforce the old patristic invectives against "pagan superstition" and to raise rational objections to the scientific basis of astrology. The result was not an abandonment of astrology, but rather the elicitation of careful defenses from would-be astrologers. Astrology's demise came much later, in the seventeenth century, with the advent of new scientific modes of thought and astrology's increasing vulgarization in the press.[98] Its detractors in the later Middle Ages kept open the debate on its acceptance, but theirs was not the last word.[99]

PIERRE D'AILLY'S DEFENSE
OF ASTROLOGY

Writing in the 1380s, Pierre d'Ailly addressed the question of whether divination by the stars was licit.[100] He took a rather conservative stand in this early work, condemning most uses of judicial astrology and following closely the arguments of Augustine, Thomas Aquinas, and Nicole Oresme.[101] In a series of treatises dating from 1410 until his death in 1420, however, d'Ailly defended astrology and its use in predicting religious change. He defined his own position as that of a "moderate" on astrology, somewhere in between those who would grant too little and those who would grant too much power to the stars.[102] The history of the science dictated a defense of even a moderate position, so d'Ailly in his writings carried on a dialogue, sometimes specific, sometimes indirect, with astrology's past critics. His defense of astrology addressed both astrology's agreement with Christian teachings and its legitimacy as a science. He dealt with the old issue of free will and fate as well as the newer rational arguments of Nicole Oresme and Henry of Langenstein. Following his own dual interests in theology and natural science, d'Ailly was most concerned with setting up the limits of astrology vis-à-vis theology, which would allow him to demonstrate, as he called it, "the concordance of astrology and theology."

In order to defend astrology from the charge that it was inconsistent with Christianity, d'Ailly first had to prune away those aspects of the science blatantly repugnant to the faith. This he accomplished by drawing upon the old characterization of astrology as part natural and part superstitious.[103] Dismissing the "superstitious" aspects of the science as the additions of wrongheaded astrologers allowed d'Ailly to concentrate his attention on the more defensible parts of astrology. In his 1414 *Vigintiloquium* or *Concordantia astronomie cum theologia*

(Concordance of astrology with theology), d'Ailly listed the following components of superstitious or false astrology:

1. The belief that all future events precede by fatal necessity from the stars;
2. The mingling of superstitious magic arts with astrology;
3. The placing of free will and matters solely under divine or supernatural control within astrology's power.[104]

The second of these objections probably was a reference to the practice of engraving stones with astrological images, against which d'Ailly's pupil Jean Gerson addressed a treatise.[105] The first and third addressed the old contention that
astrology denied human free will. The third showed as well a newer concern
with the relationship between astrological and miraculous causality, in exempting from stellar influence matters that are "solely under divine or supernatural
power." D'Ailly did not believe that the stars could signify future events without
also being their causes. Astrologers thus could predict future events because the
stars caused them. But d'Ailly still had to show how astrology could work without necessitating its effects or impinging on divine control.

On the problem of human free will, d'Ailly followed the basic solution outlined by Thomas Aquinas and Roger Bacon. In the early treatise *De falsis prophetis II*, d'Ailly relied on Aquinas's arguments about the stars. That is, he
granted the stars a direct influence on the body and an indirect influence on the
soul via the senses. Through the exercise of free will, a man might choose to
follow or not to follow the impulses of the body. That few men did resist the
flesh made many astrological predictions valid.[106] D'Ailly's later treatises used a
similar line of reasoning, now drawn from Roger Bacon. D'Ailly expounded
Bacon's views at length in his major defense of astrology, the 1410 treatise *De
legibus et sectis contra superstitiosos astronomos* (On the laws and the sects, against the
superstitious astrologers), where he argued that the heavens act upon the body
without compelling the will. The soul, since it is joined to the body, is strongly
aroused in turn, although it is not compelled to follow the body's desires.[107] Like
Bacon and Aquinas, d'Ailly found greater validity in predictions about the
masses than in forecasts about individuals. And like Bacon, he was interested in
the astrological causes of religions. Here, too, Bacon's ideas were of help. Since
group movements were often started by one famous and powerful man, the
stars' ability indirectly to arouse the soul could incline the masses to follow such
a leader.[108]

D'Ailly's concern with finding astrological causes for religions and sects
meant that he ran the danger of granting the stars control over matters supernatural or divine, the third on his list of superstitious errors. Indeed, his primary
concern in defending astrology seems to have been to show that astral causality
did not stand in the way of God's omnipotence. He devoted several treatises
in the years 1410–14 to delineating astrology's jurisdiction in relation to

religion.[109] In the first of these, the *De legibus et sectis*, d'Ailly followed Roger Bacon in associating each of six principal religions or sects with the conjunction of Jupiter and another planet and in linking changes in religions with planetary conjunctions occurring once every 240 years. But he found Bacon guilty of superstitious error when he tried to show that astrology had foretold the rise of Christianity.

The key to establishing a proper relationship between astrology and religion was, for d'Ailly, a clear distinction of natural from supernatural causality.[110] All natural things were under celestial control, he argued. Thus under the stars' jurisdiction were four of Bacon's six sects, those of the Chaldeans, the Egyptians, the Muslims, and the sect of Antichrist. The Mosaic law and the Christian religion d'Ailly classed as supernatural and thus exempt from astrological causation. Yet Christianity and Judaism had their natural aspects as well, he argued. One could properly attribute, for example, Christ's excellent physical complexion to a favorable disposition of the heavens at his birth.[111]

D'Ailly refined his distinction between supernatural and natural causes in later treatises so as to allow astrology even greater sway. In the *Concordantia astronomie cum hystorica narratione* (Concordance of astrology with historical narration) of 1414, d'Ailly put the schema quite simply: God arranged "to work naturally, with causes, except where a miraculous operation intervenes."[112] Thus astrological causality would apply to all earthly events save miracles.[113] D'Ailly further examined the limits of astrology vis-à-vis Christianity in the *Apologetica defensio astronomice veritatis* (Apologetic defense of astrological truth) also of 1414. Specifically, he questioned the role of the planets in the development in utero of Christ and the Virgin Mary. D'Ailly began with the cautious observation that the Christian faith did not compel one to exclude any stellar influence in Mary's birth, "just as it does not compel one to say that the sun did not warm her."[114] He concluded in citing Jesus' remark that he had come not to destroy the law, but to fulfill it (Matthew 5:17). Christ's words applied as well to the law of nature as to the Mosaic law, according to d'Ailly. He would have Christ and the Virgin subject to stellar influence in all matters save those "which Christian authority teaches were done by divine influence, by special privilege not of nature but of grace."[115] By reserving for God a supernatural causality beyond that of the stars, d'Ailly placed astrology among the undeniable laws of nature and gave it a scope reaching as far as the human aspects of Christ.[116]

D'Ailly showed his familiarity with attacks on astrology's legitimacy as a science from early on. In the treatise *De falsis prophetis II*, largely hostile to astrology, d'Ailly paraphrased great chunks of Augustine's *City of God* and Nicole Oresme's *Tractatus contra astrologos*.[117] The 1410 treatise *De legibus et sectis* borrowed heavily from Oresme's treatise on commensurability.[118] And in the two *Apologetice defensiones astronomice veritatis* (Apologetic defenses of astrological truth) of 1414, d'Ailly addressed specifically the criticisms of Nicole Oresme and Henry of Langenstein. In general d'Ailly responded to two sorts of assertions about astrology as a science. The first sort of attack pointed to the inaccuracy of astrological predictions and the uncertainty of human knowledge about the heavenly movements. The second kind of criticism was leveled at the rules used by the astrologers in making their predictions.

When replying to the charge of astrology's inaccuracy, d'Ailly took the line of defense that such assertions proved "the difficulty of knowing rather than the impossibility of the knowledge [*scientia*]."[119] That astrologers were frequently wrong in no way served to invalidate the theory behind their art, according to d'Ailly. In the *De legibus et sectis*, he acknowledged the great uncertainty in our knowledge of the heavenly movements. Following Oresme, d'Ailly showed that it was impossible to say with certainty that the heavenly movements were commensurable.[120] Further, he took great pains to demonstrate our lack of knowledge of even the sun's path. Even the length of the solar year was not certain.[121] For d'Ailly such reasoning did not undermine astrology, but rather indicated the peril involved in attempting to make particular judgments using the stars. "Since in the part of astronomy dealing with motions there are so many difficulties and uncertainties," he wrote, ". . . by necessity there must be many and greater ones in judicial astrology, since it depends upon the former and presupposes many uncertain things beyond it." It showed human pride, he added, to wish to try to attain knowledge of such future contingent events, which knowledge one should believe God had reserved for himself alone.[122]

Having ruled out making specific, particular predictions as an error of superstitious astrologers, d'Ailly affirmed the usefulness of astrology in foreseeing more general changes. In the second *Apologetica defensio astronomice veritatis* of 1414, he asserted that one could have cognition of the heavenly movements and their influences sufficient for making some judgments, even though this cognition was not exact.[123] In the *Vigintiloquium*, he offered examples of cases in which it would be reasonable to use astrology to conjecture about the future. For example, one might predict a good crop based upon a favorable heavenly disposition at the time of planting. Other future events relied on human free will. Even there, d'Ailly felt one could look to astrological judgments because the stars could have a great effect on "those brutish men who do not resist evil inclinations through virtues."[124] Nor did the contradictions among astrologers casting people's nativities shake d'Ailly's faith in astrology's premises: "I say that astrological judgments about nativities, more than all other judgments of the stars, are very difficult and uncertain." The "more experienced astrologers," he added, had sufficiently acknowledged that to be the case.[125] Such contradictions were to be expected because astrology was so much more difficult than the other sciences.[126]

In the criticisms of astrology drawn up by Nicole Oresme and Henry of Langenstein, d'Ailly found the additional charge that astrology's rules were contradictory, groundless, and seemingly made up. He answered these assertions in the two *Apologetice defensiones* of 1414. In the first of these treatises, d'Ailly dealt with the contention of Henry of Langenstein that it was foolish to believe that the planets ruled in turn over the nine months of human gestation. Relying on the best medical opinion, Henry had explained the fetus's development by the natural heat of the body and not by astrology.[127] D'Ailly, however, saw no reason to doubt that God had endowed the planets with a special virtue by which they would rule over successive months of human development.[128] After all, astronomers, philosophers, theologians, and physicians would all agree that the planets oversee in turn the hours of the day. D'Ailly confirmed this theory by the fact that ancient philosophers had named each day of the week after the planet ruling over its first hour.[129]

The antiquity of the astrologers' precepts also served as the basis for d'Ailly's response to Oresme's criticisms. Oresme found their rules to be utter fiction, as, for example, the assertion that some degrees of the zodiac were masculine and some feminine.[130] D'Ailly declared it preposterous that so many authors would have made up rules for the express purpose of deceiving posterity. He proposed the following sources for the astrologers' rules. First, he alleged that many of the rules were worked out by experience. As an example, he cited the couplings of the planets with the seven "climates" or regions of the earth. The validity of this association was obvious, according to d'Ailly, from the characters of the lands and their peoples. On the other hand, there were undoubtedly heavenly conditions of which the ancient astrologers would have had no direct experience, d'Ailly added. Further, the Flood would have destroyed most of the ancient writings that recorded the heavens' effects. Hence for much of astrological knowledge, he proposed as source revelation, not directly from God, but by the mediating impression of the heavens.[131] In other words, astrological knowledge had emanated from the stars themselves. Finally, d'Ailly hypothesized that it was "not incredible" that some astrological truths had been revealed directly by God to the holy patriarchs and prophets. This solution would work, he added, both to the exaltation of astronomical truth and to the honor and glory of God.[132]

D'Ailly's description of the sources of astrological knowledge involved a deliberate twisting of Oresme's words. In the *Tractatus contra astronomos*, Oresme had brought forth and rebutted the argument that astrology had been revealed to the patriarchs, charging that astrologers falsely attributed books to Solomon, Enoch, and other ancients in order to bolster the science.[133] D'Ailly knew both the argument and the rebuttal from Oresme, for he had inserted both passages into his *De falsis prophetis II*, written in the 1380s. There he followed both Oresme's intention and his order of exposition, concurring with his source's attack on astrology. In the second *Apologetica defensio*, however, d'Ailly ignored Oresme's counterargument and used the assertion that astrology had been revealed to the patriarchs to defend the science against Oresme's other attacks. In keeping with his steadily increasing estimation of astrology's worth, d'Ailly now disregarded arguments to the contrary and maintained that astrological science was a revelation from God.

D'Ailly's defense of astrology shows him familiar with all of the most important participants in the medieval debate on the stars. It also reveals his dependence on many of the older authors. In casting himself as a moderate on astrology, he relied upon the old Isidoran distinction between the "superstitious" and "natural" parts of astronomy, which enabled him to cast out the more objectionable aspects of astrology in order better to defend the remainder. When he addressed the issue of free will, he borrowed the arguments of Thomas Aquinas and Roger Bacon, which allowed him to state that the stars could incline one to action without necessity. From Bacon, too, he borrowed his conceptions of astrology's effects on religions and the implications of the planets' conjunctions. He appropriated arguments from Nicole Oresme's treatises against judicial astrology and on the commensurability or incommensurability of the heavenly motions. And, finally, he took caution from Oresme, Henry of Langenstein, and Augustine.

What set d'Ailly apart from these other authors is not so much a difference in his ideas about astrology as his concerns in defending it. D'Ailly's goal, above all, was to establish astrology as a "natural theology," which could be used to interpret prophecy as well as to validate the chronology of religious history. Hence, he ran up against issues other than the usual theological problem of reconciling astrology and free will. The key feature of d'Ailly's defense, then, was his distinction of natural and supernatural causality, and his placing astrology squarely in the camp of the former. Reserving a supernatural causality for the production of miracles meant that d'Ailly could affirm astrology without in any way denying God's omnipotence. And the wide scope given to natural causality in d'Ailly's schema meant that he could use astrology to investigate, not with certainty, but conjecturally at least, certain claims at prophecy. In 1414 he would use astrology to approximate the arrival of Antichrist. For d'Ailly, his Christian astrology provided a way of peering into the future without impinging on God's own powers.

The concern with God's omnipotence was a central feature of d'Ailly's theology, as well. Although, as he argued, God had voluntarily agreed to act only in certain ways in this world (God's so-called *potentia ordinata*), in absolute terms God's power was limited only by his inability to produce a contradiction (the *potentia absoluta*). Astral causality might well be perceived as restrictive of God's omnipotence; indeed one of the motivations behind the condemnations issued by Bishop Tempier in 1277 seems to have been a concern that Aristotelian physics impinged upon the freedom of God.[134] Although apparently unknown to d'Ailly, one of his contemporaries, Coluccio Salutati, attacked astrology on precisely this point. Among other objections to the science, Salutati argued that God could and did act outside the normal channels of causation. "Most astrologers," he noted with obvious contempt, "have thought that God leads a leisurely life and administers everything through the heavens."[135] Scholars in the fourteenth and fifteenth century paid great attention to the issue of God's omnipotence. Understandably, to defend astrology, d'Ailly had to insure a mechanism for God's free action on earth.

In truth, d'Ailly's defense of astrology answered the criticisms leveled at the science about as well as any other had, that is to say, poorly. As long as people were endowed with the free will to resist the stars' influences and God with the ability to act supernaturally, astrological predictions were uncertain conjectures at best. If one acted on the assumption that natural causality usually obtained and that men usually followed their bodily passions, however, astrological judgments were surer, but man's free will and God's omnipotence were questionable. The interesting fact is that for d'Ailly, and for the majority of his contemporaries, this defense of astrology was good enough. If one did not examine its implications too closely, it gave adequate credit to the powers of stars, man, and God.

There seem to be two cognitive processes at work in the acceptance of this sort of defense of astrology. First, there is the movement by which, with d'Ailly's increasing knowledge of and immersion in astrology, the science came to appear more and more reasonable. In a study of modern-day "witches" and astrologers, T. M. Luhrmann has noted the "interpretive drift," whereby when someone becomes a specialist in a given area, he finds its practice progressively more sensible.[136] Second, once committed to the truth of astrology, d'Ailly sought to avoid instances where his belief in astrology would conflict with his other deeply held values, such as his Christian faith. The psychologist Leon Festinger has discussed the intellectual stress (which he terms "cognitive dissonance") caused by similar conflicts in the case of failed prophecies. People act to reduce "cognitive dissonance," Festinger argues, not by abandoning one of the conflicting sets of beliefs, but by a series of patchy, short-term responses aimed at avoiding clashes rather than at maintaining any sort of real consistency. Luhrmann sees a like process at work among her modern witches; Geoffrey Lloyd has argued similarly that in ancient Greece scientific forms of thought frequently accompanied, rather than replaced, magical interpretations.[137] It would seem that d'Ailly's defense of astrology fits the same pattern; by minimizing their apparent conflict, he does not address the issue that Christianity and astrology are, in the end, irreconcilable.

CHAPTER THREE

The Making of an Astrologer:
The Development of Pierre d'Ailly's
Thought on the Stars

ALTHOUGH the bulk of Pierre d'Ailly's astrological writings date from the final decade of his life, the years 1410 to 1420, these treatises did not mark his first examination of the science. Throughout his career, d'Ailly pondered the issue of the stars' effects on mankind. Mentions of astrology appear in his letters, his sermons, and his philosophical and religious works, as well as in the treatises he devoted specifically to the topic. D'Ailly's attitudes about the stars changed significantly over the course of his lifetime. In his writings we can follow this development from a cautious, Thomistic position on astrology to an enthusiastic acceptance of the science's possibilities. In the early years of his career, d'Ailly's opinion of astrology was essentially that of his Parisian schoolmasters. Late in his life, however, he acquired a remarkable expertise in astrology, studying the major astrological sources and propagating his own views widely.

D'Ailly's eventual embrace of astrology was fueled by two factors: concern about the events surrounding him and his reading of key astrological texts. The existence of the Great Schism in the church fed apocalyptic fears and prompted him to consider ways of predicting when the End would come. Among the authors he encountered in this search was the Englishman Roger Bacon. Astrological theories in Bacon's works intrigued and, eventually, came to obsess d'Ailly. He developed, expanded, and refined these notions in his astrological writings from the years after 1410. With the approach of the Council of Constance in 1414, d'Ailly immersed himself in the study of the stars, and he found in astrology the hopeful prediction that the apocalypse was not at hand.

Pierre d'Ailly was obviously proud of his numerous writings, and he took great pains to collect and keep them, even embarking upon a systematic rereading of the entire body late in his life.[1] He constantly referred to his earlier writings, and many of his later treatises include the date and place of composition. Hence, it is relatively easy to establish the chronological sequence of d'Ailly's major writings on astrology. For the earlier works, we must rely upon conjecture for the dates. Even so, we can set out a rough chronology of these works, enough to chart the major developments in his thought. All told, the list of works mentioning astrology is impressive, and a quick glance at its contents leaves one with the overall impression that whenever time allowed and external events did not demand all his attention, d'Ailly returned to the question of the stars. (For a chronological list of d'Ailly's works dealing with astrology, see appendix 2.)

THE DEVELOPMENT OF D'AILLY'S
THOUGHT ON ASTROLOGY

It is possible to discern three rough periods in d'Ailly's thinking on the stars. In the first, from the years approximately 1375 to 1380, his writings indicate a definite interest in the stars, without, however, very much specialized knowledge about astrology. The works from the year 1410 on, by contrast, reveal d'Ailly to be accepting of astrology and its premises, with increasing sophistication in the science and faith in its capabilities. In between these two periods, the story is less clear. The years 1381–1409 were full of activity for d'Ailly, and he consequently had less time for reading and composing. The writings we have from these years point to his heightened interest in astrology as well as greater knowledge of the stars, but no real devotion to the science as yet. A coalescence of personal concerns, external happenings, and specific readings seems to have produced d'Ailly's enthusiastic use of astrology after 1410.

Pierre d'Ailly's earliest thinking on the stars would quite naturally have followed from what he learned as a student at the University of Paris. As we have seen, the university's teaching on the stars was ambivalent.[2] The prevailing Aristotelian cosmology favored viewing the heavens as causes of all that happened below. On the other hand, Christian scholars found this to be an assault on human free will and blatantly repugnant. Most followed some form of compromise, such as that articulated by Thomas Aquinas. The stars were granted a direct influence over the body and thereby an indirect influence on the soul. They thus might incline a man to action without necessitating his behavior. At Paris, there had been a clear concern to preserve man's freedom of will. Bishop Etienne Tempier's condemnations in 1277 had come down hard on astrological determinism. His sentiments would be rearticulated by the university in 1398, when the faculty of theology declared it an error to hold "that our intellectual cogitations and interior volitions are caused directly by the heavens, and that they can be known by any magic tradition, and that it is licit to make judgments with any certitude about them thereby."[3] And despite the popularity of astrology at the court, Nicole Oresme, who had taught at d'Ailly's own College of Navarre, had expressed serious doubts about the science's usefulness.

D'Ailly's earliest writings on the stars express the customary ambivalence. Accepting of the cosmological framework of astrology, he cast a disapproving eye on the more fatalistic components of the science. Such is the view of astrology set forth in 1375 in the *Principium in cursum Biblie*, the brief introduction required before the young theology student began his series of lectures on the

Bible. In d'Ailly's *Principium*, he seeks to answer the scriptural question "What new doctrine is this?" (Mark 1:27). In so doing, he imagines himself consulting the schools of the philosophers, the mathematicians, and the jurists before settling upon that of the theologians. In the school of the mathematicians, d'Ailly finds some charlatans "usurping the name of mathematicians," who have rendered nearly the whole school "disgraced and suspect."[4] Among their number are the *genethliaci* or interpreters of birth horoscopes. Passing over these "pseudo-mathematicians," d'Ailly moves on to the true ones, who study arithmetic, music, geometry, and "the influences of the heavens in astronomy [*astrologia*]."[5] Even among the true mathematicians, d'Ailly finds some so foolish as to "dare to attribute Christ's doctrine, life, works, and miracles to natural causes from celestial influences."[6] One such is the greatest authority on astrology for the Middle Ages, Abu Ma'shar: "that fantastic Albumasar" ("ille fantasticus Albumasar"). Another, Avicenna, d'Ailly terms simply "that Muhammadan."[7] Concerning the Magi of Scripture, d'Ailly repeats the standard explanation that a "marvelous star" ("stella mirabilis") announced Christ's birth to them, but that thenceforth no one was permitted to interpret anyone's nativity from the stars.[8] In short, d'Ailly presents a traditional theologian's view of astrology; "true" mathematicians rightly study the influences of the heavens, but those who push that study too far are abruptly dismissed. His brief discussion here gives no evidence of any real expertise in astronomy or astrology.[9]

Two works almost certainly dating from this early period do show d'Ailly's command of astronomical theory. In a set of *Questiones* on Sacrobosco's *Sphera*, he handled such topics as the precession of the equinoxes, epicycles and deferents, and solar eclipses. (As mentioned earlier, Sacrobosco's work was the basic elementary textbook of astronomy for the later Middle Ages.) In his commentary on Aristotle's *Meteorology*, d'Ailly succinctly discoursed on the nature and causes of comets. In both works, the influence of the heavens on the sublunar world is accepted as a given. The Sacrobosco commentary follows the standard division of astral science (*astrologia*) into the study of heavenly motions (*astronomia*) and the study of their effects (*astrologia*), with the present work, d'Ailly notes, being devoted to the former.[10] Even so, a description of the planets would not be complete without mention of their individual properties, nor does d'Ailly pass over the signs of the zodiac without going into their characteristics. Hence, in this commentary on Sacrobosco, we find statements typical of standard astrology primers like that of Alchabitius. For example, d'Ailly relates, "The sun by its aspects is fortunate, but is evil by its conjunction with some planets in some signs. It is masculine and diurnal, and it makes warmth and dryness."[11] As for the zodiac, d'Ailly lists standard characteristics for signs as well as for degrees within the signs:

> But again, some degrees of these signs are called lucid, others dark, others smoky. And the cause is that if the ascendant in the nativity of any child should be in a lucid

degree as well as the moon, then such child should be beautiful and full of light, if in a dark degree, less beautiful and ugly, if in a smoky degree, then he must be intermediate.[12]

Such astral influence clearly had its limit. In his Sacrobosco commentary, d'Ailly uses common sense and basic astronomy to argue against a possible example of the stars' effects on religion. He considers the case of the darkening of the sun at the Crucifixion. That eclipse, he explains, must have been "not natural, but rather miraculous."[13] A solar eclipse can occur only when the sun and moon are in conjunction and not in opposition. The Crucifixion, however, occurred during Passover, which happens at the time of the full moon. Thus the eclipse in question would have taken place when the luminaries were in opposition, which contradicts the rule, as d'Ailly notes. The sun's darkening was therefore a miraculous and not a natural occurrence.

In d'Ailly's commentary on Aristotle's *Meteorology*, he accepts as a given the causative force of comets, adding to Aristotle an explanation of why comets had such ill effects as the death of kings. Comets cause windy and dry weather, according to Aristotle. This drying up of the earth leads to the heat that is the beginning of wrath. "Wrath is the cause of quarrels; quarrels, of battles. Now battles are causes of the death of princes and other fighters."[14] This explanation of comets' effects is an excellent example of the "safest" sort of astrological speculation. Relying totally on one of the four elemental "qualities" (heat), d'Ailly sees comets as having an indirect influence on men (wrath as caused by heat). The most specific effect, the death of a prince, is subsumed under the more general effect of battles caused by wrath. These early works, then, perhaps composed while d'Ailly was lecturing in the faculty of arts at Paris, show nothing more than the view of the stars common to the Aristotelian cosmology of the times. Their astrological content is nothing remarkable; indeed it would rather be surprising if d'Ailly had denied stellar causality altogether.

D'Ailly also nods to astral causes in his treatise on the soul (*De anima*). Here, he allows the heavens only an indirect influence on human behavior. The soul, according to d'Ailly's treatise, contains various powers, any of which can be inclined to action by a *habitus*, although the actual use of the power depends on the will. Now a person can be moved to thinking about a *habitus* not solely by the will, but also by a host of other causes, such as the influence of the heavens.[15] Hence the stars are limited to causing a person to think about possibly performing a certain action. The *De anima*, written in the years 1377–81, reflects the concerns of the student in theology. (D'Ailly received the doctorate in theology in 1381.) While the heavens must be considered to have causative force, the stress in the *De anima* is on the preeminence of the will.

D'Ailly's fullest treatment of astrology in these early years came in his commentary on Boethius's *Consolation of Philosophy*. The *Consolation* contains a long discussion of Providence and freedom of the will. D'Ailly takes up this theme in his commentary, asking "whether everything that is done is disposed according to the order of fatal necessity."[16] Since fate, as d'Ailly points out, was commonly equated with the force of the stars, the article contains a thorough discussion of the heavens' effects on man. D'Ailly presents a model of the theological ambivalence about the stars. While affirming the theory behind astrological predictions, he urges caution and restraint in practicing the science.

D'Ailly denies quite simply that the stars can necessitate human behavior; "faith and all the teachings of the saints reprove this type of fate."[17] On the other hand, he affirms the existence of a "signifying" fate and an "inclining" fate, by which the stars incline men to action and signify future events.[18] Indeed, d'Ailly adds, "one should not deny that it is possible for us to have some sort of knowledge of future events from the disposition of the heavens."[19] He does not allow, and nor do "wise astrologers," that such disposition would necessitate the will.[20] Instead he explains how the stars might incline the will by acting upon the body, for the intellect depends upon the corporeal virtues. Thence an astrologer might make predictions from the disposition of the stars, "as from a remote cause."[21] Indeed, d'Ailly concludes that "to consult the science of the stars is useful and also licit for Christians."[22]

D'Ailly pulls back from this acceptance of astrology, however. After all, holy doctors had reproved the study of astrology. The reason, d'Ailly finds, is because of astrology's clientele. Those excessively devoted to the science might overlook more important issues, such as the salvation of their own souls. "The saints do not wish that astrology be studied while moral philosophy or the science which pertains to the salvation of the soul is left behind."[23] Not without reason had the philosophers, such as Cicero, reproved such curiosity as a vice.[24] Furthermore, astrologers run a far more dangerous risk. Citing John of Salisbury, d'Ailly cautions that philosophers can attribute too much to the authority of nature to the detriment of their faith. "Not on account of its teachings," then, "but rather on account of its students," do holy men condemn astrology, d'Ailly concludes.[25] In sum, then, although he accepts the premises of astrology, d'Ailly urges against excesses in its practice.

His Boethius commentary, like the other early writings on the stars, expresses the common ambivalence toward the stars, flavored by the hostility of the Parisian condemnations. As a student in the university, d'Ailly learned both the fundamentals of astronomy and the theological objections to astrology. In works such as Sacrobosco's *Sphera*, it was taken for granted that the heavens exerted an influence on earth. Yet Christian authors worried about the possible lack of free will in a cosmos where all events were determined by the stars. In his early works, d'Ailly followed both these conflicting teachings inasmuch as it was possible: he accepted the theory behind astrology, yet reproved its practice.

Works composed in the 1380s demonstrate d'Ailly's continued interest in and increasing knowledge about astrology, while retaining the overall negative tone of the writings from the earlier period. These works are distinguished not merely by d'Ailly's greater knowledge of the stars but also by a concern with astrology's relationship to prophecy. In a sermon preached for Advent in 1385, he considered whether astrology might be useful in predicting Christ's arrival for the Final Judgment. He examined astrological portents of Christ's birth in another Advent sermon. And the treatise *De falsis prophetis II* included a section discussing whether divination by the stars is licit. His reluctance in these works to concede astrology's power to predict religious change stands in marked contrast to his later views on astrology.

In *De falsis prophetis II*, d'Ailly sets out to determine how one is to distinguish true from false prophecy.[26] After discoursing on the nature of hypocrisy and the different degrees of prophecy, he devotes several pages to the question of astrology's merits. Here, he finds judicial astrology to be for the most part "false, superstitious, and useless."[27] On the other hand, throughout the treatise d'Ailly gives many indications that he accepts the notion of some celestial influence below, namely on the body. Hence he allows that a great many of the astrologers' predictions are true, since so few men resist their bodily passions.[28] Such predictions are not prophecy, however, which by definition exceeds the force of natural reason.[29] Indeed, d'Ailly carefully points out that even the astrologers acknowledge the superiority of prophecy to their art:

> In the first proposition of the *Centiloquium*, Ptolemy says that some persons, on account of a force dominant in their souls, although they have no great knowledge of astrology, have a better cognition of future events and are nearer to the truth than the astrologer. About which Haly says in his Commentary . . . [that] this way [of foretelling the future], when it is pure, is called "divine" by the philosophers.[30]

When d'Ailly discusses prophetic dreams (*somnia*), he mentions the stars as a cause whereby the soul might be disposed to receive such dreams. He even gives a specific configuration of the heavenly bodies which could aid prophecy.[31] Quoting pseudo-Haly's commentary on the *Centiloquium* again, he states, "The stars are to souls as the elements are to bodies." D'Ailly explains that just as certain elements might dominate a particular body, "so the soul acquires diverse properties, according to the lordship of the stars dominant in its infusion."[32] Seeking, apparently, to avoid imputing a direct *astral* influence on the soul, d'Ailly elaborates. The *mover* of each celestial orb makes an impression on souls, just as the stars themselves direct the bodies of men, "so, for example, that if Mercury is strong in the hour of the beginning of anyone's generation, or the hour of his nativity, . . . the mover of the orb of Mercury will make [his] soul . . . of strong memory; and the virtue of the same orb of Mercury directs his body to this end."[33] The notion of celestial "intelligences" or "movers" directing the spheres of the heavens was not uncommon, but d'Ailly's acceptance of their "impressions" on the soul here was perhaps risky for him. Such thinking was specifically condemned in 1277 by Bishop Tempier and would be censured

again in 1398.[34] It seems clear that d'Ailly was here trying to avoid having the
stars themselves affect the soul; for example, he quoted Thomas Aquinas's conclusion that a corporeal substance could have no effect on an incorporeal being
such as the soul.[35] He apparently sensed, however, that his statement about the
celestial "movers" went a little too far, for he did not repeat this reference to
their influencing the soul anywhere in his later writings on the stars.

In answering whether divination by the stars is licit, d'Ailly accepts some
celestial influence on earth while taking an overwhelmingly negative view of
astrology. He considers three issues with respect to the stars: first, the virtues and
effects of the stars; second, their meaning in foretelling the future; and third,
what parts of astronomy are licit. Under the first head, he finds that the stars do
indeed have a causative force here below; to foretell future effects, the stars must
be their causes and not merely signs.[36] In the second part, d'Ailly specifies what
sorts of future events can be known by means of the heavenly bodies. Because
the stars designate the future by being causes of future events, then it follows that
they can predict any happening of which they are a cause. D'Ailly removes two
types of events from astral causality: those effects which happen by accident and
those which proceed from human free will.[37] On the other hand, he admits that
the stars can indirectly move the will by their influence on the bodily passions.[38]
Hence, he posits the following response to his question: astrology is licit when
used to predict natural events such as the weather and illicit when used to predict
fortuitous events or to foresee the acts of men with any certainty.[39]

So far d'Ailly has merely rehearsed the views of Thomas Aquinas. He continues, however, in a far more negative vein, to investigate more fully which parts
of astrology might be licit. In short, he finds three goals of astral science: first, to
discover the motions, sizes, and positions of the heavenly bodies; second, to lead
the student to the contemplation of higher things; and third, to make predictions
about future events.[40] Praising the first two of these ends, d'Ailly goes to great
length to condemn the third. He reproduces almost in its entirety Augustine's
attack on astrology from *The City of God* and repeats a good portion of Nicole
Oresme's *Tractatus contra iudiciarios astronomos* (Treatise against judicial astronomers). These two mark the most thoroughgoing attacks on astrology of the
entire Middle Ages.[41] It might be argued that in following Aquinas, d'Ailly left
open the possibility of making some general predictions using astrology. He had,
as noted, admitted that some of the astrologers' forecasts were true. That his
opinion is, overall, negative seems apparent in his closing quotation from John
of Salisbury: "I can recall no one who remained in this error for a long time from
whom the hand of the Lord did not exact a worthy revenge."[42]

In December 1385, Pierre d'Ailly preached a sermon dealing with the various
advents of the Lord: in the flesh, in the minds of people, to his physical suffering
(the Crucifixion), and at the Last Judgment. His major theme was whether the
time of Christ's final advent was knowable, and in the course of the discussion
he had occasion to deal with astrology. This sermon shows again the curious

combination of ideas found in the treatise *De falsis prophetis II.* D'Ailly seems at once to accept the validity of astrological prediction and harshly to condemn its use.

D'Ailly speaks of "gentiles" who had some foreknowledge, "although shadowy," of Christ's birth: Hermes (Trismegistus), the Sibyl, Ovid, and "the ancient Chaldeans."[43] According to d'Ailly, the astrologer Albumasar had followed these Chaldeans in describing "a certain Virgin and that son of hers, whom many peoples call Jesus."[44] But that is not all. "Before the advent of this boy, i.e., Christ," d'Ailly adds, there had been a conjunction of Saturn and Jupiter signifying both the birth of a boy to a virgin mother and that this boy would become the greatest of prophets.[45] He offers no explanation of this astrological digression. Presumably it belongs to that category of writing which gave Christians delight in their faith by showing that even those without access to revelation agreed with some Christian teachings. Such is the impression left by d'Ailly's opening remarks. The holy fathers were not alone in expecting Christ's advent, d'Ailly states, "for hardened gentility . . . also pierced the mist now and then and dreamed the same."[46]

When d'Ailly discusses the fourth and final advent of Christ, for the Last Judgment, he takes up astrological themes again, asking whether the time of this last advent can be foreknown by us with any certainty. Replying, not surprisingly, in the negative, d'Ailly argues that the consummation of the world, just like its creation, will be a supernatural event and therefore cannot be known naturally—"neither by the reasoning of the philosophers nor by the speculations of the astrologers."[47] He goes on to mention one such astrological speculation that had been reported (and attacked) by Arnold of Villanova. This theory began with the supposition of a rate of precession of the equinoxes of one degree per 100 years. From this figure, one could conclude that in 36,000 years the fixed stars would return to their original positions. According to d'Ailly's sermon, some astrologers held that this entire revolution would be necessary "for the universal perfection of the world."[48]

As does his source, d'Ailly rebuts the notion that this "Great Year" of the fixed stars will coincide with the time of the world's end. We cannot readily assume that the present world will last 36,000 years, he cautions, "because God did not bind his power and wisdom with natural causes." In order to clarify, d'Ailly notes, "He is empowered to speed up the motion of the orbs as much as pleases him and complete this sort of revolution in the briefest time, so that he would complete the revolutions of one hundred years in one year or one-half."[49] That such a speeding of the heavens is indeed to happen near the end of the world d'Ailly affirms by reference both to Scripture and to the Sibyl.[50] D'Ailly's quotation of this passage from Arnold of Villanova's treatise is telling. In his view God is both above and bound by the laws of nature, and the Creator's power to override natural causes reminds d'Ailly's listeners of the fragility of any astrological prediction.

In another Advent sermon from the same period, d'Ailly again mentioned the astrological prefiguration of Christ's birth. In this sermon he stressed the supernatural aspects of the Nativity, noting God's ability "to form a baby boy in a virgin womb."[51] That the heavens had testified to this birth d'Ailly confirmed by reference to Albumasar and the pseudo-Ovidian *De vetula*. But in keeping with his emphasis on the unusual circumstances of the Nativity, he implied that the conjunction before Christ's birth was not an ordinary phenomenon and termed it "mirabilis quedam coniunctio."[52] Indeed, d'Ailly conflates this conjunction with the star seen by the Magi, asserting that the Magi learned of Christ's birth from both "a marvelous new star in the heavens . . . and an otherwise unheard of conjunction."[53] He asserts that both of these astral phenomena were not natural occurrences but rather supernatural signs placed in the heavens by God. Hence, although there was a conjunction before Christ, it was, he appears to state, not one deriving from the ordinary heavenly movements. If the stars foretold the rise of Christianity, it was only because God had placed unusual and extraordinary signs in the heavens.

The two sermons for Advent and the treatise *De falsis prophetis II*, likely all from the 1380s, point to d'Ailly's continuing ambivalent attitude about the stars in these middle years. There is a sense, however, in which the tension between the acceptance and rejection of astrology is heightened. On the one hand, d'Ailly goes beyond positing a general astral influence on men, for he cites some specific astrological effects: a conjunction signifying the Incarnation and a configuration of the heavens disposing to prophecy. On the other hand, his condemnations of judicial astrology seem a bit harsher than those of his earliest writings. In the Boethius commentary, as noted earlier, he called astrology "useful" and "licit," if kept within moderate bounds. This work resounds with the moderate voice of Thomas Aquinas. In the *De falsis prophetis II*, judicial astrology is "false" and "superstitious," and the dominant voices are the strident rebukes of Augustine and Nicole Oresme.

One reason for the harsher condemnation in these middle years is that d'Ailly addresses the specific question of astrology's relation to religious change, whereas in his earlier writings he had discussed the stars in a more general fashion. He wanted to be careful to exempt religion from astral causes. But it is possible also that this more critical view reflected tensions within d'Ailly's own mind. He was now clearly intrigued by Albumasar's astrological explanation of the Virgin birth, whereas he had censured "ille fantasticus Albumasar" for the same doctrine in his 1375 *Principium* on the Bible. The ambivalence in these works of the 1380s points to a d'Ailly tempted, and troubled, by that *curiositas* which the medieval theologians reproved. Increasingly attracted by the science of the stars, he seems to "protest too much."

D'Ailly's heightened interest in astrology in these years was an outgrowth of his concerns about the Great Schism. Like many of his contemporaries, he interpreted the events in the church in apocalyptic terms. Astrology held out the promise that it could predict the apocalypse. D'Ailly rejected astrology in these years, but he clearly was intrigued by its claims. Later, when he did use the stars to look at the end of the world, it was the theories he had encountered in the 1380s to which he returned.

In the final decade of his life, d'Ailly's interest in the stars became virtually an obsession, as the crisis in the church came to a head in the councils of Pisa and Constance. He produced at least sixteen works dealing directly or indirectly with astrology in the years 1410 to 1420. In these treatises d'Ailly defended the use of astrology with increasing openness and enthusiasm. His particular passion was for the doctrine of the great conjunctions, periodic alignments of Saturn and Jupiter that were held to have great significance here below. He read avidly about this theory, and we can follow his growth in expertise through his writings. Finally, these late works are united by their concern with the stars' effects on religion. D'Ailly now believed in an astral influence on religious behavior. And, in a shift from his earlier teachings, he now sought to use astrology to predict the advent of Antichrist.

D'Ailly laid down the elements of his defense of astrology in his 1410 treatise *De legibus et sectis contra superstitiosos astronomos* (On the laws and the sects, against the superstitious astrologers).[54] He followed the Thomistic explanation of the stars' effects and taught, as he had in his early writings, that the stars incline without necessitating the will. He exempted God from the heavens' rule by retaining the notion of a supernatural causality independent of the stars, at work in the case of miracles and other preternatural events. In later writings d'Ailly answered modern detractors of astrology: Henry of Langenstein (or Hesse), Nicole Oresme, and his own former pupil Jean Gerson.[55] He gave increasingly open support to astrology, as indicated even in the titles he assigned his writings. In 1410 d'Ailly's defense of astrology was presented as an attack on "superstitious astrologers." In 1414 he wrote in defense "of astrological truth." In 1419 he offered "a defensive apology of astrology." D'Ailly did not hesitate to make his views public. In 1414 he praised astrology in a letter to the pope; astrological themes also appeared in a sermon preached at the Council of Constance in 1416.[56] This sermon offers an instructive contrast with that preached in 1385, when d'Ailly dismissed the possibility of any astrological knowledge about the Last Judgment. In 1416, addressing a similar question, he urged holding a middle ground between those who believed in an astral prefiguration of Christ's birth and those who "in these mysteries of faith . . . thoroughly reject astronomical judgments."[57]

The focus of most of these later works was the astrological doctrine of the great conjunctions, those conjunctions of Saturn and Jupiter which occurred every 240 and every 960 years and were said to cause, among other things, changes in governments and in religion.[58] D'Ailly described this theory in the treatise *De legibus et sectis* of 1410 and applied the doctrine of the great conjunctions to biblical chronology and human history in 1414.[59] Not satisfied with chronicling past conjunctions and their effects, d'Ailly exploited the predictive value of this doctrine as well. In the *Concordantia astronomie cum hystorica narratione* (The concordance of astrology with historical narration), he extended his study of conjunctions into the future and made a provisional guess at the time of Antichrist's arrival.[60] He examined the same event in the light of other astrological theories in his *Elucidarium* (Elucidation) from the same fruitful year of 1414.[61] In 1418, he returned to the doctrine of the great conjunctions in a treatise relating the predictions of the book of Revelation to current events.[62] In the years after 1410, and particularly in the months leading up to the Council of Constance, d'Ailly was greatly concerned with the stars' influence on religion. He studied the great conjunctions for some insight into their role.

In the years 1414 and 1415 Pierre d'Ailly read deeply in astrological sources and developed a real expertise on the doctrine of the great conjunctions. In 1410 he had described the theory in the rudimentary form reported by Roger Bacon, mentioning in passing that Bacon's source was the Arab astrologer Albumasar.[63] By May 1414 d'Ailly had read some of Albumasar and had consulted the works of Alchabitius and Leopold of Austria for information on conjunctions.[64] In September of the same year d'Ailly revised his earlier thought on the great conjunctions, based upon a reading of John of Ashenden, Johannes de Muris, Abraham ibn Ezra (Avenesra), and Henry Bate of Malines. He altered the dates he had previously given for these conjunctions and noted the discrepancies between theory and actual celestial events.[65] The list of his sources, now quite impressive, included nearly all of the major Arab and modern writers on astrology.[66] Not yet satisfied, d'Ailly soon penned another revision of his table of conjunctions, this version giving both their times and their locations down to the minute (one-sixtieth of a degree).[67] Having now situated the dates and times of the great conjunctions to the best of his ability, d'Ailly apparently turned to questions of interpretation. In January 1415 he tried to sort out the numerous and conflicting theories relating various signs of the zodiac to different regions of the earth.[68] D'Ailly's expertise on the great conjunctions was well known into the sixteenth century. And in 1465, when the celebrated astrologer Regiomontanus gave Jacob von Speier a reading list on the great conjunctions, d'Ailly's name appeared next to those of Albumasar and Messahalla.[69]

D'Ailly's mature writings on the stars, finally, continue to evince a persistent concern with the stars' effects on religion. In the *De legibus et sectis* in 1410, he presented two conflicting views about the astrological origins of religions. As a compromise solution, d'Ailly suggested that religions were under the sway of the heavens inasmuch as they were natural. Hence a supernatural doctrine, such as Christianity, would be exempt from astral control. Even within this treatise, d'Ailly showed some signs of broadening the stars' jurisdiction, for he allowed stellar influence on the human aspects of Christ's life.[70] After 1410, d'Ailly continued his examination of these themes. He urged the use of astrology as an aid to interpreting prophecy in 1414 in the *Concordantia astronomie cum theologia* (The concordance of astrology with theology, also called the *Vigintiloquium*).[71] He reaffirmed the likelihood of astral influence on the births even of Mary and of Christ.[72] And he used astrological theories, along with the book of Revelation and other apocalyptic texts, to approximate the arrival of Antichrist.[73]

D'Ailly's writings testify to the changes and developments in his thinking on the stars. His earliest works offer little more than the prevailing attitude of his Parisian masters; the later writings would make astrological speculation a natural component of theology. Even given this dramatic shift, d'Ailly held many beliefs about astrology throughout his life without change. Both the old and the young man, for example, followed Aquinas in saying that the stars incline without necessitating.[74] The traditional ambivalence about the stars left room for a broad spectrum of beliefs, from near total determinism to an almost complete denial of celestial causality. By staying within those limits, d'Ailly could even present his own writings as consistent, after a fashion.[75] Clearly, however, his ideas changed between the years of 1385 and 1410. Looking at d'Ailly's own readings and at the events of the world around him can help us to pinpoint the reasons behind this shift.

THE SOURCES FOR D'AILLY'S THOUGHT ON THE STARS

As noted earlier, much of d'Ailly's thinking on astrology remained unchanged throughout his career—for example, his consistent use of Thomas Aquinas's explanation of the stars' effects on human behavior. One feature, however, clearly distinguishes his early hostile writings from the later favorable ones, and that feature is d'Ailly's enthusiastic embrace of the doctrine of the great conjunctions. The story of d'Ailly's acceptance of astrology is, in effect, the history of his knowledge about that theory.

D'Ailly first spelled out the theory of the great conjunctions in the treatise *De legibus et sectis* of 1410. He lifted much of that work from Roger Bacon's *Opus maius*, one of the major sources for d'Ailly's *Imago mundi* from the same year.[76] Although d'Ailly, like Bacon, cited Albumasar as a source on the great conjunctions, it seems clear that d'Ailly had not yet read the Arab astrologer's work, for he slavishly followed Bacon's exposition. Indeed, we have no positive evidence of his having read Albumasar until his *Concordantia astronomie cum hystorica narratione* of 1414. There he cites a theory found in Albumasar and not in Bacon.[77] By this time d'Ailly had already produced two works dealing with conjunctions of Saturn and Jupiter. Clearly, then, his source for the doctrine of the great conjunctions was Roger Bacon's *Opus maius*.

When did d'Ailly read the *Opus maius*? Bacon's name is conspicuously absent from d'Ailly's early writings on astrology, as, indeed, it is from the *De legibus et sectis,* where he refers to Bacon as "a certain English doctor" ("quidam doctor anglicus"). In 1378 d'Ailly composed his *Epistola ad novos Hebraeos* (Letter to the new Hebrews) to contest Roger Bacon's attack on Jerome's translation of the Bible. In that work he acknowledges that he has not read Bacon's writings himself; nor does he name his adversary.[78] In a companion piece written more than a decade later, around 1390, d'Ailly clearly has read Bacon and acknowledges that there is some truth in what he says.[79] Max Lieberman has speculated that this contact with Bacon's critical thought led d'Ailly to peruse his other works, in particular the *Opus maius,* which inspired much of d'Ailly's later writings.[80] He seems to have encountered the *Opus maius* even earlier than 1390, however. Recall that in two sermons for Advent he presented astrological interpretations of Christ's birth. Although d'Ailly did not cite Roger Bacon, the *Opus maius* is the most likely source for at least some of that information.

In his 1385 Advent sermon (and also in his other undated Advent sermon), d'Ailly mentioned that Albumasar, following the teachings of the "ancient Chaldeans," had described a virgin and her son, "whom many peoples call Jesus."[81] The "virgin" that Albumasar described was a particular constellation which rose with the first "face," that is, the first ten degrees of the zodiacal sign of Virgo. Because Albumasar described a figure of a virgin and child called "Iesu," medieval Christian readers took this text to describe Mary and the infant Jesus.[82] This supposedly astrological description of the Incarnation was one of the most celebrated passages on astrology in the Middle Ages, and d'Ailly could have taken it from several sources besides Albumasar himself. For example, he could have found it quoted in Albertus Magnus's *Speculum astronomiae,* or he could have read a paraphrase in Bacon's *Opus maius* or in the pseudo-Ovidian *De vetula.*[83] Bacon appears to be his most likely source, however, for the references d'Ailly makes to both Albumasar and the "ancient Chaldeans" appear together only in the *Opus maius.*

What follows in d'Ailly's sermon was apparently also drawn from Roger Bacon. Before Christ's birth, d'Ailly relates, "by about six years and a few days and hours, there was a greatest conjunction of Saturn and Jupiter near the beginning of Aries, over which Mercury, lord of Virgo, presided, all of which signified clearly a boy to be born of a virgin, who would be the greatest of prophets."[84] This conjunction appears nowhere in Albumasar or in Albertus Magnus, although mention is made of one such in the *De vetula.*[85] Roger Bacon quoted this passage from the pseudo-Ovidian work and added, "If we revolve the motions of Saturn and Jupiter to that time, we find that they were conjoined according to their mean motions six years, five days, and three hours before Christ's nativity . . . in Aries, ten degrees, fifty-six minutes, and fifty-two seconds."[86] D'Ailly's conjunction appears to correspond to that described in the *Opus maius.* Furthermore, in Bacon d'Ailly would have found that Mercury was dominant over Virgo and that a religion under that planet's dominance would be one of a prophet born to a virgin.[87] Thus in d'Ailly's 1385 sermon for

Advent, we find a clear borrowing from Roger Bacon and his earliest reference to a conjunction. Significantly, this sermon was the writing in which, as noted previously, d'Ailly seemed particularly intrigued by astrology.

This early acquaintance with Bacon's *Opus maius* makes the question of the date of d'Ailly's *De falsis prophetis II* especially interesting. In that treatise, most likely a work of the 1380s, d'Ailly presented an extremely negative view of astrology. It shows no evidence of d'Ailly's having read Bacon's views on the stars. Indeed, the bulk of the astrological material in the treatise comes directly from the writings of Thomas Aquinas, Augustine, and Nicole Oresme. If the treatise *De falsis prophetis II* does, in fact, antedate d'Ailly's 1385 sermon, one might argue that the Advent sermon reveals a softening of d'Ailly's position on astrology, a change attributable perhaps to his reading of Roger Bacon. Unfortunately, there is simply not enough evidence to substantiate this surmise.

One could just as plausibly argue for a later date of composition. In 1389, d'Ailly's friend Philippe de Mézières penned the *Songe du vieil pelerin*, an inventory of the troubles of his own age.[88] Mézières's allegory incorporated large parts of Nicole Oresme's *Livre des divinacions*, including much of the same material found in d'Ailly's *De falsis prophetis II*, where it came from the Latin version of Oresme's work (the *Tractatus contra astrologos*). The two friends were accustomed to sharing ideas. D'Ailly had written his *Epistola ad novos Hebraeos* at Mézières's request. It is not improbable that the two men would have read Oresme and followed his teachings on astrology at about the same time.[89] If so, the treatise *De falsis prophetis II* might represent a reaction against d'Ailly's 1385 encounter with Roger Bacon. Whatever the case, d'Ailly was unquestionably occupied by the relationship between astrology and religious changes in the first decade of the Great Schism.

D'Ailly returned to the *Opus maius* in the years around 1410. After the disastrous Council of Pisa in 1409, d'Ailly retired to his bishopric of Cambrai for a period of reading and composition.[90] A number of works composed in this period (from July 1409 to the summer of 1411) bear the distinct stamp of the *Opus maius*. In August 1410 d'Ailly completed the *Imago mundi*, a geographical work drawing heavily on Bacon.[91] Other cosmographical works followed, as well as a treatise on calendar reform also greatly indebted to the *Opus maius*.[92] Thus steeped in the English Franciscan's ideas, d'Ailly presented a tentative defense of Bacon's doctrine of the great conjunctions in the *De legibus et sectis*, completed December 24, 1410. Whatever the extent of d'Ailly's acquaintance with Bacon's work in 1385, when he followed the Englishman in describing the conjunction before the Nativity, his later interest in astrology seems to be directly linked to his scrutiny of the *Opus maius* in 1410.

To a great extent, d'Ailly's astrological writings from 1410 until his death in 1420 represent a development of Bacon's themes. In the *De legibus et sectis*, he presented much of Bacon's work and accepted the theory of the great conjunctions as basically true. In 1414 d'Ailly followed Bacon's theory to correlate conjunctions with biblical and profane history in the *Vigintiloquium* and in the *Concordantia astronomie cum hystorica narratione*. In the same year d'Ailly traveled throughout Germany as papal legate. He came across sources that showed him just how simplistic Bacon's version of this doctrine had been.[93] He issued an *Elucidarium* of his earlier works. True to his original source, however, d'Ailly seemed to urge greater accuracy in dates and not a revision of the theory itself.[94] His last writing on the great conjunctions, from 1418, returned to Bacon's (and Albumasar's) version, as presented in the *Concordantia astronomie cum hystorica narratione*. D'Ailly had now read most of the major authorities on the great conjunctions, and he incorporated many of their writings into his treatises of 1414 and 1415.[95] He clearly was unable, however, to distance himself from the English doctor's teachings.

D'Ailly's writings, then, reveal the *Opus maius* to have been the major influence in his changing thought on the stars. Yet they also illustrate his methods as a scholar. In the 1380s, d'Ailly was drawn to Bacon's ideas as part of a program of reading on the apocalypse, as illustrated in his 1385 Advent sermon. In 1410, he scrutinized the *Opus maius* itself, with the result being the *Imago mundi* and the *De legibus et sectis*. Intrigued by the theory of the great conjunctions, d'Ailly read Bacon's source (Albumasar's *De magnis coniunctionibus*) and other astrological textbooks, such as the works of Leopold of Austria and Alchabitius. Next, he read more technical works of astrology, such as those of John of Ashenden and Henry Bate of Malines. The fruits of this research appeared in the *Elucidarium* and the treatise *Pro declaratione decem dictarum figurarum*. And he put forth his own intricately calculated list of conjunctions in the *De figura inceptionis mundi*. Finally, in 1418, d'Ailly returned to the question of the apocalypse and offered his own astrological prediction of that event in the *De persecutionibus ecclesie*. His studies had come full circle.

THOUGHT AND ACTION: THE MAKING OF AN ASTROLOGER

It seems obvious that d'Ailly was inspired to study astrology by his reading of Bacon's thoughts on the great conjunctions. It is less apparent, however, why Bacon's views struck such a responsive chord in 1410, since d'Ailly had apparently first read them in 1385. The events surrounding his life provide some insight. D'Ailly was a public figure of enormous stature, and his political fortune rose and fell with the happenings in the French court and in the church. Not surprisingly, these events had an impact on his thought. The central preoccupation in d'Ailly's public life and in his thought became the Great Schism. Like so many of his contemporaries, d'Ailly saw in the Schism a sign of the approaching end of the world. With time, he discarded that view of the crisis, and astrology helped him to confirm his new interpretation.[96]

D'Ailly's early works echoed the prevailing attitude about the stars he had encountered at the University of Paris. As a student in theology from 1368 to 1381, he quite naturally would have followed his masters' teachings. Even as a master in the faculty of arts, astronomy rather than astrology would have been his subject, as witnessed in his writings on Sacrobosco's *Sphere* and Aristotle's *Meteorology*. At Paris, astrology was under the aegis of the medical faculty, and there was a special college of astrology and medicine newly founded in the 1360s.[97] There was a sense, too, that a theologian should concern himself with theology, and not with the stars. "We know," d'Ailly would write in 1414, "that some have objected to us that it would be more fitting to our profession and similarly to our age to be occupied with theological rather than those mathematical studies."[98] His early caution on the stars provides no surprise.

When we move to the works of the 1380s, a new theme becomes apparent. In both the Advent sermon of 1385 and the treatise *De falsis prophetis II*, d'Ailly expresses his concern that the Schism in the church might signal the beginning of the apocalypse. "Concerning this bitter sedition of the present Schism," he writes in 1385, "it is altogether to be feared that . . . it be that horrendous division and schismatic persecution after which the furious persecution of Antichrist is quickly to come."[99] In the *De falsis prophetis II*, he says bluntly that these times "seem to be near the end of the world."[100] In both these works d'Ailly examines means of predicting the future, astrology being one. His disinclination to delve any further into astrology at this point can be imputed in part to prudence and in part to his belief that prophecy was a surer guide to the apocalypse. Even if d'Ailly had desired to read more on the stars, it is hard to see where he would have found the time. In 1384 he was named head of the College of Navarre; in 1389 he became chancellor of the university and almoner to the king. D'Ailly had his own career to look after, and in the troubled years that followed, it took all the political skill and acumen he could muster.

As noted previously, the political situation in France was unstable and volatile. In 1380 the French king Charles V had died, leaving his minor son, Charles VI, to succeed him. Charles's minority and subsequent insanity left France under the domination of his warring uncles. Further, the election of the Aragonese Benedict XIII to succeed Clement VII in 1394 provoked disagreement in the French court over continued support of the Avignonese papacy. D'Ailly, sent by the king to greet the new pope, received a string of honors from Benedict culminating with his appointment as bishop of Cambrai in 1397. These papal favors caused d'Ailly trouble at home, however.

By the time d'Ailly resigned his post as chancellor of the University of Paris in 1395, he was perceived as being closer to Benedict than to the king; his position at the factionalized court was slipping. His acceptance of the bishopric of Cambrai two years later proved to be a political nightmare. The diocese was divided in its allegiance between the two papacies, and the bishop of this French-speaking province was also a prince of the Holy Roman Empire. Furthermore, d'Ailly's appointment had enraged Philippe le Hardi, duke of Burgundy and count of Flanders, who had wanted one of his own cronies installed at Cambrai. The cathedral chapter was hostile; there were designs on d'Ailly's life. It was more than a year before d'Ailly was able to take effective control of the bishopric.[101]

By 1410, when d'Ailly retreated to Cambrai and reread Roger Bacon, the situation was grim indeed. The Schism had now endured more than thirty years. The duke of Orléans, d'Ailly's last supporter at court, had been assassinated in 1407. Following the assassination, there was a roundup of partisans of Benedict in Paris.[102] D'Ailly's only ally in the city was now his former pupil and successor to the post of university chancellor, Jean Gerson. The general council which d'Ailly himself had recommended to end the Great Schism had met in Pisa the previous year with disastrous results. There were now three popes. Perhaps as a distraction, d'Ailly undertook the writing of the *Imago mundi*, following the *Opus maius*. Clearly intrigued by what he read about the stars, he pursued his reading of Bacon in the *De legibus et sectis*. Astrology's ability to predict religious change played into d'Ailly's intense curiosity about the End. He sought to know whether the church would survive the Schism or if the crisis indeed would mark the advent of Antichrist. According to d'Ailly's 1410 interpretation of Bacon, all religions save Christianity and Judaism were under the control of the stars, including the "diabolical sect" of Antichrist.[103] One thus might be able to use astrology to predict his advent.

In the years between 1410 and 1414, d'Ailly showed himself increasingly to be disturbed by the events of his times. In 1412 he felt compelled to write an *Apologia* in answer to Boniface Ferrer's charges that he had displayed covetousness, overweening ambition, and greed in his handling of the Schism.[104] During the same period he devoted himself to spiritual writings, a number of which he produced in the period before the Council of Constance. Max Lieberman has suggested that d'Ailly's turn to spirituality was a direct reaction to his troubled times. But, as Lieberman also argues, the spiritual peace that should follow upon mystical studies was distinctly foreign to d'Ailly's nature, and, accordingly, he returned to astrology.[105]

In 1414, d'Ailly again had cause to hope, and be worried, about the future. On the urging of the German emperor Sigismund, the Pisan pontiff at last had summoned another general council to meet in Constance in November. That year saw a flurry of astrological writings from d'Ailly's pen, as he sought to determine what the church's future would be: the *Vigintiloquium*, the *Concordantia astronomie cum hystorica narratione*, the *Elucidarium*, the two *Apologetice defensiones astronomice veritatis*, and probably the *De figura inceptionis mundi*. Following Bacon's doctrines, he could at last gain some measure of hope: Antichrist would not arrive until the distant year of 1789.[106]

In 1418 d'Ailly retired to Avignon as papal legate for the new Pope Martin V. The Council of Constance had come successfully to a close. The Schism was over, and d'Ailly had played a leading role in the restoration of peace to the church. He turned to the book of Revelation, seeking, apparently, to fit the events of his own lifetime into John's vision of the final things. Gone, now, was the fear that the End was imminent; gone, any sense of urgency or anxiety. Antichrist's advent, d'Ailly wrote, "even now does not seem near."[107] Astrology appears in this treatise as a valid tool for speculating on the apocalypse. D'Ailly ignores the caveats that plagued him in 1414 and recapitulates his prediction from the *Concordantia astronomie cum hystorica narratione*: Antichrist will come in the year 1789, "if the world shall last that long."[108]

The *De persecutionibus ecclesie* was d'Ailly's last long treatise. After that work we have only an exchange of letters with Gerson, the subject being again astrology. He confidently urged his former pupil not to take an excessively harsh stand against the science. Just as he would call astrologers "superstitious" who grant too much power to the stars, d'Ailly wrote, so those who depress astrology's power "against philosophical reasoning" are "superstitious theologians."[109] D'Ailly wrote Gerson in November 1419. He died the following August.

At the end of his life, then, we find d'Ailly occupied with the question that intrigued him throughout his career. In his youthful days at Paris he followed the university's teaching on the stars, finding in the writings of Thomas Aquinas the seeds for his later defense of astrology. He became expert in astronomy, perhaps lecturing on Sacrobosco's *Sphere* and Aristotle's *Meteorology*. Following the start of the Great Schism, d'Ailly's interest in the stars took a new turn, as he pondered whether astrology might be used to predict the End, which now seemed so imminent. Finally, in the months preceding the Council of Constance, d'Ailly consulted the stars to conjecture about the church's future. In the end, his calculations gave him hope, as he found a distant date for the apocalypse using the rational speculations of the astrologers.

CHAPTER FOUR

Astrology and the Narration
of History

IN THE FINAL decade of his life, Pierre d'Ailly devoted considerable attention to astrological theories relating changes on earth to conjunctions of Saturn and Jupiter. In a series of treatises, he attempted to demonstrate and to clarify the effects of these conjunctions throughout history. Armed with an increasingly sophisticated knowledge of the stars, d'Ailly constructed a thoroughgoing astrological history that encompassed the course of events from Creation to Last Judgment. He encountered many inaccuracies and inconsistencies in his star-based narration, yet remained committed to his system, believing always that more knowledge would resolve the conflicts. For d'Ailly, history proved the truth of astrology, and astrology revealed the broad patterns of history. A person equipped with a thorough knowledge of both could use the stars to predict the future course of events.

Broadly speaking, d'Ailly's forays into history writing had three interrelated goals. First, he sought to establish astrology's place as a science useful to the fields of chronology and history. Second, he hoped that the evidence of his astrological history would confirm the use of the stars to examine the future and the present. Third, d'Ailly attempted to combine his astrological narrative with traditional Christian ways of understanding history. This last was perhaps his most difficult task, for the Christian and astrological views of time were in apparent conflict. D'Ailly blended these two types of history by employing two astrological theories: one insisting upon periodic changes dictated by the conjunctions of Saturn and Jupiter, and the other deriving from the order of the planets the information that there would be only six major religions in the course of the world's history. By combining these two doctrines, d'Ailly was able to extend his astrological studies to the end of time in the Christian schema, that is, to the investigation of the apocalypse.

ASTROLOGY, THEOLOGY, AND HISTORY

In the years 1414 and 1415, d'Ailly composed a series of treatises investigating the stars' role in history. In these works he pursued the supposed correlation between conjunctions of Saturn and Jupiter and events on earth. He began this task by following the theory of the great conjunctions that he had encountered in Roger Bacon's *Opus maius*. Bacon, in turn, had borrowed from the ninth-century Arab astrologer Albumasar (Abu Ma'shar), so that d'Ailly's teaching on the great conjunctions was much the same as that in Albumasar's *De magnis coniunctionibus*, the major source on conjunctions in the Middle Ages.[1] As did his sources, d'Ailly focused on those conjunctions of Saturn and Jupiter occurring at intervals of roughly 240 and 960 years, that is, on the greater conjunction, involving the shift of the series of conjunctions to a new triplicity, and the greatest conjunction, marking the return to the first of the sign of Aries.

In a related theory, also important to d'Ailly's conception of history, Albumasar and his followers taught that six principal religions (called laws or sects) had existed or would come to exist in this world: Judaism (signified by the conjunction of Jupiter and Saturn), pagan religion (Jupiter and Mars), the worship of stars and idols (Jupiter and the sun), the Saracen religion (Jupiter and Venus), Christianity (Jupiter and Mercury), and a lying sect (Jupiter and the moon).[2] D'Ailly followed the Christianized version of this theory, as outlined by Roger Bacon in the *Opus maius*.[3] According to Bacon's redaction, Jupiter is associated with matters of faith because it rules over the ninth of the twelve mundane houses, signifying religion. Therefore, conjunctions of the other planets with Jupiter have great significance for sects and religions. Since six planets can be joined with Jupiter, it follows that the world will see six major religions, each with the characteristics of one of the six planets. Thus, Bacon states, the conjunction of Jupiter with Saturn signifies Judaism, older than and prior to all other religions, just as Saturn is the farthest planet. The conjunction with Mars relates to the Chaldean law; that with the sun, to the Egyptians' religion. Jupiter and Venus together represent the sect of the Saracens, while Christianity constitutes a Mercurial religion. Finally, the conjunction of Jupiter and the moon signifies a sect characterized by magic and lying. This is the sect of Antichrist, and it will be the last of all sects, just as the moon's is the last of the celestial spheres. Finally, Bacon argued that greater conjunctions of Saturn and Jupiter had foretold the rise of these religions. It is important to note that the six conjunctions that signify the religions are not taken to be causative or even to refer to actual planetary conjunctions. No one conjunction of Jupiter and Mars, for example, caused the appearance of the Chaldean religion. Rather, Jupiter's coupling with the other planets indicates what religions are to appear on earth; individual Saturn–Jupiter conjunctions show when these religions will arise.

Throughout his series of historical treatises, d'Ailly's plan was deceptively simple, to relate events on earth to the changes foretold by the greatest and greater conjunctions. In the 1414 *Concordantia astronomie cum theologia* (Concordance of astrology with theology, also known as the *Vigintiloquium*), d'Ailly attempted to use this method to settle a vexing theological and historical question: how many years were there between Creation and the birth of Christ?

The question of the world's age had a history stretching back to the early Christian centuries. Drawing upon a Jewish tradition linking the six days of

Creation to six millennia of world history, authors such as Julius Africanus (ca. 160–ca. 240) had sought to compose a chronology extending from Creation through the Last Judgment. Crucial to the endeavors of such Christian chronographers was the historical information contained in the Old Testament, but herein also lay a major obstacle. There was a sharp disagreement between the Hebrew Bible and the Greek translation known as the Septuagint regarding the number of years from Creation to the birth of Christ, with the Septuagint enumerating about 5,500 years and the Hebrew version yielding somewhere around 4,000.[4]

This discrepancy survived into the Middle Ages. The standard solution for the early medieval centuries was that outlined in Eusebius's *Canons*, a work designed to synchronize the various religious and secular chronologies of the ancient world. Eusebius (269–339) followed the Septuagint's dating and placed Creation at 5,228 years before the beginning of Christ's ministry, that is, 5,199 years before the birth of Christ.[5] Eusebius's *Canons* were translated into Latin by Jerome in the fourth century, yet through his translation of the Vulgate Jerome passed on the dating of the Hebrew Bible as well. Isidore of Seville popularized the *Canons'* Creation date, whereas Bede, in the eighth century, put forth a new figure based on the Hebrew chronology preserved in the Vulgate. His new figure of 3,952 years was also frequently copied and posed a challenge to the Isidoran number.[6] There the uncertainty remained with tradition supporting both sides. In the thirteenth century, Vincent of Beauvais in his *Speculum historiale* followed the smaller number, yet cited both figures, noting that any certitude in this matter was scarcely possible.[7]

D'Ailly's treatise represented an attempt to settle this question by adducing an additional sort of information: the times of the greatest conjunctions of Saturn and Jupiter. It was a project, as he wrote, suggested to him by a friend skilled in both astrology and theology. The theory was that one might use a list of Saturn-Jupiter conjunctions to verify or correct the chronologies of the theologians. After all, such conjunctions were supposed to announce major changes in kingdoms and religions. Presumably, the dates of important conjunctions could be expected to precede the various eras of world history. The evidence provided by the stars, then, might help historians choose between sets of dates. In fact, d'Ailly implied, such disagreement would never have happened if chroniclers had only bothered to consider the great conjunctions.[8]

Before applying this theory to biblical history, d'Ailly felt compelled to offer a defense of astrology and to show how the science might be useful to theologians. He began cautiously. Astrology must be purged of superstitious errors, he argued, such as the attribution of fatal necessity to stellar causality and the assertion of the stars' superiority over free will and divine power.[9] D'Ailly admitted, however, that the stars could incline men to certain behavior without necessitating their effects.[10] Having laid down this conservative framework, he proceeded with more abandon. For example, he asserted that Augustine's flat denial of astrological necessity would meet with the approval of (pseudo) Ptolemy in his statement that "the wise man rules the stars."[11] D'Ailly was quoting from Augustine's scathing attack on astrology in *The City of God*. It takes a certain aplomb to imply that Ptolemy, the great astrologer, and Augustine, the science's great critic, were essentially in agreement about the stars.

D'Ailly insisted that astrology was useful to theology, and he recommended that it be "adapted" to certain types of prophecies.[12] For example, he argued, astrology can illuminate predictions of abundant or scarce harvests, or even of acts proceeding from human free will. Naturally, in both cases the astrologers can offer only conjectures, for neither the weather nor human actions can be forecast with certainty. But to judge from d'Ailly's examples, the latitude to be given to astrological conjecture is very wide indeed.

According to d'Ailly, astrology had relevance to theological prophecies about the Flood and the Nativity of Christ. Although Noah had advance knowledge of the Flood by prophetic revelation, d'Ailly states, "it seems probable that some astronomical constellation could have presignified that effect and somehow could have been a partial cause of it."[13] Later, d'Ailly cites a greatest conjunction of Saturn and Jupiter, said to have occurred two years before the Flood.[14] Similarly, he mentions a conjunction six years before Christ's birth said to signify the future "Mercurial" law (Christianity). "Although the Incarnation and Nativity of the blessed Christ were in many ways miraculous and supernatural," d'Ailly argues, "nonetheless, Nature, just as a handmaid to her Lord and Creator, was able to cooperate in these events and, through the virtue of the heavens and the stars, to work together with the natural virtue of his virgin mother."[15] Lest he appear to give too much power to the stars, d'Ailly notes that God could have augmented the good influences and suspended any evil stemming from Christ's natal horoscope.[16] Despite his caveats, however, we are left with the strong impression that astrology could have predicted both the Flood and the birth of Christ.

Following his introductory remarks about the stars, d'Ailly abruptly turned to the question of the age of the world and left astrology to the side. In the *Vigintiloquium*'s prologue, d'Ailly had cited three conflicting figures for the number of years from Creation to the Nativity: 6,000, 5,228 (correctly attributed to Eusebius), and 5,199 (wrongly ascribed to Bede).[17] Seeking a solution, d'Ailly first noted that there had been varying opinions about the length of the year.[18] Yet even if the conflicting figures were all reduced to years of 365¼ days, d'Ailly concluded, they would still disagree.[19] Further inquiry into authors such as Eusebius, Isidore, Orosius, Bede, and Vincent of Beauvais presented d'Ailly with even more conflicting estimates of the world's age.[20] Most authors, he noted, seemed to follow the translators of the Septuagint in estimating somewhere around 5,200 years between Adam and Christ.

D'Ailly suggested that astrological considerations might yield a more exact figure for the age of the world. For known history after the Flood, one need only consult the Alfonsine tables. From the tables d'Ailly took the information that there had been 3,101 years, 10 months, 12 days, and 30 minutes of a day (i.e., one-half day) from the Flood to the birth of Christ.[21] This dating of the Flood to 3102 B.C. corresponds to the beginning of the Kaliyuga, our current era, in Hindu astronomy. This date entered the West via Sassanian Persia. It is the Flood date given, for example, by Albumasar, in the astronomical tables of al-Khwarizmi, and in the Alfonsine tables.[22]

In d'Ailly's mind, the Indo-Alfonsine figure, based on calculations of planetary motions, is more accurate than any of the church sources, and he accepts it to be the truest estimate of the number of years between the Flood and the Nativity. To know the number of years before the Flood, d'Ailly suggests an examination of the great conjunctions. His first instinct is to calculate backward from the time of a known conjunction, but he realizes that this method brings him up against the same problem. Being unsure of the number of years before the Flood, he does not know how far back to reckon or which conjunction to call the first from the beginning of time.[23] Yet he believes it expedient to calculate the times of the greatest conjunctions of Saturn and Jupiter because great historical changes will be found around these times.[24] In other words, establishing a list of conjunctions by working backward could help verify an existing chronology but could not establish a Creation date.

Having rejected what he calls the a posteriori method of figuring the world's age by conjunctions, d'Ailly proceeds to an a priori determination. He acknowledges that he can offer only a conjecture, for his calculations require drawing up a horoscope with the location of all of the planets at the moment of their creation on the fourth day.[25] D'Ailly was by no means the first to interest himself in the horoscope of Creation (or *thema mundi*). The subject comes up not infrequently in astrological literature—and in art—though with no great uniformity in solutions.[26] It was an astrological commonplace that the degree of the ecliptic known as a planet's exaltation corresponded to the position in which the planet was created. With such reasoning Julius Firmicus Maternus in the fourth century had sketched out a *thema mundi*, which was taken up in Macrobius's widely read *Commentary on the Dream of Scipio*. Firmicus Maternus and those following in his footsteps assigned Aries to the prominent position of the midheaven (*medium celum*) in the world's horoscope and had the sun in the sign of Leo.[27] Christian tradition, however, dictated that the world was created when the sun was at or near the vernal equinox (Aries 0°), so that the Creation, the Annunciation, and the Crucifixion might all fall (in theory) on the same date. Hence, the sun would have to be in Aries, and not in Leo as Firmicus Maternus would have it.

D'Ailly's horoscope, accordingly, differs substantially from Firmicus Maternus's. Drawing, as he claims, on Albumasar, he presents a horoscope that also has Aries in the midheaven, but with drastically different planetary positions. The planets, he says, are in their exaltations.[28] The positions correspond more closely, though not exactly, to the exaltations given by Alchabitius than to those of Firmicus Maternus. His endeavor has also led d'Ailly to consider such vexing theological and astronomical questions as where in the sky the sun was created and whether the moon was created full, new, or crescent.[29] (D'Ailly's *thema mundi* appears in figure 4.)

Based upon this horoscope, d'Ailly follows "the opinion of certain astrologers" in enumerating a series of Saturn-Jupiter conjunctions, finding the world's *first greatest conjunction to have occurred in the 320th year of the world*.[30] It is not clear who d'Ailly's unnamed sources are or if these same "astrologers" are responsible for his Creation horoscope as well. Whatever the case, their expertise was far from reliable. Given the planets' original positions at Creation, in the year 320 Saturn would have been in Libra and Jupiter in Leo, hardly in conjunction.[31] D'Ailly had at least appropriately noted that he would not be proceeding upon "accurate and precise calculations."[32]

Following the lead of his unnamed sources, d'Ailly continued to list greatest conjunctions of Saturn and Jupiter, noting that the third of these would have occurred in the 2,240th year of the world. "Arab astronomers," d'Ailly interjects, had said that there was a conjunction of Saturn and Jupiter in Aries 2 years before the Flood. D'Ailly equates this with his third greatest conjunction. Hence he concludes that there must have been 2,242 years before the Flood, a number that, he is pleased to note, agrees with the Septuagint and church teaching. Finally, adding the Alfonsine tables' 3,101 years from the Flood to Christ, d'Ailly obtains a total of 5,343 years from Adam to Christ.[33]

He concludes with the appropriately humble offering that his treatise provides no definite answers but only a beginning. He suggests, however, that the church, with the authority of a general council or the pope, "commit several skilled astrologers and theologians to this task," so as to make a definite determination of the world's age. Thereby would be removed what he terms the "scandal" of the disagreement of various church sources.[34] D'Ailly's use of the term "scandal" is probably an overstatement. We have no evidence that the question of the world's age caused the church any great embarrassment. It was, however, a concern of both astrologers and a number of mathematically minded theologians of the later Middle Ages. Perhaps they were drawn to this question by the writings of Arabic astrologers such as Albumasar and Messahalla. Besides d'Ailly's *Vigintiloquium*, there were astronomical attempts to calculate the world's age by John of Ashenden, Walter of Odington, and Jean de Bruges, all of them citing Arabic examples.[35]

On May 10, 1414, d'Ailly completed the second treatise of his series, the *Concordantia astronomie cum hystorica narratione* (The concordance of astrology with historical narration). Having established the world's age in the *Vigintiloquium*, he now chronicled all of human history, from Creation to Last Judgment, pausing here and there to relate great events to the heavenly movements. In particular, he looked at two astrological phenomena. For most of history up until the birth of Christ, d'Ailly relied upon the greatest conjunctions of Saturn

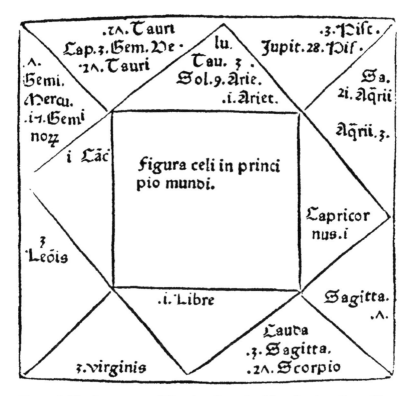

Figure 4. The horoscope of Creation from the *Vigintiloquium*. From Pierre d'Ailly, *Tractatus de imagine mundi et varia ejusdem auctoris et Joannis Gersonis opuscula* (Louvain: Johann de Westphalia, ca. 1483), fol. [bb5r]. The figure is slightly in error. Jupiter should be at Sagittarius 15° and Mars at Scorpio 28°. Cf. Vienna, Österreichische Nationalbibliothek, MS 5138, fol. 48r–v.

and Jupiter, whose times he had established in the *Vigintiloquium*. But the long period between these conjunctions (960 years) rendered them less useful for explaining recent history, and d'Ailly therefore turned to a second guiding principle. Albumasar, again, had taught that changes in mores, sects, and kingdoms were linked to the completion of ten revolutions by Saturn. Since Saturn took roughly 30 years to circle the zodiac, every 300 years one could expect great upheavals in religions or dynasties.[36]

D'Ailly's narrative is dry and often tedious, and he records many events for which he gives no astrological explanation. The astrological portions of the work are of interest, however, for d'Ailly displays an extremely liberal sense of chronology and causality in his attempts to correlate stellar with earthly events.[37] For example, the first greatest conjunction of Saturn and Jupiter is said to have happened in the 320th year of the world. However, d'Ailly relates that conjunction backward to the murder of Abel (which he places in the year 100) and to the evils introduced to the world by Cain.[38] For the second greatest conjunction, in the year 1280, d'Ailly finds no corresponding changes save the birth of Methuselah, "a little before" that conjunction.[39] Of the first three conjunctions, only the third (in the year 2240) seems to have borne a real relation to the events in d'Ailly's narrative. That conjunction preceded the Flood by two years and was its partial cause.[40]

With the fourth greatest conjunction in 3200, d'Ailly returns to his rather free sense of correspondences, and notes that it was *preceded* by the discovery of magic arts by Zoroaster[41] and the birth of Abraham.[42] That conjunction also followed changes in the kingdoms of Assyria, Egypt, and Greece.[43] The effects of this conjunction were apparently felt entirely before its occurrence. Moving farther, d'Ailly's scheme becomes even looser, for he finds not only the lives of Joseph, Moses, and the ancient Greek heroes and philosophers but also the fall of Troy to have happened *between* the fourth and fifth conjunctions.[44] Never at a loss, d'Ailly suggests an alternative astrological explanation. These events can be said to correspond to the intervening greater conjunctions and to the revolutions of Saturn.[45]

The fifth greatest conjunction is consigned to the year 4160, more than 100 years after the fall of Troy. D'Ailly tells of many changes in kingdoms both before and after this conjunction, such as the demise of the kingdom of Athens; the reign of Saul, first king of the Hebrews; and the founding of Carthage.[46] Rome's foundation occurred rather inconveniently in the year 4434,[47] and d'Ailly offers no astrological explanation for this event, the Babylonian Captivity of the Jews, or the reign of Alexander the Great.[48] Indeed, he neglects the stars entirely until he comes to the sixth greatest conjunction in the year 5120, approximately 225 years before Christ. Close before and after this conjunction, d'Ailly finds "great and marvelous changes," chiefly the translation of the Septuagint, the First Punic War (264–241 B.C.), and other Roman wars in the East.[49] Thence the narrative becomes quite dense, and d'Ailly carries his history up to the reign of Vespasian without again mentioning astrology.

Having brought his history this far, d'Ailly recapitulates briefly, then introduces the theory of the ten revolutions of Saturn. He finds this astrological doctrine a convenient one, for, according to Albumasar, the effects of the tenth revolution could be delayed or made earlier by a proximate conjunction or by the planet dominating the affected kingdom.[50] Hence, following Albumasar, he is able to relate this astrological phenomenon to Alexander the Great, Jesus, Mani, and Muhammad.[51] D'Ailly is careful to note that the revolutions of Saturn did not cause Christianity, but he does attribute to their influence the accession of Augustus and the persecution of Christians under Nero.[52]

D'Ailly uses the ten revolutions of Saturn as an organizing principle for the remainder of his chronicle, noting events at approximately 300-year intervals. Once again, his sense of chronology is rather loose, though justified now by

Albumasar's explanation of Saturn's effects. The ten revolutions that ended in 1189, for example, had significance over the pontificate of Innocent III (1198–1216), the Crusaders' capture of Constantinople in 1204, the Fourth Lateran Council (1215), the reign of Frederick I (1152–90), the capture of Edessa in 1144, the murder of Thomas Becket in 1170, the birth of the mendicant orders, and the rise of the Tartars.

Finally, d'Ailly follows his astrological guideposts beyond the events of his own lifetime and into the future. With the eighth greatest conjunction (1693 A.D.) and the completion of ten revolutions of Saturn in 1789, he predicts great changes in laws and sects, probably the advent of Antichrist.[53]

After wider reading in astrological sources, d'Ailly put forth an *Elucidarium* (Elucidation), completed September 24, 1414, in which he corrects the *Vigintiloquium* and the *Concordantia* in several places. In particular it was a reading of the fourteenth-century astrologer John of Ashenden's *Summa* of astrology that chastened d'Ailly about his earlier treatises, and he incorporated large sections of the Englishman's work into the *Elucidarium*.[54] The treatise has a rather disjointed feel, as d'Ailly jumps from topic to topic, but the work's overall design is clear. In the first half, d'Ailly retracts statements from the earlier works, essentially demolishing the astrological framework he erected in those treatises. In the second half, he sets up a new and remarkably similar system for correlating the stars and world events.[55] One has the distinct impression that, faced with difficulties in his first foray into astrology, d'Ailly rushed headlong into another equally simplistic schema.

In the *Elucidarium*, d'Ailly has to make a number of corrections to the list of eight greatest conjunctions he established in the *Vigintiloquium*. First, following John of Ashenden, he finds that the majority of astrologers agree that the Flood was signified by a conjunction of Saturn and Jupiter in the middle of Cancer 279 years before the deluge's onset. Thus, he notes, it would be astronomically impossible for there also to have been a greatest conjunction of Saturn and Jupiter 2 years before the Flood, as he had argued in the *Vigintiloquium*.[56] Accordingly, he has to abandon his original astrological calculation of the age of the world. Since the dates for the greatest conjunctions established in the first treatise were wrong, d'Ailly is forced to alter them based on a known conjunction, occurring 5 years before Christ's birth.[57] (See table 1.) In other words, he abandons the attempt to derive his conjunctions from the positions of the planets at Creation, in effect admitting defeat in his attempt to offer an astrologically based Creation date.

But this is not the only problem with d'Ailly's list. Precisely speaking, as he acknowledges, these conjunctions do not occur every 960 years, as he had originally calculated, but rather every 953 years or so. That is, neither Saturn nor Jupiter completes its course in exactly the 12 or the 30 years implied by the 960-year period. However, d'Ailly considers 960 years a close enough approximation for his purposes.[58] Given the example of the *Concordantia*, it is easy to see

TABLE 1

Greatest Conjunctions of Saturn and Jupiter from d'Ailly's
Major Historical Works

Conjunctions in the Vigintiloquium *and the* Concordantia *(Creation at 5343 B.C.)*	Conjunctions in the Elucidarium *(Creation at 5343 B.C.)*
1. A.M. 320 = radix for d'Ailly's calculations	1. A.M. 538
2. A.M. 1280	2. A.M. 1498
3. A.M. 2240	3. A.M. 2458
4. A.M. 3200	4. A.M. 3418
5. A.M. 4160	5. A.M. 4378
6. A.M. 5120 = ca. 225 B.C.	6. A.M. 5338 = radix for d'Ailly's calculations, 5 years and 320 days before the Nativity
7. A.M. 6080 = A.D. 735	7. A.D. 955
8. A.M. 7040 = A.D. 1695	8. A.D. 1915

Note: A.M. = anno mundi.

why. Unable to come to a neat answer, d'Ailly sees divine purpose behind this confusion: "Perhaps God wished this imprecision so that he might hide the secrets of the heavens from the unworthy, lest [these secrets] become contempt-ible to [the unworthy] on account of too great familiarity."[59]

D'Ailly had to face a far more serious obstacle to his system. The schema of great, greater, and greatest conjunctions recurring at regular intervals was based on the planets' approximate and not true longitudes. Ancient and medieval as-tronomers distinguished between a planet's true position and its "mean *motus*" (*medius motus*), an imaginary point that moved about the ecliptic with a uniform angular velocity. A planet's true position represented its actual location on its epicycle, measured by a line drawn through the centers of the earth and the planet itself. The *medius motus*, on the other hand, is also a point on the ecliptic lying on a line drawn from the earth's center, but that line must be parallel to the line running from the equant point to the center of the epicycle.[60] To determine a true planetary longitude using, say, the Alfonsine tables, one began by calculat-ing the planet's *medius motus* by reference to some epoch position (*radix*). This was the simplest step in a rather lengthy calculation.[61] A conjunction according to "mean *motus*" (a "mean conjunction"), then, represented the moment when the two planets had the same *medius motus*, not the same true longitude. The corresponding "true" conjunction might happen earlier or later, and in a differ-ent part of the zodiac, than the mean conjunction.

Of all d'Ailly's major sources, he admits, only Albumasar gave consideration to mean conjunctions; the rest examined true conjunctions only. D'Ailly was unwilling to dismiss Albumasar, both because of his great authority and because, as he explains, his results were close to those of the other authors.[62] Probably also, d'Ailly did not wish to abandon a schema that was both simple and regular. To calculate a series of mean conjunctions one had only to begin with one at a known epoch (*radix*) and perform a simple sum based upon the known time and increase in longitude between successive conjunctions. To calculate a true conjunction required a long, iterative process of estimation, calculation, and sometimes recalculation of both planets' positions.[63] To work from mean conjunctions was immeasurably easier; it may be that d'Ailly himself was unwilling or unable to calculate the dates and locations of true conjunctions.[64] Hence, he accepted both systems, which forced him to consider whether any given conjunction was true or mean if he wished to calculate the times of other conjunctions from it.[65] This solution gave d'Ailly twice as many conjunctions to relate to earthly changes, for he readily gave heed to the effects of both mean and true conjunctions.[66] All in all such a confusion worked to his advantage. With so many variables to consider, an incorrect astrological prediction was more likely the fault of the astrologer and not due to the falsity of the science.[67]

Finally, d'Ailly had to acknowledge a third weakness in the *Vigintiloquium* and the *Concordantia*. The teaching about the ten revolutions of Saturn was highly questionable. D'Ailly had examined such doubts in the *Concordantia* but had rather blithely dismissed them.[68] Now d'Ailly returned to these criticisms, once again finding Albumasar to be almost alone in judging from the revolutions of Saturn. The major objection to this doctrine is that it is not clear where to begin calculating ten revolutions. Even if astrologers begin from "some great Saturnine effect," d'Ailly admits, that base is weak. Hence, he says, astrologers commonly, and he now as well, prefer to use what they call the revolutions of the *magnus orbis*.[69]

The doctrine of the *magnus orbis*, as d'Ailly expounds it, is both simple to grasp and easy to apply to astrological judgments.[70] According to this schema, a new *magnus orbis* begins every 360 years, with each *orbis* under the domination of one planet and one sign of the zodiac. This period also has its origins in Hindu astrology, filtered, like the doctrine of the great conjunctions, through the authors of Sassanian Persia.[71] According to d'Ailly, Leopold of Austria, Albumasar, and others had related these *orbes* to great changes in the world.[72] The appeal of this doctrine to d'Ailly is obvious. It offered the same type of schema he believed that he had found with the greatest conjunctions and the revolutions of Saturn, with none of the complications. A new *magnus orbis* began every 360 years regardless of the planetary motions. There was no need to consult astrological tables to compute this happening, and the arithmetic involved was mercifully easy. Further, prediction was simplified by the association of each *orbis* with a planet and a sign. The changes to occur during that *orbis* would be in keeping with the characteristics of the ruling planet and sign. The only major problem was when to begin the cycle of the *magni orbes*. Even here there was a ready answer. Astrologers commonly agreed, d'Ailly notes, that with the conjunction signifying the Flood there began a new *orbis* under Saturn and Cancer.[73] Hence, d'Ailly had an epoch for his calculations.

The theory of the *magni orbes*, then, forms one of the cornerstones of the alternative system that d'Ailly erects in the second half of the *Elucidarium*. He also relies on the greatest conjunctions of Saturn and Jupiter, now redated from the *Vigintiloquium*,[74] and their greater conjunctions, those involving a change to a new triplicity.[75] Using the admittedly inaccurate approximations of 960 and 240 years between successive (mean) greatest and greater conjunctions, he establishes tables of all such conjunctions from Creation well into the future. (The last conjunction computed is for the year A.D. 3072.)[76] Likewise, he lists all of the *magni orbes* up to the year 1659.[77] Finally, he includes a curious section based, as d'Ailly says, rather on "imaginary conjecture" than on "certain reasoning."[78] According to the proponents of this opinion, the entire duration of the world can be determined by drawing a parallel between the four ages of man; the four cardinal signs and their four dominating planets; and the four figures of Adam, Moses, Jesus, and Antichrist. There will be as many years from the Nativity to the end of the world as there were from Adam to Christ.

D'Ailly leaves it largely up to the reader to connect his new system with earthly events. Presumably, the reader is to follow the lead of Abraham ibn Ezra (Avenesra), whose relation of greater conjunctions and terrestial changes d'Ailly paraphrases at length.[79] There remain a few problems, however. Avenesra's greater conjunctions are twenty years off from those given by d'Ailly in his new numbering.[80] Further, d'Ailly relies here on mean conjunctions, an approach he has already questioned. Indeed, among the examples he gives relating greater conjunctions to earthly events, the most notable effects follow true conjunctions, not mean ones.[81] Finally, even though he calculates his greater conjunctions according to mean motions, few of them are in fact the first in a triplicity. Thus they are not, strictly speaking, greater conjunctions, as d'Ailly himself acknowledges.[82] How this fact affects their significance d'Ailly does not offer.

What, then, is the overall tone of this work? It seems as if d'Ailly might be heading for a dismissal of astrology.[83] He has practically repudiated the *Vigintiloquium*, finding his estimate of the world's age, although plausible, to have no sound basis in astrology. And the foundations of his *Concordantia*, the theories of the greatest conjunctions and the revolutions of Saturn, have undergone serious revisions. In the place of these treatises d'Ailly has sketched a system that avoids their major pitfalls but which has problems of its own.

That the work does not express a profound pessimism about astrology is indicated as much by d'Ailly's statements therein as by his own later writings. He viewed the *Elucidarium* as a correction and complement to the first two treatises. Its purpose was to supplement and not to supplant.[84] D'Ailly was also explicit about the *Elucidarium*'s implications for the conclusions of the *Concordantia*. They could be "saved," d'Ailly said, by considering true conjunctions, the *magni orbes*, and the greater (mean) conjunctions.[85] In other words, if one read deeply enough in the book of the heavens, one would indeed find these events inscribed therein. As d'Ailly would argue in his *Apologetica defensio astronomice veritatis* (Apologetic defense of astrological truth) composed later that year, these obstacles proved the difficulty of the knowledge rather than the impossibility of the science.[86] D'Ailly's faith in astrology, then, remained unshaken, and the *Elucidarium* represented an earnest attempt to find a true and accurate astrological theory.

D'Ailly carried on this quest in three additional treatises composed in the latter months of 1414 and in January 1415.[87] In *Pro declaratione decem dictarum figurarum* (To explain the ten said horoscopes), he described a number of conjunctions of Saturn and Jupiter, drawn from the *Elucidarium*. D'Ailly added little beyond the text of the *Elucidarium* in this listing of significant conjunctions, but he did note yet another new astrological theory, apparently drawn from Albumasar. If a given conjunction of Saturn and Jupiter had significance for a particular religion or people, every time the two planets were again joined in the original sign—that is, every sixty years—there would be some change to that religion. For an example, d'Ailly cited the conjunction signifying the birth of Muhammad. That conjunction had taken place in Scorpio in the year 571; sixty years later a conjunction of Saturn and Jupiter in the same sign signified the prophet's death.[88]

D'Ailly produced yet another catalogue of conjunctions in a treatise also dating from the final months of 1414, *De figura inceptionis mundi et coniunctionibus mediis sequentibus* (On the horoscope of creation and the following mean conjunctions). This work contains d'Ailly's most detailed calculation of Saturn-Jupiter conjunctions, specifying times to the second and locations down to the fraction of a degree. The dates and positions of these conjunctions are quite different from those in the *Elucidarium*, however. D'Ailly begins his treatise with yet another horoscope and date for Creation. The new Creation date is 5,328 years and 243 days before the birth of Christ, obtained by adding Albumasar's estimate of 2,226 years to the Flood to the 3,102 years between the Flood and Christ found in the Alfonsine tables. The 243 additional days are to insure that the sun will enter Aries at the moment of Creation.[89] This difference, however, does not explain the discrepancy from the conjunctions in the *Elucidarium*, for d'Ailly did not derive his new list from the planetary positions at the world's beginning. Rather, he calculated them from an epoch conjunction—in fact, the same epoch conjunction he had used in the *Elucidarium*, the conjunction some 5 years and 320 days before the Nativity.

What explains the changes from the old list? D'Ailly is working with two new values. He has now set the time between any two Saturn-Jupiter conjunctions at 7,253 days and 14 hours (ca. 19.85 years).[90] D'Ailly was also using a new and larger figure for the gain in longitude between successive conjunctions, for the time required for the pattern to return to the initial point is now closer to 800 years than to the usual 960. The "gain" specified in Arabic sources of 242° 25' pointed to a figure of 960 years between greatest conjunctions. The "gain" given in the Alfonsine tables was 242° 58', and it yielded a period of 800 years between greatest conjunctions. D'Ailly apparently based his calculations in the *De figura inceptionis mundi* on the Alfonsine "gain."[91] These new values mark an implicit criticism of d'Ailly's earlier work on the great conjunctions. He passes over this point in silence, yet he must have hoped that excessive accuracy in his calculations now could make up for the inaccuracies in the theory of the great conjunctions as he had applied it so far.

D'Ailly also noted three important true conjunctions: one in 841 B.C. in Sagittarius, one in 36 B.C. in Leo, and one in A.D. 748 in Sagittarius. The interesting feature of these conjunctions was that they signified effects that would happen centuries later. The reason behind the delay was that "slower" (or less frequent) conjunctions, according to d'Ailly, produce their effects more slowly, "first some preambles like flowers, and later other complete [effects] like fruits."[92] Hence the conjunction of 841 B.C., while marking the commencement of the kingdom of the Romans, had special significance about the rise of Christianity. The conjunction of 36 B.C., meanwhile, pointed to the birth of Muhammad. And the conjunction that had occurred in the year 748 signified the appearance of a new prophet and new religion to appear around the year 1600.[93] With the extremely long time given for a conjunction's effects to begin, an astrological prediction was given almost a limitless chance to come true.

In January 1415, d'Ailly composed yet another treatise on the stars' effects on earth, the *De concordia discordantium astronomorum* (On the concordance of discordant astrologers). He devoted this work to an investigation of the relationship between the four triplicities and various regions on earth, a topic on which astrologers were nowhere in agreement. During the course of the discussion, d'Ailly brought forth yet another astrological program of history. The four triplicities ruled the earth in turns, according to the pattern of Saturn-Jupiter conjunctions. Hence, d'Ailly, following the astrologer Henry Bate of Malines, noted that the dominant kingdom during each of these periods was one signified by the reigning triplicity. For example, Troy was destroyed under the earthy triplicity. That triplicity points to the strength of a western region, and Troy fell to western invaders from Greece.[94] It was an extremely loose theory, yet it provided d'Ailly with further evidence that history flowed along the patterns written in the stars. By 1418, d'Ailly had convinced himself of the validity of the theory of the great conjunctions. He returned to the greatest conjunctions of his original dating and the revolutions of Saturn to predict, as he did in the *Concordantia*, the advent of Antichrist for 1789.[95]

HISTORY AS THE PROOF
OF ASTROLOGY

Why did d'Ailly write his historical treatises? He showed no great interest in history before and after their composition. Rather, he ventured into historiography to garner evidence—evidence that would demonstrate both the truth and the usefulness of astrology. As we have seen, this evidence could be rather flimsy. But for d'Ailly, as for the other authors who sought to introduce astronomical data into biblical chronology, the amazing correspondences between astronomy and Scripture underscored the truth of them both. Astrology strengthened d'Ailly's faith and suggested that God had arranged history according to the patterns of the heavens. In the words of his 1416 sermon for All Saints' Day, God was "the supreme astrologer."[96]

By d'Ailly's own testimony, these works related more to astrology than to history. At the end of the *Elucidarium*, d'Ailly offered a small prayer of thanksgiving for the completion of his work. There he named three historical treatises as belonging to a group of eight "useful enough" treatises which "we have collected . . . from the sayings of wise men."[97] This series also included the *Imago mundi* and its epilogue, works on the calendar and the moon's cycle, and the treatise *De legibus et sectis contra superstitiosos astronomos* (On the laws and the sects, against the superstitious astrologers). Thus, d'Ailly himself saw his historical treatises as part of a larger opus moving from cosmology to astronomy to astrology.[98] Whereas in the treatise *De legibus et sectis* he provided a defense of astrology and cleansed the science of "superstitious errors," in d'Ailly's historical works he offered empirical evidence of astrology's truth and gave examples of its relevance to other disciplines.

Certainly this goal is apparent in the *Vigintiloquium*, where the apparent disharmony of the two parts of the work reveals the author's true purpose. Recall that d'Ailly opened the treatise with a discussion of astrology's application to theology but rather quickly took up the question of the world's age, pursuing that topic for most of the work. He made no overt attempt to relate the two parts, and, indeed, there is very little astrology in the latter section of the treatise. Although d'Ailly's calculation of the age of the world had an astronomical basis, his solution rests not on the *significance* of the greatest conjunctions but rather on the report of an Arab astrologer that one occurred two years before the Flood.

The two parts are presented as a coherent whole, however, and one must believe that they both work to demonstrate the concord between astrology and theology promised by the work's title. Thus, in the second section, d'Ailly offers a concrete example of the premise developed in the first: all truth must agree.[99] Accordingly, he used astrological methods and data to resolve the theological dispute over the world's age. His figure of 2,242 years before the Flood, as he points out, verifies the calculation according to the Septuagint, which the church follows.[100] That the two figures agreed suggested that the guiding hand of truth was behind both methods. In the *Vigintiloquium*, then, d'Ailly draws upon history (in the form of chronology) to offer a cautious example of astrology's usefulness. In the first part of the treatise he argued that astrology could predict major changes in the world, citing the Flood as an example. In the second part, he relied upon that conclusion to correlate a particular conjunction with that epoch-making event.

Finally, the purpose of these treatises is made manifest by the sheer number of examples d'Ailly gives connecting stellar and earthly events. The two conjunctions discussed in the *Vigintiloquium* pertained to crucial moments in Christian history: the Flood and the Incarnation. In the *Concordantia*, d'Ailly linked every greatest conjunction with some important biblical or secular happening. In the prologue he stated that the major purpose of this work was to declare more specifically what he had asserted in general terms in the *Vigintiloquium*: "That if we note the times of the greatest conjunctions and compare them with history, we will find great and marvelous changes to have happened in this world around those times."[101]

Even in the *Elucidarium*, after having to concede so many points, d'Ailly still found conjunctions to have signified many major historical events. The Flood, for example, was preceded by some 279 years by a conjunction of Saturn and Jupiter in Cancer 14°.[102] The Black Death was the result of what d'Ailly believed to have been a conjunction of the three superior planets in Aquarius 19° on March 21, 1345.[103] A conjunction of Saturn, Jupiter, and the sun in Scorpio 8° in 1365 explained the Great Schism, for the unnamed astrologers whom d'Ailly cites noted that Scorpio was an "enemy of religion."[104] Other notable conjunctions signaled the rise of Muhammad, the kingdom of the Franks, and the mendicant orders.[105] Even later, he added three important conjunctions to this list in the *De figura inceptionis mundi* and explored ways to chart their influences in the *De concordia discordantium astronomorum*.

Taken as a body, d'Ailly's historical treatises constitute a remarkable attempt to justify astrology by demonstrating its past effectiveness. D'Ailly argues his case forcefully but, at times, carelessly. First of all, there is d'Ailly's rather rough sense of chronology, particularly evident in the *Concordantia*. There, the greatest conjunctions relate as often to the events preceding them as they do to future things. Even in the more cautious *Elucidarium*, the carefully verified conjunction signifying the Flood comes a rather distant 279 years before that happening. And with all of d'Ailly's examples of world changes following closely on heavenly events, he gives few instances of accurate astrological predictions.[106] What appears clearly in these treatises is d'Ailly's willingness to overlook most obstacles to an acceptance of astrology. This trait is especially noticeable in the *Elucidarium*, where he abandons one astrological schema only to rush headlong into another. D'Ailly in these treatises is fully committed to astrology and desirous to give the science a firm empirical backing. Telling is his treatment of his sources in the *Vigintiloquium*: given a choice between following theologians or astrologers for the years from the Flood to Christ, without hesitation he chooses the figure from the Alfonsine tables over estimates by Bede, Isidore, and Vincent of Beauvais.[107]

What is perhaps most revealing of d'Ailly's true purpose is what he does not do in these treatises. His great concern is not with the rules for *interpreting* specific conjunctions. He gives no indication of having read any of the bulk of

Albumasar's *De magnis coniunctionibus*, in which the author lists rules for determining precisely what effects a given conjunction will produce. (The usual procedure was to scrutinize a chart drawn up for the moment of the sun's entry into Aries in the year of the conjunction's occurrence.) Rather, d'Ailly was at pains to establish as accurately as possible *when* such epochal celestial events would occur and that they in fact did usher in great changes on earth. Hence his enthusiasm not just for conjunction theory but also for the doctrines of Saturn's ten revolutions and the *magnus orbis*.

The urgency behind d'Ailly's demonstration of the stars' role in history reflects nothing less than his desire to find in astrology the patterns of human destiny. Through astrology, he sought to understand the crisis of his own age (the Schism) and to establish its place in the chronology of the apocalypse set out by Scripture.[108] Only by understanding precisely the rhythm of the stars could he hope to know if his own lifetime was to see the beginning of a new astrologically determined era, the world's final one, or if that change belonged to a more distant time.

THE WRITING OF AN ASTROLOGICAL HISTORY

For a would-be historian in the Middle Ages, astrology was not the only, or even the primary, means for finding a pattern to the world's history. The historian could avail himself of any of a number of such schemas. There was, for example, a long tradition of dividing all of history into seven ages, sometimes equated with the seven ages of the human life cycle and the seven days of Creation week. According to this view, the Incarnation had begun the world's sixth age, and the Last Judgment would usher in its final one. Another periodization, based on Daniel 7, described history as a succession of four great empires, with that of the Romans being the last. Yet another view categorized history in terms of great moments in the story of salvation: before the law (Adam to Moses), under the law (Moses to Christ), and under grace (after Christ). Still another division assumed history fell into twelve hours, with Christ's advent coming at the eleventh hour.[109] The drawback to all of these periodizations was that they left little room for contemporary history. Anyone writing after Christ's birth could look forward only to the end of time. There was no space to periodize modern history. The rhythm of the great conjunctions offered a way of understanding those times other than in apocalyptic terms.

In 1892 Friedrich von Bezold published an important article about astrological history in the Middle Ages.[110] He stressed the contrast between astrological and traditional sorts of periodization and argued that astrology provided the opportunity for an enlightenment in historical writing parallel, as he said, to modern positivism in historiography.[111] Bezold particularly praised the theory of the great conjunctions outlined by Albumasar. Stressing the growing prestige of astrology from the thirteenth century on, he gave a number of examples of its use in historiography, including that of Pierre d'Ailly. He noted that he had found less astrological historical writing than he expected in the fifteenth century, perhaps because of the humanists' interest in ancient models of history. But he concluded that the astrological schema retained a strong hold on the historical imagination into the sixteenth century and beyond.[112]

Bezold's essay was and remains noteworthy because he emphasized an aspect of medieval historiography long neglected (even today). Most important, he identified the capacity of astrology to offer a scientific alternative or addition to the traditional Christian view of history. Like many apologists, however, Bezold overstated his case.[113] Although many European authors cited conjunctions or astrological events, very few made the heavenly movements into an organizing principle for their historical narratives. Even fewer carried their astrological histories as far back as Creation or as far forward as the distant future. Even if one considers earlier Arabic models, the astrological history composed by Masha'allah (Messahalla) extended only into the tenth century and not to the end of time.[114] Indeed, it appears that the Middle Ages' most systematic example of astrological historical writing may have come not from the hand of a historian or a professional astrologer but rather from that of the theologian Pierre d'Ailly.

Although there were excellent Arabic models of astrological history writing, it would seem that the use of astrology in historical periodization does not occur until the fourteenth century in the Latin West. In 1303, Peter of Abano in his *Conciliator* related greatest conjunctions of Saturn and Jupiter to the appearance of certain major historical figures.[115] The first real use of conjunction theory as organizing principle appears to have come several years later in Giovanni Villani's chronicle of Florence (d. 1348). Villani looked backward from the conjunction of Jupiter and Saturn in Aquarius in March 1345. He found great events linked to other conjunctions after periods of 20, 240, and especially 960 years. He carried this astrological schema backward 960 years from 1345, finding a new epoch to have begun then with the barbarian invasions and the fall of Rome.[116] Villani did not carry his astrology as far back as the birth of Christ. Bezold suggests that this is because of the memory of Cecco d'Ascoli, burned as a heretic in Florence in 1327. Cecco had erected a horoscope of Christ's birth and had asserted that one could predict from it Jesus' great wisdom, his birth in a stable, and his death on the Cross.[117]

After Villani, the next known example of astrological historical writing is that of Pierre d'Ailly.[118] In comparison with the offerings of Peter of Abano and Giovanni Villani, the great scope of d'Ailly's model is immediately apparent. D'Ailly considered not simply the greatest conjunctions, but also the revolutions of Saturn and, in the *Elucidarium*, the greater conjunctions of Saturn and Jupiter. He did not draw back from the astrological interpretation of Christianity but rather relied on his distinction of natural and supernatural causality (and his exemption of Christianity from the former) to save him from an error such as Cecco d'Ascoli's.[119] Finally, he carried his astrological arrangement of events throughout the entire course of history, from Adam to Antichrist. The most comprehensive astrological history after d'Ailly, that of the abbot Johannes Trithemius in the early sixteenth century, extended only as far as 1525.[120]

There remains unanswered, then, a question only partly raised by Bezold. Why are there not more examples of astrological history writing in the fifteenth century, and why does the most comprehensive such history appear to be that of Pierre d'Ailly? The most probable answer has several parts. First, most chronicles were written for a purpose—for the glorification of their subjects, say, or for the moral edification of their readers. Given such goals, it is easy to see why the historian might incorporate a specific astrological event into his narrative, to illustrate the auspicious astrological beginnings of a city or a reign, for example. But to organize a history totally along astrological lines might detract from these goals. In the same way that astrology was feared to deny a man's free will, a history dictated by a succession of conjunctions might detract from the achievements of a particular group of people. Fate, rather than character, might appear to lie behind a people's accomplishments. To paint the entire scope of history as astrologically caused neither glorified the heroes nor vilified the wrongdoers.

Second, many chroniclers may not have known much astrology. While doubtless most were familiar with its principles and its practitioners, they may not have read the types of astrological sources that would inform them about the theory of the great conjunctions.[121] Salimbene of Parma, for example, could cite prognostications of Michael Scot, yet did not write an astrologically based chronicle.[122] In Italy, where astrology had the greatest hold in the schools, humanist ideals dictated a history written along classical lines.[123] Finally, of those with the knowledge necessary to compose such a history, the practicing astrologers, one must assume, would have had little interest in such a task. They were far more occupied with creating predictions and almanacs, or in the practice of medicine. Only a few men, then, could have had both the knowledge and the interest to produce a systematic astrological history.

Why d'Ailly should have come to be the foremost creator of an astrological history is really of no surprise, given his own preoccupations. His models for historical writing were theological ones: the Bible, the works of Petrus Comestor and Vincent of Beauvais. Like the Bible, the *Speculum historiale* of Vincent of Beauvais began with Creation and extended through the Last Judgment. What sort of schema could incorporate both the grand history of salvation and the pressing concerns of 1414? For d'Ailly the explanation had to include astrology. Drawn to the theory of the great conjunctions by Roger Bacon and by a desire to understand the present and future fate of Christianity, he delved into astrology in order to understand the history of his own times and into history in order to justify astrology. For Albumasar's conjunctions to have predictive value, they had to hold true for the past. What d'Ailly was doing was following a suggestion he could have read in Roger Bacon: look back and compare astrological tables with historical events; then you can forecast similar effects from similar heavenly configurations.[124] To understand his own times it was not enough for d'Ailly to know that he was living in the world's eleventh hour. He needed to know how many minutes were left. Only if astrology and theology were in harmony could the periods of the astrologers and the ages of the world come together to yield an answer.

GOD'S TIME AND THE
ASTROLOGERS' TIME

In 1960, an important study by Jacques Le Goff examined the tensions between merchants' and the church's view of time in the Middle Ages.[125] Following Le Goff's suggestive work, Tullio Gregory has argued for an essential conflict between Christian and astrological views of time.[126] Thus the biblical interpretation saw history as punctuated by a series of unique divine interventions; astrological time was divided into periods between regularly occurring celestial figures. If the heavens acted as God's agents, then astrology would eliminate God's direct intervention in history.[127] According to Gregory, the universal belief in celestial causality (sanctioned as part of Aristotelian cosmology) meant that the conflict between the two views of time was not only necessary but also bound to finish with the victory of the astrological model. Hence, Christian authors like Aquinas and Bonaventure would take up astrological themes with a great awareness of the difficulties involved. But in the end, the broad sweep of history would seem free from human free will and entirely subject to the control of the stars.[128] Gregory gives the impression that the balance tipped steadily toward the astrological view of time from the twelfth and thirteenth centuries on, even with the conflict and attacks from the Christian side.[129] Yet despite the encumbrance of man's free will, Gregory sees astrological time as, in the end, a liberating and secularizing force. Through astrological elections and interrogations, man could seize a favorable moment for some enterprise and thereby become master of his own fate.[130]

Taking up Gregory's ideas, Krzystof Pomian has recently reexamined the relationship between astrological and Christian views of time.[131] He terms the two models "chronosophies," by which he means that they integrate past, present, and future into one coherent image that explains the purpose behind men's actions.[132] The biblical view, which Pomian calls the "theology of history," was essentially the notion of history expressed by Augustine in *The City of God*. This history was linear, universal, and directed by Providence toward a certain end. According to the astrological model, Pomian claims, history was cyclical and dictated by recurring movements of the stars. Astrology rendered intelligible both local and universal events and thus constituted a "naturalistic theology of history."[133] According to Pomian, these two chronosophies are logically inconsistent and mutually incompatible. Nonetheless there were several attempts to synthesize them in the thirteenth through fifteenth centuries, the most interesting, Pomian says, being Pierre d'Ailly's. By subordinating the naturalistic to the theocentric theology of history, Pomian argues, d'Ailly came up with one grand scheme ranging from history to prophecy. But this work was a historiographical dead end, and a new chronosophy replaced astrology after the sixteenth century.[134]

Both Pomian and Gregory make the important point that the predictive science of astrology gave rise to a view of history that could challenge the traditional Christian interpretation. D'Ailly's historical treatises bespeak both the existence of this conflict and the desire to bring the two systems into closer harmony. To insist too greatly on the conflict between the two, as Pomian does, however, is misleading.[135] D'Ailly was both unable to deny the theory of celestial influences and committed to the thesis that all truth must agree. If the primary purpose of his historical treatises was to confirm astrology, they were also written in part to show the concordance of astrological and historical truth. Thus d'Ailly could offer, in both the *Vigintiloquium* and the *Elucidarium*, an astrological explanation of the longevity of the Old Testament patriarchs. Astrologers listed the minimum, median, and maximum life-spans for persons born under each planet's influence. Some of these periods were quite long, which explained the extreme ages reported by Scripture.[136]

It is misleading, also, to stress the cyclical as opposed to linear nature of astrological history. The Arabic astrologers who developed Hindu notions of a "world year" erected systems that implied both a beginning and an end for the world.[137] As Gregory himself points out, one of the early translators of Arabic astrological sources, Hermann of Carinthia, had foreseen the necessity of the end of the world in the stars.[138] Particularly for d'Ailly, as well, astrological time was both cyclical and linear. On the one hand was the rhythm of conjunctions, revolutions, and *orbes*. But on the other hand, there were the six great religions based on the conjunction of Jupiter with each of the other planets. If the cyclical array of the great conjunctions allowed d'Ailly to relate astrology to the past, the combination of these two theories enabled him to extend his scheme into the future. That there were to be only six great religions gave him confidence that there would be no sect after Jesus save that of Antichrist.[139] And since Christ's birth had been signified by a conjunction of Saturn and Jupiter, d'Ailly felt reasonably optimistic about using astrology to foresee that final sect.

The difficulty for d'Ailly lay in fitting the two systems together so as to have their major points coincide. For example, the greatest conjunctions were clearly meant to represent major historical watersheds, yet for the epochal event of Christian history, the Incarnation, there was no corresponding conjunction save one occurring 225 years before.[140] And when d'Ailly redated the greatest conjunctions in the *Elucidarium*, he had a conjunction 5 years before the birth of

Christ but now no greatest conjunction closely foretelling the Flood.[141] Further, d'Ailly was unable to mesh his astrological history with the traditional Christian periodizations. For example, his conjunctions do not neatly point to six ages of world history, although he did cite Augustine on the six ages.[142] He also had to overlook the common view of history as a series of *translationes imperii* or shifts in power to successive peoples.[143] Similarly, although d'Ailly noted the existence of four great empires in history (headed by Babylon, Carthage, Macedon, and Rome), he attempted no astrological explanation of this fact.[144] D'Ailly's inability to mesh his astrological history with the usual Christian views of time doubtless explains the appeal of the strange theory he reported in the *Elucidarium*. Combining astrology, history, and the four ages of man, d'Ailly at last found an astrological system that highlighted the major points of Christian history.

Although d'Ailly could not make the epochs of Christian and astrological time coincide neatly, his star-based history did integrate the two into a single time-line. In the *Vigintiloquium*, the *Concordantia*, and the *Elucidarium*, he dated nearly all events in terms of years from Creation, an event whose date, in turn, he obtained using astrology. And the framework of his historical narrative, the list of the greatest conjunctions of Saturn and Jupiter, derived from computations backward and forward from one given conjunction. In composing an astrological history, d'Ailly presumed that both Christian and astrological time could be measured using a single scale, that they were, in effect, in harmony. But his effort was not merely an assertion of astrology's truth; it was also an attempt to solve a genuine problem for medieval chroniclers.

To put all of history onto a single time-line was no simple task. Aside from the problems in biblical chronology mentioned earlier, a historian in the Middle Ages would have to correlate dates based on regnal years, dates from the founding of Rome (*ab urbe condita*), dates expressed in terms of the fifteen-year Indiction cycle, and dates in the more recent anno domini form, all the while encountering inaccuracies, confusion, and omissions. Speaking of the medieval advances in chronology, historian Bernard Guenée has termed Bernard Gui's fourteenth-century catalogue of all the kings of France (itself still inaccurate) a remarkable achievement.[145] No wonder some historians have assumed that the logical numbering of years B.C. and A.D. was an invention of the modern era.[146] It was not. A. D. von den Brincken has recently demonstrated that such a dating system has its roots in thirteenth-century attempts to present a workable chronology for dating historical events. The expression of ancient dates in terms of years before the Incarnation (B.C.) represented an attempt to do for biblical chronology what medieval lists of popes and emperors had done for the years after Christ, that is, to provide a convenient and simple way to correlate the various dating systems in current use.[147]

To understand fully what d'Ailly was doing when he composed his historical treatises, one must look again at his lists of Saturn-Jupiter conjunctions. Here was a time-line running from Creation into the distant future, anchored at one verifiable point by a known conjunction, and punctuated at 240- and 960-year intervals. Like the lists of pontiffs and emperors of the twelfth- and thirteenth-century chronographers, d'Ailly's conjunctions provided yet another scale against which one might measure historical dates. In his final list in the *De figura inceptionis mundi*, he located all conjunctions and events in terms of years before and after the Incarnation. As in the case of the Roman historian Varro, d'Ailly seems to have been aiming at some mixture of technical chronology and astrological history.[148] What his treatises demonstrated was that, besides the chronologies of biblical history, regnal lists, and papal successions, the data provided by the stars deserved an equal, indeed even at times an exalted, place.

D'Ailly's historical treatises lie at the heart of his astrological writings, just as the theory of the great conjunctions lies at the center of his teachings on astrology. These treatises form the· bulk of the astrological works produced in 1414–15, from which time most of d'Ailly's texts on the stars are dated. Taken as a body they reveal three basic goals: to use historical events to prove the truth of astrology, to establish astrology as a viable vehicle for historical composition, and to synthesize the astrological and Christian chronologies. That d'Ailly failed fully to realize these goals is all too apparent to modern eyes, yet in his own mind he had smoothed out the inconsistencies reasonably enough. He was willing to overlook these problems in the belief that astrology was a valid science.

As Keith Thomas has noted, a large part of the appeal of astrology in the early modern period is explained by the fact that it was impervious to an internal attack.[149] To the believer in astrology, an incorrect prediction indicated a faulty calculation or a failure to consider all the celestial influences; prognostications could also be wrong due to the unpredictable actions of man's free will or of God. Never did astrological theory itself come into doubt. Similarly, the obstacles that d'Ailly encountered in his harmonizations of astrology and history served, in his mind, to demonstrate the complexity, not the falsity, of the science. Indeed, a major attraction of astrology lay in the intricacy of the calculations that it entailed. Alexander Murray has described the reverence for mathematics in the later Middle Ages and the position of astrology at the pinnacle of the mathematical arts.[150] D'Ailly felt this sort of awe for calculations and rejoiced in the opportunity to apply his arithmetic to theological problems.

In the end, though, he composed these treatises as the foundation of a system, the "natural theology" that he dubbed astrology in the *Vigintiloquium*. D'Ailly turned to history because only by explaining the past could he hope to interpret the present and the future. If the stars underlay the events of biblical times, then their rhythm must also dictate the pattern for later years. The motivation behind d'Ailly's forays into astrological history was in the end his concern about his own and future times. He composed most of his historical treatises in 1414, the year in which he preached the opening sermon of the Council of Constance. Over and over, d'Ailly's writings express this fear: "That this is that Great Schism which should be the preamble to Antichrist's advent."[151] The astrological forecast for 1414 was grim: "the retrogradation of Jupiter in the first house in the beginning of the year signifies destruction of religion and that peace is still not established in the church."[152] Would this turmoil mark the beginning of the apocalypse? Perhaps astrology, its truth borne out by history, could provide an answer.

CHAPTER FIVE

The Great Schism and the
Coming of the Apocalypse

WITH THE OUTBREAK of the Great Schism in 1378, Pierre d'Ailly and many of his contemporaries assumed that the apocalypse was at hand. They based this dismal conclusion both on their reading of Scripture and on a long medieval tradition of speculation about the end of time. Beginning with the church fathers, theologians had elaborated upon the biblical descriptions of the apocalypse. They described the characters and careers of the major antagonists in Christianity's final battles, while artists made familiar scenes of Antichrist and the Last Judgment. Many writers of the thirteenth and early fourteenth centuries went so far as to suggest times when these events might happen and to interpret the apocalypse in political terms. By d'Ailly's lifetime, apocalyptic utterances and lamentations of the sorry state of affairs were commonplace. The existence of the Schism, however, seemed to fulfill many of the prophecies about the end of the world and gave such predictions a new sense of relevance and urgency.

The connection between the Schism and the end of time in d'Ailly's thought has not gone unnoticed. In 1980, Louis Pascoe published an article stressing the importance of apocalyptic notions to d'Ailly's view of the crisis. Spurred by the Schism, Pascoe argued, d'Ailly came to see all of history as a series of persecutions of the church. Although he proposed practical, realistic solutions to the Schism, he followed the events around him with profound fear and anxiety. D'Ailly's urgent sense that the final persecution was near, Pascoe concluded, must be seen as the intellectual and spiritual background for his ideas about the Schism.[1]

Although Pascoe quite rightly emphasized the relationship between the Great Schism and d'Ailly's apocalyptic notions, he failed to note both that his opinions about the End changed over time and the importance of astrology to this development. D'Ailly originally did view the Schism as a sure sign that Antichrist's reign was imminent. In sermons and treatises from the first decade of the division, he explored questions about the end of the world and warned his audiences to prepare for the tribulation at hand. In the years after 1400, however, he reexamined his early apocalypticism. D'Ailly came to hope that a council would end the Schism and that the apocalypse thereby would be postponed. Astrology helped to confirm his predictions.

During the Schism's early years, however, d'Ailly limited himself to learning as much as possible about the torments he believed were near. He frankly doubted that any human science, including astrology, could determine when the apocalypse would occur. Nonetheless, he read broadly in treatises discussing both the timing and the circumstances of the end of the world. The medieval study of the Last Things formed a distinct area of enquiry embracing several branches of learning. When d'Ailly wished to learn about the end of the world, he could choose from a wide range of sources: biblical texts and commentaries, non-Christian prophecies (such as those of the Sibyls), more recent visions, and astrological calculations.[2] D'Ailly was acting within established medieval tradition both when he interpreted the Schism as a preamble to Antichrist's reign and, later, when he used astrology to determine when that torment would begin.

D'Ailly immersed himself in eschatological sources in the early years of the Schism. Key biblical texts provided him with information about Antichrist and his reign and with the important fact that a schism would precede this torment. Medieval commentators had constructed elaborate "biographies" of this great villain and had linked historical events to passages in Revelation. Such treatises made it possible for d'Ailly and others to conceive of Antichrist's reign within contemporary events. (For example, a number of authors portrayed the Saracens as the great enemy of the final times.) D'Ailly also knew the works of Joachim of Fiore and more recent writers on the end of time. These authors suggested that the world was nearing its final age, sometimes specifying the time of its end. Armed with this reading and faced with the fact of the Schism, d'Ailly saw only one possible interpretation in the years before 1400: to pray and fear, for the end is at hand.

THE BIBLE ON THE END OF TIME

Scripture was by far the most important source of information about the apocalypse for d'Ailly and his contemporaries. Passages in Daniel and Revelation spelled out, albeit in enigmatic form, God's plan for the world's end. Commentaries on these two books were key vehicles for eschatological speculation in the Middle Ages. Other scattered texts provided information about Antichrist, who, although mentioned by name only twice in Scripture, was taken to be the object of numerous other references. The words of Daniel and Revelation are obviously metaphorical and invite various interpretations. Not surprisingly these texts were read and analyzed throughout the Middle Ages.

The book of Revelation presents a series of visions culminating in that of the Last Judgment and the New Jerusalem. By Pierre d'Ailly's time, the reader could draw on a number of commentators relating these visions to historical events. First, there was the vision of the book with seven seals, with new and untoward torments unleashed with the opening of each successive seal (Rev. 5–10). Next appeared a woman in travail, the dragon who sought to devour her offspring, and the beast with seven heads and ten horns (Rev. 12–13), the beast whose number, for "him that hath understanding," is 666 (Rev. 13:18). Further visions described seven last plagues, the whore of Babylon, the binding of the devil for a thousand years, and the final onslaught of evil prior to Judgment Day. Already in the eighth century Bede had devised a theory of history linking the seven seals of Revelation to seven periods in the history of the church.[3]

If Revelation provided the basic outline of the apocalypse, Daniel gave the reader a clue as to calculating its timing. In Daniel 12, Daniel asks when the End will come and is told, "from the time that the daily sacrifice shall be taken away, and the abomination that maketh desolate set up, there shall be a thousand two hundred and ninety days" (Dan. 12:11). Further, he hears, "Blessed is he that waiteth, and cometh to the thousand three hundred and five and thirty days" (Dan. 12:12). In Jerome's commentary on Daniel these figures were interpreted as relating to the time of Antichrist (1290) and the Last Judgment (1335), respectively. The rigors of interpretation demanded that the subtraction be made, leaving Jerome with the conclusion that there would be a period of 45 days between the end of Antichrist's reign and the Last Judgment.[4] The figure of 1,290 days, together with the obscure phrase "a time, times, and an half" of Daniel 12:7, yielded the information that Antichrist's reign would endure three and a half years. (The equation was thus: "time" equals one year, "times" equals two years, and "an half" equals a half a year; likewise 1,290 days make approximately three and a half years.)

Many other biblical texts dealt with eschatological themes, of which three were of particular relevance for d'Ailly. In 2 Thessalonians occurs a passage traditionally taken in the Middle Ages to refer to Antichrist. Thence came the important information that Antichrist's advent would be preceded by a schism, in the words: "for that day shall not come, except there come a falling away first [nisi venerit discessio primo], and that man of sin be revealed, the son of perdition" (2 Thess. 2:3). In the standard medieval commentary on the Bible since the twelfth century, the glossa ordinaria, the discessio of 2 Thessalonians was interpreted as signifying either the destruction of the Roman Empire or a division in the Roman church. When the Schism began in 1378, d'Ailly and many contemporaries quite naturally viewed it in terms of the second of these alternative interpretations of Paul's epistle. In d'Ailly's reading, the passage referred to "a schismatic subtraction of obedience from the Roman church or empire."[5]

Matthew 24 contained more information assimilated into the medieval view of Antichrist. Thence came the important warning that his persecution would be accompanied by false prophets, who would deceive many "insomuch that, if it were possible, they shall deceive the very elect" (Matt. 24:24). The identification of such pseudoprophets formed the subject of a pair of treatises by Pierre d'Ailly. Further, this text gave medieval exegetes the rationale for identifying minions of Antichrist among their own contemporaries.

One additional passage carried inestimable weight in medieval discussions of the End, and d'Ailly cited it on numerous occasions. This was the stern admonition in Acts 1:7 that the time of the apocalypse was beyond human ken: "It is not for you to know the times or the seasons, which the Father hath put in his own power" ("Non est vestrum nosse tempora vel momenta, quae Pater posuit in sua potestate"). Despite this warning, and scores of others from Saint Augustine on, the temptation to calculate the time of the apocalypse was irresistible. Yet the weight of Scripture was such that each would-be calculator, indeed each writer on the End, had to bow to what Robert Lerner has dubbed the "uncertainty principle" of Acts 1.[6]

THE APOCALYPSE IN THE MEDIEVAL IMAGINATION

Athough the Bible provided the bulk of the material for speculations about the end of the world, a number of medieval traditions not stemming directly from Scripture helped shape d'Ailly's view of the apocalypse as well. Among the most important and enduring of these notions was the belief that the course of the world would in some way parallel the account of Creation week in Genesis, such that each day of Creation would represent one millennium in the history of the world ("One day is with the Lord as a thousand years," according to 2 Peter 3:8). Hence, the end of the world was anticipated in its 6,000th year, with the eternal repose of the blessed representing God's rest on the seventh day. This belief, Richard Landes has argued convincingly, lay behind every attempt to construct a world chronology in the early Middle Ages, for, if one established the date of the world's beginning, then by simple arithmetic one could calculate the time of its end. In particular, Landes urges, the acceptance of the new dates for Creation set by Eusebius and, later, Bede was directly motivated by the desire to show an agitated populace that the world's 6,000th year was, in fact, not at hand.[7]

This belief survived in the later Middle Ages, albeit in somewhat changed form, dictated by the fact that the world had passed its 6,000th year (according to Eusebius's calculation) some time around the year 800.[8] Hence some later authors argue for the world's end in its 7,000th year, others for any time before its 6,500th year. The theory no doubt held a certain appeal in its simplicity; further it seemed to mesh with the seven ages of history laid down in Augustine's *City of God*.[9] As late as 1456 Felix Hemmerlin used the parallel with Creation week to predict the end of the world for 1492. Adding 5,508 years for the time from Adam to Christ to the current year of 1456, Hemmerlin reasoned that the world had only 36 years left before it reached the fatal age of 7,000.[10]

Another major source of d'Ailly's eschatological lore was an anonymous seventh-century treatise ascribed to Methodius, which purported to be a vision received by that bishop and martyr.[11] Pseudo-Methodius, drawing upon the fourth-century Tiburtine Sibyl, popularized the figure of the Last World Emperor, whose reign was to usher in a period of peace before the final onslaught

of Antichrist and his hosts, Gog and Magog. Important not merely for its characterization of this final emperor, pseudo-Methodius's *Revelationes* also meshed
political elements with apocalyptic. For pseudo-Methodius, the Saracens were
the great enemy of the last times, whom the Last World Emperor, "the king of
the Greeks or Romans," would subdue, before himself failing in the face of
Antichrist's hosts.[12] Particularly in the time of the First Crusade, the parallel
with present events seemed irresistible, and the prophecies about this emperor
abounded.[13] Through the figure of the Last World Emperor, political prophecy
became an established part of medieval eschatology.

JOACHIM OF FIORE AND OTHER
PROPHETS OF THE END

If d'Ailly found general information about the end of time in writings from the
early Middle Ages, when he turned to more recent works, he read that the
apocalypse was near and learned of actual estimates of the time of its occurrence.
Sir Richard Southern has argued that the thirteenth century witnessed an important shift in thinking about the future. Before that time, people were mainly
curious about the outcome of contemporary events. Hildegard of Bingen (d.
1179), for example, received far more questions about current happenings than
she did about the end of the world. By the mid-thirteenth century, however, the
scene had changed. People were thenceforth concerned that the end of the
world was at hand.[14]

In confirmation of Southern's thesis, there are scores of apocalyptic predictions from the later Middle Ages, many of which were known to d'Ailly. Not
simply the works of visionaries or propagandists, many of these treatises show
that the timing of the apocalypse had become a question for serious study. These
writings display the blend of disciplines that Southern has described as composing the science of the future: Scripture, non-Christian prophecies, recent visions, and astrology.[15] Some of these prognostications were clearly intended as
discourses for learned consumption; others circulated openly as prophecies, with
the dates of their predictions changed again and again with each successive disappointment.[16]

By far the central figure in later medieval prophecy, and an important source
in d'Ailly's mind as well, was the Calabrian abbot Joachim of Fiore (d. 1202).
Joachim did not view himself as a prophet, but rather as having been given a gift
for understanding Scripture.[17] Joachim's special insight was his understanding of
history as composed of patterns of twos, threes, and sevens. By one such group-
ing, he viewed all of history as composed of seven ages. His own times, Joachim
felt, were near the end of the sixth age; the seventh age would be as if a Sabbath,
a time of repose within human history after Antichrist's persecutions yet before
the Last Judgment. Joachim anticipated that within two generations (by ca.

1260), Antichrist would arrive, bringing the sixth age to an end and ushering in the seventh age of peace. Superimposed upon the pattern of seven ages was that of the three *status*: those of the Father, the Son, and the Holy Spirit. The third *status* corresponded to the seventh age; its men would live in peace and have greater understanding of things divine than those of the first two periods.[18]

In later years this notion of Joachim's was held to be dangerous. In 1254, Gerard of Borgo San Donnino, a fanatic follower of Joachim, intimated that the third *status* would involve the overthrow of the Old and New Testaments and, thereby, of all established institutions; henceforth the only authority would be certain of Joachim's writings known as the Eternal Evangel. Needless to say, Gerard was condemned, and while Joachim's writings were not censured along with him, the doctrine of the third *status* remained suspect.[19] There were, however, many other aspects of Joachim's work, and the abbot's writings had an enormous influence. Prompted in large part by spurious writings circulated by Joachim's adherents, men came to see Joachim as having prophesied the mendicant orders.[20] And the failed prediction for the year 1260 seems only to have increased expectations for the future appearance of such a crisis, for Joachim's name appears appended to apocalyptic prophecies throughout the later Middle Ages. Henceforth, Joachim's teachings would figure in any discussion of the apocalypse.

Another important source on d'Ailly's eschatological reading list was a treatise composed by the physician, religious thinker, and mystic Arnold of Villanova in 1297, *De tempore adventus Antichristi* (On the time of the advent of Antichrist).[21] Arnold made several remarkable claims, and his treatise understandably caused something of an academic storm in Paris. First, he asserted that the injunction of Acts 1:7 ("Non est vestrum nosse tempora . . .") applied only to the actual end of the world and not to the arrival of Antichrist.[22] Hence, one might make a specific prediction about Antichrist's advent, leaving in doubt only the amount of time left between that event and the Last Judgment. Second, Arnold assumed that one could apply to the prophecies of Daniel the equation of days and years implied by Ezechiel 4:6 ("I have appointed thee each day for a year." The Latin is more straightforward: "Diem pro anno . . . dedi tibi."). Counting 1,290 years "from the time when the Jewish people lost possession of their land," Arnold concluded that Antichrist would come some time around the year 1378.[23]

Arnold dismissed entirely what he called the astrologers' argument that the world must last 36,000 years, the time necessary for the complete revolution of the eighth sphere (assuming a rate of precession of one degree per 100 years). The end of the world, just like its beginning, would be a supernatural event. Even if the complete revolution of the eighth sphere were necessary, he added, God could speed up the heavens so as to accomplish this in a much shorter time.[24] In 1299, Arnold submitted his treatise to the faculty of theology at Paris, where it was condemned, but it was upheld by Boniface VIII, whom he served as physician.[25] In 1305, still engrossed by the topic and now claiming divine inspiration, Arnold proclaimed that Antichrist had been born already and was then three years old.[26]

Controversy surrounded Arnold's writings, and a number of authors produced works in reaction: John of Paris, Peter of Auvergne, Nicholas of Lyra, Guido Terrena, and Henry of Harclay.[27] At least one Parisian reply to Arnold's treatise was also well known to d'Ailly, that of John of Paris (Joannes Quidort Parisiensis). He composed his *Tractatus de antichristo* in 1300, proposing to review all that was accessible in Scripture and by human conjecture about Antichrist. John strongly contested the notion that one could have any certain knowledge about the time of Antichrist's arrival, although he believed that we might conjecture indeterminately about this event.[28] In fact, John offered his own prediction that the world would end within 200 years, based upon the notion that the world would endure only six millennia. John believed the world to be in its 6,300th year at the time of his writing; by the principle of synecdoche, however, any quantity of years up to 6,500 would stand for 6,000, and any quantity over 6,500 would represent 7,000. Hence for the world to end after its sixth millennium, by John's reasoning, it was necessary only for it not to surpass the age of 6,500, which would happen within two centuries.[29] John's argument is an excellent example of the persistence of belief in the theory about the six millennia despite the world's having surpassed its predicted end. John further believed that astrological arguments could bolster his prediction by giving "scientific proof" ("scientiffica probatio") of the world's age.[30] By positing the latitude of the sun's apogee (*aux*) at Creation and comparing it with its position in Ptolemy's day, John determined that the world had endured 5,100 years to the year A.D. 130, and thus approximately 5,000 years to Christ, precisely the figure he used for his prediction.[31]

Although John was happy to use astronomy to prove his theories, he had little tolerance for the astrological prediction cited by Arnold of Villanova. Arnold had attacked this notion of the world's enduring 36,000 years by reference to God's omnipotence, his ability to speed up the heavens at will. To this argument John added astronomical reasons as well. The theory, he noted, was based on the notion that the eighth sphere moves at the rate of one degree per 100 years, hence requiring 36,000 years to make a complete circuit. John pointed out that there were other figures for this motion than the one degree per century given by Ptolemy, so that the figure of 36,000 years must be doubted. Besides, "Thabit" had shown that rather than moving in a circular orbit, the eighth sphere oscillated back and forth with the motion called trepidation.[32] Hence even the premise of the astrologers' theory was wrong. John apparently did not see, or chose not to notice, that these arguments undermined his own astrological proof as well, for it too depended upon Ptolemy's rate of precession. As for Arnold's prediction about Antichrist, John found it wrong on all accounts. Arnold had erred, John argued, in applying the "day for a year" equation to Daniel; his interpretation did not fit the text at all. And even if Arnold's exegesis were correct, he was mistaken in his computation, which should yield the year 1366 and not 1378 for Antichrist's appearance.[33]

By the latter part of the fourteenth century apocalyptic prophecies were taken seriously by learned and ignorant alike. Many people held the outbreak of plague in 1348 to be a sure sign of the End.[34] This was a period in which men anxiously compiled anthologies of prophetic texts, such as that of Henry of Kirkestede (d. after 1381).[35] These prophetic anthologies reflected an intense desire for knowledge about an End that seemed increasingly close. So, too, did learned eschatological treatises find the apocalypse impending.[36]

Perhaps the most influential such works were those composed by Jean de Roquetaillade (John of Rupescissa). In his commentary on the *Oraculum Cyrilli* of 1345–49 he predicted that a Hohenstaufen Antichrist and an Antipope would stand in opposition to the king of France and a true pope in France around 1365. After many tribulations, the millennium (Joachim's third *status*) would begin in 1415, upon which would follow the onslaught of Gog and Magog and the Last Judgment. Similar predictions followed in Roquetaillade's famous *Vade mecum in tribulatione* (Handbook for times of tribulation, translated roughly) of 1356. In succeeding years Roquetaillade's vogue was enormous, and he was believed to have foretold correctly plague, schism, and disaster to France.[37]

THE GREAT SCHISM AS PREAMBLE
TO THE APOCALYPSE

By the time d'Ailly entered the University of Paris in 1363 or 1364, many scholars were obsessed by the study of the Last Things. As countless historians have remarked, their apocalyptic notions found a focus in and drew inspiration from the appearance of the Great Schism.[38] Its seemingly insoluble pitting of church against church appeared to embody the *discessio* of 2 Thessalonians as well as Joachimist predictions of an Antipope. Following its outbreak, there was a pervasive sense that the End was at hand. Rumor had it that Antichrist had already been born. Such, for example, was the message of a widely circulated letter, purportedly from the grandmaster of the Hospitallers on Rhodes, bearing the date 1385 in its earliest versions.[39] Similar beliefs found expression in the writings of Nicolas of Clamanges, Pierre d'Ailly's pupil and friend. "I estimate scarcely three years to be left," Nicolas wrote, "until the time of that day most greatly to be feared."[40] Apocalyptic fears were not found just among the clergy, for in his poems Eustache Deschamps (1346–1406) repeatedly discussed signs of the End.[41]

In this atmosphere of eschatological uncertainty and confusion in the church hierarchy, burgeoning numbers of people claimed to have received private prophetic revelations. Prophetic anthologies from these years reflect anxiety about the Schism's presumed link to the apocalypse.[42] Many of the prophecies that circulated were blatantly partisan, such as the visions reported by one Marie Robine beginning in 1387, which continually supported the Avignon papacy.[43] Church authorities had a variety of reactions to this outpouring of eschatological speculation. According to André Vauchez, university theologians jealously guarded their right to be the sole interpreters of Scripture, while at the same time they acknowledged the possibility of other revelations outside of Scripture, such as Marie Robine's and Joachim's. The burning issue, as attested by works by d'Ailly, Gerson, and Henry of Langenstein, thus became how to determine which of the many claimants to special revelation were in fact true prophets. Although theologians urged caution in accepting any purported revelations, such determinations could easily fall along partisan lines.[44]

Some authors used the apocalyptic interpretation of the Schism to make a political statement. One such, writing under the name of Telesphorus of Cosenza, claimed to have received inspiration in the years 1356–65 about the Schism and the Last Things. His book, which took on its final form in the years 1378–90, owed much to Roquetaillade's political eschatology. Telesphorus predicted tribulations to come, in which an evil third Frederick would oppose the French king, to be followed by a time of great peace under a French world emperor. Only then would the final Antichrist arrive, upon whose defeat a seventh age of peace would ensue.[45] Telesphorus's treatise was immensely popular, providing an interpretation of events that seemed to make sense only in an eschatological context, and holding out the promise of better times ahead. Apparently disliking its partisan tone, Henry of Langenstein composed a scathing attack on Telesphorus and other would-be prophets in 1392.[46]

The Schism itself had a profoundly unsettling effect on some, as witnessed by an anecdote reported by Henry of Langenstein. A French monk named William, a learned and holy man and friend of Henry's, began to have visions about the Schism. He believed that the Holy Spirit had informed him that the Schism was to endure only a short time more. Based on these revelations, William set a time for the Schism's end, only to be disappointed in his prediction, for the Schism still was ongoing ten years after William's prognostication. Assuming he had misinterpreted the revelation he thought he had received, William recalculated the time for the Schism's end, only to be wrong again. Whence, deeply troubled, William abandoned the monastery, and "having left his religious habit there, he wandered about in the woods next to the monastery in a vile secular tunic."[47] The Schism was such a dramatic event that it led William to have visions, the validity of which was clearly questioned by Henry. That a man who appeared to be "of great sanctity and perfect religion" ("apparens magnae sanctitatis et religionis perfectae") would come to leave his monastery shows the intense spiritual agony that the Schism and William's frustrated hopes for its end could produce.

Probably the best example of apocalyptic from the time of the Schism, and one certainly familiar to Pierre d'Ailly, is embodied in the life and works of Saint

Vincent Ferrer (1350–1419).[48] Dominican confessor and chaplain to Benedict XIII in Avignon, Vincent had a vision in 1398 of Christ standing between Saint Francis and Saint Dominic, ordering Vincent to go forth and preach. In 1399 Benedict granted Vincent license to travel and preach wherever he chose, free from the jurisdiction of the local clergy. Thus began an extraordinary twenty-year mission of preaching, which took Vincent throughout most of Europe. Vincent's sermons were enormously popular; huge crowds came to hear him preach, and a small band of flagellants attached themselves to him and accompanied him on his travels. His message was simple: the imminence of the Last Judgment and the necessity of repentance.

Vincent set out his complete view of the End in a letter to Benedict XIII written July 28, 1412.[49] No one could know with certainty the time of Antichrist's advent, Vincent wrote. One could ascertain, however, that the time between Antichrist's death and the Last Judgment would be short, that is, the forty-five day period deduced from Daniel 12.[50] Because of this time limit, Vincent rejected the notion set forth by Joachim and other authors that there would be an era of peace on earth following Antichrist's death. The only hopeful note in Vincent's message was for those stalwart souls who could survive Antichrist's persecutions and repent for their sins. Further, Vincent told Benedict that Antichrist's reign would begin very soon.[51] He based this conclusion in part on the existence of the Schism, which he equated with the *discessio* of 2 Thessalonians.

Vincent also adduced some rather disturbing pieces of evidence for the imminence of the End. Some nine years previously, when he was preaching in Lombardy, a messenger from a group of Tuscan hermits came to the preacher. These hermits had received divine revelation telling them that Antichrist had already been born. If these revelations were true, Vincent added, Antichrist would already be nine years old, and the time of his reign would indeed be close at hand.[52] In Piedmont, he had heard of another revelation, this time from a Venetian merchant. When he was overseas, the merchant related, in a Franciscan convent hearing vespers, two young novices were obviously taken up in some kind of rapture. "Together, and in a terrible fashion, they cried out, 'Today at this hour Antichrist has been born, the world's destroyer.'" Carefully questioning the merchant, Vincent determined that this vision too had occurred nine years previously.[53] Throughout his travels, he added, people would come up to him and tell him visions they had had about Antichrist and the end of the world, all agreeing with the stories he had heard previously.[54] Thus armed, Vincent felt quite comfortable telling the pope that, in all likelihood, Antichrist was already nine years old.

In his sermons Vincent frequently stressed apocalyptic themes, for example, in a sermon on Antichrist given in 1404. That Antichrist would come soon, we could be sure, he preached, by the existence of the Schism, the *discessio* of Scripture.[55] This sentiment must have been quite common, for Vincent felt little need to offer further proof of the imminence of the End. Rather, he expounded upon the details of Antichrist's career and his means of seducing his converts. Many people would be won over by gifts of money and privileges, such as letting men have several wives.[56] False miracles worked by Antichrist and his minions would deceive another group. More followers would come over to Antichrist's camp following public disputations, in which Antichrist and his demons would appear to better their Christian adversaries. A final group would convert after being tortured. Suddenly and terrifyingly, Vincent's words shift to the second person: "First, he will take all of your temporal goods away from you. *Item*, he will kill children and friends in the presence of their parents. *Item*, from hour to hour, from day to day, he will tear away from you one member after another, not all at once, but continuing over a long time."[57] Having thoroughly frightened his audience with this vivid description of their impending torments, Vincent offered the one remedy available in these times: to pray morning and night.[58']

Vincent's sermons drew extremely large crowds, and one can well imagine that he struck a responsive chord in his audience. His views typify a common reaction to contemporary events, one that saw in the Schism a preamble to Antichrist's persecution. The events of the later fourteenth century were extremely unsettling, and the Schism in particular seemed to have significance far beyond its quotidian manifestations. It was an event that appeared to fulfill many prophecies, from the Bible through Roquetaillade. To many learned observers, it was best interpreted in eschatological terms, and this view of the Schism gained currency through sermons and popular prophecies. The Great Schism began just three years before Pierre d'Ailly received his doctorate in theology. His inaugural disputations in 1381 all centered on questions of church government.[59] Thenceforth, the Schism, the way to bring about its end, and its eschatological significance became central preoccupations in his thought.

PIERRE D'AILLY'S APOCALYPTICISM
IN THE YEARS BEFORE 1400

Like many of his contemporaries, d'Ailly saw in the Great Schism a sign of the approach of the apocalypse. In his sermons and in two treatises he composed on false prophets, d'Ailly examined apocalyptic themes and warned of the imminence of Antichrist's reign. He did not simply use eschatological language for the preacher's aim of rousing his audience. Rather, d'Ailly's works show that he engaged in a program of reading about the End, a program that came to include some astrological authors. D'Ailly's early writings evince a serious and sincere apocalypticism. Indeed, we have reason to suspect that d'Ailly anticipated Antichrist's arrival in the year 1400. In later years, he would change his attitudes, doubting and arguing openly against the notion that the end of the world was at hand. Yet in his early works, we see a persistent fear that the apocalypse was impending.

Eschatological themes appear as early as 1379, in a sermon d'Ailly preached for the feast day of Saint Dominic.[60] The history of the church, he explains, can be seen as a series of wars waged by the devil against the church. One such war was in the time of the martyrs; another involved heretics, against whom Saint Dominic had fought. Now a third persecution has arisen, that of the schismatics, "against whom our Dominic is fighting in heaven."[61] D'Ailly interprets the Schism in this sermon as one of a series of persecutions that the church must suffer, a series, as Saint Augustine says, that has no definite number. He tells his audience, "Even if the schismatics are defeated or brought to concord . . . (which I do not know when it will be; God knows), the persecutions of the city of God will not thereby cease."[62]

The Schism began on September 20, 1378, with the election of Pope Clement VII. D'Ailly preached his sermon on Saint Dominic in August 1379, with the Schism less than a year old. Had d'Ailly and other members of the University of Paris begun to see the Schism as the *discessio* of 2 Thessalonians? There is no evidence in this sermon to show that d'Ailly was concerned that the apocalypse was at hand, but he was interested in questions of eschatology. Although our preacher insists on an indefinite number of persecutions for the church, for example, he makes a point of stating that they will be worse near the end of the world.[63] Citing Acts ("Non est vestrum . . ."), d'Ailly emphasizes that it is vain to attempt to calculate the time of the apocalypse. He mentions predictions cited by Augustine, that the world would end four hundred, five hundred, or a thousand years after Christ, all of which are patently false. He calls the belief that the world will endure seven thousand years (mentioned by William of Auvergne) superstitious.[64] We cannot be sure if such statements indicate apocalyptic sentiments in d'Ailly or the members of his audience. What is apparent, already in 1379, is a vague link between the Schism and eschatology in d'Ailly's thought.

In 1380 there was a great stir at the University of Paris. Mathias of Janov, a member of the Bohemian nation, composed a book about Antichrist. He also taught publicly that Antichrist had already been born and that he had seduced all of the members of the university, "so that now they taught nothing that was sound or true Catholic doctrine."[65] Whatever the effects of Mathias's unusual message, in a sermon most likely preached in the same year, Pierre d'Ailly shows an increased concern with apocalyptic matters. In his sermon on Saint Francis, d'Ailly follows Hildegard of Bingen's exposition of the apocalypse.[66] The Schism now appears in the context of the descriptions of Revelation.

According to Hildegard, d'Ailly preaches, the church is to suffer seven persecutions, foretold by the seven angels and trumpets of Revelation 8–10. Three of these persecutions have already passed, and we are now in the fourth. This is the "womanish time" ("muliebre tempus"), according to Hildegard, in which cupidity, sensual delight, and vanity have arisen, so that the church and the clerics today are worse than the Jews, heretics, and pagans of the previous three persecutions.[67] Saint Francis himself is the eagle of Revelation 8:13, who flies through the heavens crying, "Woe, woe, woe, to the inhabiters of the earth," with each "woe" symbolizing one of the three tribulations yet to come. Now Hildegard had foretold that the "womanish time" would end with a schism. Hence, d'Ailly tells his audience, we can conclude that the fifth persecution either has now begun or is soon to happen.[68] D'Ailly does not discuss the future beyond the present schism, except to mention in a general way that Hildegard expected a horrendous tribulation to follow the schism, and he warns his listeners to prepare for it.[69]

Already in 1380, then, d'Ailly had begun to see the Schism in eschatological terms. To understand the events around him, he turned to Revelation and to the prophets, Hildegard and Joachim, whom he also cited.[70] D'Ailly's exposition of Hildegard is lengthy; indeed her ideas form the inspiration of the sermon. It would seem that d'Ailly based his sermon on an actual reading of the holy woman's works and not on secondhand knowledge of her ideas. In other words, he was interested enough in apocalyptic matters to seek out and read Hildegard's writings. It is impossible to tell, however, just how imminent the apocalypse was in d'Ailly's mind. Clearly he sensed an extraordinary crisis in his own times, which he placed on the brink of the final three horrible persecutions of Revelation. But if these three trials were of the same duration as the first three, the church would have another thousand years and more left to survive.

If d'Ailly's sermon on Saint Francis expressed a vague notion that his were no ordinary times, his sermon on Saint Bernard, from roughly the same period, proposed that the Schism be seen as an immediate preamble to Antichrist.[71] D'Ailly pointed out that both Saint Bernard and Saint Gregory the Great had seen in the evils of their own lifetimes the signs of the end of the world. He cited the passages from Augustine and William of Auvergne that he had used in the sermon on Saint Dominic to prove the absurdity of attempting to calculate the time of the apocalypse. He even brought out the familiar "Non est vestrum" of Acts 1.[72] Then d'Ailly made a rather surprising shift. Although we cannot and should not define the time of the apocalypse with any certainty, d'Ailly preached, "nonetheless, from the scriptures which we read and the experiences which we see, we are able [to mark], *by plausible conjectures*, the approach of Antichrist and the nearness of the end of the world."[73]

D'Ailly gave his listeners three signs that were preambles to Antichrist: an abundance of iniquity in the world, a lessening of Christian charity, and a schismatic division of the Catholic church.[74] Since the first two of these signs might always seem to be present, he chose to expound upon the third. He reminded his audience of the text of 2 Thessalonians, where it is told that the Day of Judgment will not come unless there first be a division, and the son of perdition be revealed. D'Ailly explains the text in his own words: "before the day of the Lord, it says, there will be a certain dissension or division of the church of God, as an *immediate* preamble to Antichrist."[75] Although Jerome had interpreted this text in terms of the Roman Empire, d'Ailly says, he prefers the interpretation of the *glossa*, which sees the *discessio* as a division in the Roman church. With all probability, he adds, we must believe that the present Schism is that *discessio*, "after which Antichrist's persecution is quickly to come."[76] Bringing his sermon to an end, d'Ailly urged his audience to fear and prepare for the tribulations to come. No longer vague in his apocalypticism, d'Ailly now preached that the End was at hand.

A thorough expression of d'Ailly's apocalyptic views appeared in the sermon he preached for Advent 1385. In this sermon we see that d'Ailly has furthered his readings in eschatological sources. Also, for the first time, astrological themes emerge. Although d'Ailly's sermon purported to discuss the four advents of Christ (in the flesh, in people's minds, to mortal punishment, and at the Last Judgment), he devoted the most time to exploring Christ's final coming. Could the time of this advent be foreseen with any certainty, he asked? Although this knowledge would be extremely useful to the church, d'Ailly concluded, it was not knowable by human industry or by deduction from Scripture.[77] On the other hand, d'Ailly argued, such knowledge could be revealed, both by God's absolute power and by his ordained power.[78] This distinction is important, for there is some question as to whether d'Ailly believed God would ever act according to his absolute power.[79] That knowledge about the End (beyond that given in Scripture) might be revealed by God's ordained power, however, represents a far more optimistic hope for such revelation.

The examples with which d'Ailly bolsters his conclusions give fascinating testimony to his reading at the time. He was clearly in possession of a copy of Arnold of Villanova's treatise on Antichrist, for he uses lengthy excerpts in his sermon. Almost verbatim, he repeats Arnold's version of the astrological argument about the world's 36,000 year duration. Like his source, d'Ailly counters this theory with reference to God's ability to speed up the heavens as much as he wishes.[80] He offers Arnold's example as proof that the time of the End cannot be known by human reasoning. Arnold's name comes up again when d'Ailly preaches the futility of attempting to use Scripture to deduce any certain time for the apocalypse. Since Arnold had predicted Antichrist's advent for 1375, 1376, or 1377, d'Ailly argues, and it is now 1385, the end of the world should already have happened.[81]

D'Ailly had also been reading another source dealing with the stars, for he mentions pagans who had astrological knowledge of Christ's birth. Following Roger Bacon's *Opus maius*, d'Ailly describes a conjunction of Saturn and Jupiter that preceded Christ's birth by some six years.[82] It is not clear how far d'Ailly had read in Bacon at this time. If he had read only two pages (in the modern edition) previous to the passage he cited, d'Ailly would have seen mentioned the astrologers' sect of the moon, equated by Bacon with Antichrist. If he had gone back a few more pages, he would have encountered Bacon's assertion that by using astrology we could be forewarned of Antichrist's advent.[83] If d'Ailly was aware of Bacon's promise that astrology could foresee the coming of Antichrist, he chose to ignore it. The reason behind this choice must have been his hope that such knowledge had been or would be revealed by God, and his belief that such revelation would be more certain than human reasoning. Had he disapproved entirely of using astrology to predict Antichrist's arrival, then surely he would have ridiculed, rather than praised, the astrological prefiguration of Christ's birth in the *Opus maius*.

From the words of this sermon, d'Ailly would seem to have concluded that prophecy and current events pointed toward the impending end of the world. Witness his choice of text: "Know ye that the kingdom of God is nigh at hand" (Luke 21:31). He may well have believed that the type of prediction advocated by Bacon was simply not necessary in the face of glaring testimony. The Schism once again seemed the likely preamble to Antichrist.[84] The sins of the present appeared to fulfill the words of Matthew 24 and Luke 21, which foretold the obscuration of the heavens near the end of time. "What, indeed," he asked, "is 'There shall be signs in the sun, and in the moon, and in the stars' [Luke 21:25], if not that the luminaries of the church shall grow dim? Which certainly now, alas, we behold has happened. . . . Let us fear the future judgment of greatest terror," d'Ailly warned his audience, "for it is at hand."[85]

D'Ailly must have thought it likely that revelation, too, foretold the impending apocalypse. He told his listeners that certain knowledge about the last things could be revealed to men directly by God; "so, indeed, the blessed Cyril, Abbot Joachim, and Saint Hildegard are believed to have prophesied many things about the last times."[86] D'Ailly did not explicitly say that Cyril, Joachim, and Hildegard were prophets, only that they are "believed" to have prophesied. His statement, however, implies approbation of their writings. So, too, does his quoting of prophetic verses attributed to Joachim, verses whose truthfulness d'Ailly neither defends nor questions directly. According to this prophecy, there were only fifteen years remaining before Antichrist's reign:

> When there have been completed one thousand three hundred
> And ten times ten years after the dear Virgin's giving birth,
> Then Antichrist will reign, full of the demon.[87]

D'Ailly presents this prophecy as if to withhold judgment on its truth. "It has been reported by some," he states, "that according to a revelation made to the same Joachim, there do not remain more than fifteen years until the reign of Antichrist."[88] The prophecy itself follows, and its position within the text of the sermon leaves open the question of how seriously d'Ailly took its message. Immediately after quoting the pseudo-Joachistic verses, d'Ailly seems to counter their validity by citing Acts 1 ("Non est vestrum . . .") and similar passages from Augustine. His argument continues, however, and he explains that Jesus' words were in the present tense ("Non *est* vestrum . . ."), laying open the possibility of some *future* revelation about the time of the End.[89] Although d'Ailly does not return to the pseudo-Joachim verses, he goes on to give the signs that he senses mark the nearness of the apocalypse, ending his sermon with a warning that the kingdom of God is at hand. Given the thrust of his words, it seems likely that d'Ailly suspected Joachim's verses might be true, a sentiment he wished his audience to share. Perhaps these were the thoughts d'Ailly had in mind when in 1399, at the age of forty-nine, he gave orders to have his tomb built in the cathedral of Cambrai.[90]

Around the same time as the Advent sermon, d'Ailly composed two treatises on a subject with obvious eschatological overtones: the art of recognizing false prophets. The theme was a particularly apt one, for not only did theologians anticipate an increase in the number of false prophets as the time of Antichrist approached, but they also worried about the very real problem of the numerous claims to prophecy during the Schism.[91] In the *De falsis prophetis II* (which apparently is the prior in order of composition), d'Ailly spelled out yet another version of the persecutions of the church.[92] In this interpretation, he stresses the nearness of the End. The church is to suffer three persecutions before that of Antichrist, d'Ailly states. The first was by violence, and it was carried out in the time of the martyrs. The second persecution was by fraud and occurred in the time of the heretics. The third persecution will be by hypocrisy, brought on by the false prophets of the title, who will seduce Christians first by fraud and later by force.[93]

D'Ailly is apparently taking up a theme that he first set out in the sermon on Saint Dominic in 1379, but now he intensifies the apocalyptic message. The hypocrites of the third persecution are to be precursors or preambles to Antichrist, the "son of perdition" of 2 Thessalonians.[94] A multitude of such hypocrites, then, will be a sign of the approach of Antichrist. By fraud, pretended sanctity, and hypocrisy they will seduce the populace and divide the church, until "at last there is to be obtained a general division or schism of other churches from the Roman Church, . . . which division will be the way or the preparation for Antichrist to come, about which the Apostle prophesied in 2 Thessalonians."[95] How useful it would be in these modern times to recognize such false prophets, d'Ailly continues, for even now we are experiencing such a Schism, "about which it likely is to be feared that it be a preamble to Antichrist."[96] He goes on to discuss various types of divination with the aim of identifying the features of true prophecy. The treatise ends with d'Ailly denouncing any claims of astrology to be such.[97] He appears unwilling to take up Roger Bacon's suggestions, as yet.

In the second treatise of the series, *De falsis prophetis I*, d'Ailly continued his investigation into false prophets and their deeds with an eye, it would seem, on contemporary claimants to revelation. As before, he maintained that such false prophets were the precursors of Antichrist, and he noted that Scripture and Hildegard had prophesied that their lying would eventually be recognized.[98]

They would, however, seduce many people at first, and, with that in mind, d'Ailly dwelt at length on the hypocrites' false miracles. These miracles should be considered suspect, he wrote, "especially in these times that seem to approach the end of the world."[99] There were to be many such false miracles around the time of Antichrist, while, according to Isidore, true miracles would cease.[100] Accordingly, d'Ailly warned his readers that "all those who preach or teach publicly and are not sent by God are false prophets and pseudodoctors."[101] The miracles and claims by which they sought to prove their mission must be examined diligently before being believed, lest the credulous be deceived.

In sum, in the years before 1400 Pierre d'Ailly became increasingly convinced that the apocalypse was near. He read eschatological works: sections of Augustine's *City of God*, writings by Hildegard of Bingen, the pseudo-Joachim *Super Hieremiam* and commentary on the *Oraculum cyrilli*, Saint Bernard and Saint Gregory the Great on the approaching end, Arnold of Villanova's *De antichristo*, and more. Important for his later interests, he learned of astrological attempts to calculate the time of the apocalypse, although in these years he rejected any such prognostication. It may be that he assumed such calculations were unnecessary; he may have believed the pseudo-Joachistic verses that predicted Antichrist's reign for the year 1400. Certainly he sensed that the time was ripe. In his *Oration on Matthew*, he put forth a lengthy catalogue of the sins of his age. Concluding, he told his listeners that there was little left "but for Antichrist to come!"[102]

CHAPTER SIX

Astrology and the Postponement
of the End

IN THE YEARS before 1400, Pierre d'Ailly had come increasingly to see the Great Schism as a sign of the approach of Antichrist. He warned his audiences that the End was at hand, all the while rejecting any human means of determining when the Judgment would come. In subsequent years, however, d'Ailly revised his early interpretation of events. He began to hope that human efforts, with God's assistance, could end the Schism. With the church healed and reformed, he believed, God would withdraw the torments he had prepared for the earth, and the apocalypse would be postponed. D'Ailly's later examinations of the apocalypse bear another striking difference from his early works. Whereas in the 1380s he had denied astrology's ability to predict the End, he now looked to the stars to forecast the arrival of Antichrist. Astrological calculations confirmed his new hopeful interpretation of the Schism, for they put the fiend's advent in the distant future.

D'Ailly had two excellent reasons for fighting apocalyptic beliefs in his later years. He had seen how such notions could lead to behavior that threatened the church's authority. Flagellants and other heretics had arisen during the long years of Schism, and in the church's eyes their misdeeds stemmed from apocalyptic beliefs. Further, d'Ailly had come to realize that despairing because the End was at hand could hamper any efforts to heal the Schism. If the Last Judgment were imminent, the only proper response would be to pray and repent for one's own sins. There would be little sense in trying to bring the church back to unity. If a reformation of the church could postpone the apocalypse, however, people would have a strong motivation to work together to end the Schism.

Following d'Ailly's early apocalyptic writings, there is a lacuna of several years in our knowledge of his eschatological thought. In 1385–87 he was occupied with university politics, first the affair of Jean Blanchard and then that of Juan de Monzon. He became almoner to the king and then chancellor of the university in 1389. Then in 1395–97 he was named to the bishoprics of Le Puy, Noyon, and Cambrai. Perhaps he was too busy to write much more about the end of the world, or perhaps he was awaiting the fateful year of 1400, subject of a pseudo-Joachistic apocalyptic prediction. Whatever the case, he was silent on the question of the apocalypse until after that year had passed.

Traces of d'Ailly's new interpretation of the Schism appear as early as 1403, in a treatise he composed entitled *De reformatione ecclesiae* (On the reformation of the church). In this work, d'Ailly lamented the horrendous schism in the church and cited his favorite apocalyptic authors: Joachim of Fiore and Hildegard of Bingen. Yet, in marked contrast to his early sermons and treatises, he neglected to mention the approach of Antichrist or to equate the Schism with the *discessio* of 2 Thessalonians, which was to precede the fiend's advent. Instead, he suggested that a merciful God had allowed the Schism to occur "so that . . . His church might be reformed for the better." If this improvement did not happen swiftly, d'Ailly warned, "I dare to say that although we are seeing great [torments] now, soon we will see incomparably greater ones."[1] This was his only indication that the Schism might be a preamble to the apocalypse.

ASTROLOGICAL CALCULATION AND THE POSTPONEMENT OF THE END

In the years between 1410 and 1414, d'Ailly refined his new interpretation of the Schism. He sought to confirm his revised view by a program of reading in eschatological sources and in books of astrology. In the series of astrological treatises he composed in those years, d'Ailly frequently dealt with questions touching upon the apocalypse, and he attempted his own astrological calculation of the time of Antichrist's advent. In the end, astrology and prophecy both gave him hope that the Schism would be healed and Armageddon would be postponed.

On December 24, 1410, d'Ailly completed his treatise *De legibus et sectis contra superstitiosos astronomos* (On the laws and the sects, against the superstitious astrologers). There he pitted against one another the opinions of William of Auvergne and Roger Bacon on the stars' effects on religion. Although he proposed to offer a middle ground between the two authors, d'Ailly came down much more on the side of Bacon, and his treatise offers, in essence, a defense of astrology.[2] As we have seen, Bacon had argued that the stars dictated that there would be only six major religions or sects in the world's history; further, the times of origin of these sects could be foretold using astrology. Each of these religions was associated with the conjunction of Jupiter with one of the other six planets. The sixth and final sect was to be that signified by Jupiter's conjunction with the moon, and, in Bacon's exposition, it represented Antichrist.

The eschatological implications of Bacon's theories were clearly apparent to d'Ailly. He quoted from the *Opus maius* the passage in which Bacon urged the church to include astrology in its study of the Last Things:

> I know that if the church desired to go over the sacred writing and the holy prophecies, and the prophecies of the Sibyl, and Merlin, the eagle, and Joachim, and many others, and also the histories and books of the philosophers, and if it should command that the ways of astrology be considered, there would be found a sufficient indication or, rather, certitude about the time of Antichrist.[3]

D'Ailly countered Bacon's claims with the "Non est vestrum" of Acts 1 and referred the reader back to his Advent sermon of 1385. This was the sermon in which he had followed Bacon in describing a conjunction preceding Christ's birth, yet had rejected any astrological forecasting of the apocalypse. Even if d'Ailly in 1410 disapproved of the astrological prediction of Antichrist's advent, the rationale for such a calculation was implicit in his treatise. All religions and sects that were diabolical in origin (as surely he must have considered Antichrist's), d'Ailly had written, are naturally under the control of the stars.[4]

D'Ailly's second astrological treatise, the *Concordantia astronomie cum theologia* (Concordance of astrology with theology, also known as the *Vigintiloquium*), would seem at first glance to have little to do with the end of time. In this treatise, completed early in 1414, d'Ailly sought to use astrology to determine the true age of the world. As Richard Landes has alerted us, however, this information could be vital to would-be calculators of the apocalypse.[5] D'Ailly, too, was aware of traditions giving the world only six thousand or seven thousand years until the Last Judgment. He had mentioned such notions in three of his early sermons, those for Saint Dominic and Saint Bernard, and the Advent sermon of 1385.[6] In the *Vigintiloquium*, he discussed these beliefs again, referring his reader back to the 1385 sermon.[7] While he was writing this treatise, then, d'Ailly knew that his conclusions could have eschatological importance.

For those wishing to calculate the time of the apocalypse using d'Ailly's figures, his treatise was decidedly anti-apocalyptic. According to d'Ailly, some 5,343 years had elapsed between Adam and the birth of Christ.[8] Thus, the world had reached the age of 6,000 more than seven centuries previously, and it would not approach its dreaded 7,000th year until 1657. D'Ailly did not make these subtractions in his treatise, nor do we know if he even had them in mind. But if his sentiments about the apocalypse were the same as they would be in subsequent works, d'Ailly in the *Vigintiloquium* offered a subtle argument against those who would preach the imminence of the End. Just as, according to Landes, many early medieval chronologies were intended to counter apocalyptic beliefs, so too, d'Ailly's dating could have been meant to show that the End was centuries away.

If in the *Vigintiloquium* there was a vague link between astrology and eschatology in d'Ailly's mind, in his next composition he openly offered an astrological conjecture for the timing of the apocalypse. The *Concordantia astronomie cum hystorica narratione* (The concordance of astrology with historical narration) was finished on May 10, 1414. In this work, d'Ailly sought to establish the relationship between astrological and earthly events throughout the course of history. He concluded with a section on the apocalypse. D'Ailly's treatise manifests both his suspicion that the apocalypse was not at hand and his continued reading in both astrological and eschatological sources.

In the *Concordantia* d'Ailly sought to explain much of history as the product of periodic conjunctions of Saturn and Jupiter and of each completion of ten orbits by the planet Saturn.[9] When he described the events of his own lifetime, d'Ailly did not hesitate to list the Schism as one of many evils following the triple conjunction of Saturn, Jupiter, and Mars in 1345.[10] He next situated the Schism in a list of twenty-two schisms that had befallen the church. Needless to say, the present division was the longest and worst the church had endured.[11] As in d'Ailly's 1403 *De reformatione ecclesiae*, at this point in the text one would anticipate a reference to the *discessio* of 2 Thessalonians and a prediction that the reign of Antichrist was at hand. What follows is rather surprising.

First d'Ailly laments the failure of the Council of Pisa (1409) to end the Schism and mentions that a council has been called to meet in Constance later that year. "And nevertheless," he continues, "according to some astronomers, it has been foretold from the horoscope of the present year that the retrogradation of Jupiter in the beginning of the year in the first house signifies the destruction of religion and that peace is not yet established in the church."[12] In astrological terms, when a planet's motion was retrograde (i.e., its apparent motion carried it in the reverse of its ordinary course), dire consequences followed. That the planet involved was the ordinarily benign Jupiter, signifier of religion, spelled disaster for the church. Again we might expect a dire apocalyptic prediction to follow, but d'Ailly instead offers his readers some hope. "God is the true sage who alone rules the stars," he states, in an interesting twist on the oft-quoted "The wise one rules the stars" ("Sapiens dominatur astris"). With his help, d'Ailly adds, we may be able to find some remedy to this evil. Otherwise, we must fear that the Schism is the preamble to Antichrist's advent.[13] Now, at last, he cites Hildegard, Joachim, and 2 Thessalonians.

This is a remarkable passage, and it reveals much about d'Ailly's interpretation of the state of events in 1414. He is clearly so anxious about the Schism that he has consulted the astrological forecast for the year and can recite its conclusions in some detail. On the other hand, the call for a new council in Constance appears to have given d'Ailly hope, for surely this must be the remedy with God's aid of which he writes. D'Ailly must have concluded that the success or failure of the Council of Constance would have a decisive impact on the course of history. Should the Schism come to an end, mankind will survive. Should the council fail, we must reasonably expect the imminent reign of Antichrist.

If d'Ailly sensed that the world's fate hinged upon the actions of the Council of Constance, in the concluding chapters of the *Concordantia*, he implied his belief that its outcome would be successful. Having introduced the topic of Antichrist's reign, d'Ailly went on to examine this question by means of astrology, following the theories outlined by Roger Bacon. D'Ailly noted a number of astrological phenomena pointing to the importance of the year 1789. A greatest conjunction of Saturn and Jupiter, of the sort that happens once every 960 years, would occur in the year 1692. After this conjunction, another astrologically significant period would end with the completion of ten trips through the zodiac by the planet Saturn in 1789. Further, from the years 1764 to 1789, the eighth sphere would "stand still."[14] Based on these observations, d'Ailly concluded, "if the world shall last until that time, which only God knows, then there will be many great and marvelous alterations and changes in the world, and chiefly with respect to laws and sects [i.e., religions]."[15] Whence, he continued, we can conclude with due probability that around that time will arrive Antichrist and "his damnable law or sect."[16]

Unlike so many of d'Ailly's astrological writings, this passage does not appear to be a borrowing from another author. Rather, the cardinal himself seems to have derived the date of 1789 from his own list of greatest conjunctions and revolutions of Saturn. His choice of date rested on sound theory. He incorporated in his prediction the three most important signifiers of religious change according to Albumasar (Abu Ma'shar): the greatest conjunction, occurring once every 960 years; Saturn's completion of ten revolutions, every 300 years; and the period of the eighth sphere's *accessus* and *recessus*. D'Ailly must have been struck by the coincidence of the last two of these three celestial phenomena in 1789. Albumasar had taught that great changes were likely when the eighth sphere's completion of its access or recess went along with a shift in signs for Saturn.[17]

These calculations offered, albeit conditionally, a balm to those troubled by apocalyptic anxieties. For those who might miss his point, d'Ailly himself made the necessary subtraction: "it is apparent that from this year of 1414, up to the standing still of the eighth sphere [1764], there will be 253 [should read 350] complete years."[18] Astrologically speaking, the apocalypse was not at hand. Yet d'Ailly left his prediction conditional upon God's will. Although the stars pointed to Antichrist's arrival in 1789, d'Ailly was careful to note that whether the world would endure that long was a matter for God's knowledge alone. And, one might add, it was a question whose answer appeared to hinge on the outcome of the Council of Constance.

Not surprisingly, d'Ailly added further eschatological reading to his astrological investigation of the End in the *Concordantia*. In the treatise's final chapters he gave lengthy excerpts from pseudo-Methodius's *Revelationes*. In d'Ailly's schematic presentation, pseudo-Methodius could be said to have listed eight preambles to the advent of Antichrist. It would seem that only the first three of these preambles could have happened by the time d'Ailly was writing, although he specifically relates only the first of these to contemporary events. The first of pseudo-Methodius's preambles was to be the *discessio* of 2 Thessalonians, "on which we have already briefly touched."[19] The second and third preambles listed by d'Ailly might also be considered to have happened in his lifetime. The second preamble was to happen when the sons of Ishmael and Hagar (identified with Islam) should rise up against the Roman Empire, in the world's seventh millennium. A third sign was to be that the promised land would fall to the sons of Ishmael, "that is, the Saracens." This would be a time of great corruption, which pseudo-Methodius identified with the *discessio* of 2 Thessalonians (although this was presumably fulfilled by the first preamble).[20] Pseudo-Methodius had predicted, fourthly, that there would follow a time of tribulation, in which many would deny the faith. Then, fifth, the Ishmaelites would be subjugated by a king of the Greeks or the Romans (the Last World Emperor). The sixth preamble would be a time of great peace in the world, following the defeat of the Saracens. Upon its end, the "doors of the north" would be opened, and the peoples enclosed within by Alexander the Great would come out and subject the world to another tribulation. At last would follow the death of the king of the Romans in Jerusalem and the appearance of Antichrist.[21]

D'Ailly's lengthy excerpts from pseudo-Methodius must be interpreted as a sign of his approval of the author, for they are presented without comment or rebuttal. It would seem that d'Ailly found in pseudo-Methodius, as he had in his astrological calculations, support for his revised interpretation of the eschatological meaning of the Great Schism. In d'Ailly's new view, the Schism was indeed part of a program of the Last Things, but it was no longer the *immediate* preamble to Antichrist's reign that it had been in his sermon for Saint Bernard, for example. D'Ailly's reliance on these new authorities, astrology and pseudo-Methodius, reveals his aspirations that the Council of Constance would end the Schism and even bring about a much-needed reformation in the church. With his writings in 1414 d'Ailly introduced a new sentiment in his thinking about the Schism—hope.[22]

The new interpretation of the current situation is even more apparent in a letter he wrote about a month later to John XXIII, the pope in the Pisan line to whom d'Ailly now owed his allegiance.[23] In this letter, d'Ailly summarized for the pontiff the words of a sermon preached by a certain doctor of theology in the presence of Urban V, that is, prior to the outbreak of the Schism.[24] This doctor had listed certain signs that would herald the nearness of Antichrist's persecution: the *discessio* of 2 Thessalonians, an abundance of sin in the church, tyranny of rulers, and popular commotions.[25] He had also reproved a number of wrong opinions about the apocalypse. In particular, d'Ailly's unnamed theologian had reproached those who would despair of reforming the sins of their own times, seeing them as the sure preambles to Antichrist. Although a remedy was difficult, he argued, it was not impossible. Indeed, theologians, philosophers, and astrologers had all opposed the belief that the future was immutable.[26]

D'Ailly went on to clarify his point that proper action now could save the church. His unnamed doctor of theology had preached that there was a tribulation prepared for the church, yet it could be headed off by a swift reformation of the clergy's morals. That this improvement had not occurred was all too apparent by the course of events. For after the death of Urban's successor, Gregory XI, there indeed had followed "such a horrendous tribulation of the church," that is, the Schism. As he had written before, d'Ailly told the pontiff, the cause of the Schism was certainly to be found in the sins of the clergy.[27] D'Ailly clearly meant to draw a parallel between the situation prior to the outbreak of the Schism and his own times. As the words of his anonymous preacher demonstrated, a change in the morals of the clergy could have saved the church from the tribulation of the Great Schism. Now, on the eve of the Council of Constance, d'Ailly saw the church in a similar position:

> although I am not a prophet, nor the son of a prophet [Amos 7:14], nevertheless without rash assertion I dare to say that unless in the General Council soon to be celebrated there be remedies foreseen for this damnable scandal, in order thoroughly to extirpate this schismatic division and to reform for the better the church, deformed in so many ways, we must believe with all likelihood that greater and fuller evils are still to come.[28]

Since d'Ailly goes on to refer to present happenings as "preambles to Antichrist," apocalyptic torments must be the "greater and fuller evils" of which he speaks here. As in the *Concordantia*, he implies that decisive action in Constance will ward off this persecution.

D'Ailly continues his letter with a short bibliography, a list that makes apparent the link between his astrological and eschatological interests. "In these matters," he tells the pontiff, "I do not require [you] to put your faith in a sinner, but rather in those who are said to have been inspired by the Holy Spirit," citing Joachim and Hildegard.[29] Continuing, he refers John to his own astrological writings. He notes that his conclusions were in agreement not simply with the prophets, "but also [with] astrological judgments, which are not entirely to be rejected by Catholics, just as I recall having noted in the treatise that I lately compiled, *De concordia Astronomicae veritatis et narrationis historicae.*"[30] In fact, he refers the pope to the body of his astrological writings, by name, as well as to his other works on the Schism, politely and wisely submitting them to the Holy See, and to the upcoming council, for correction or approval. The combination of sources to which d'Ailly directs the pontiff here makes it clear that d'Ailly himself saw his astrology as a means for studying the apocalypse.

A work completed on September 24 of the same fruitful year of 1414, the *Elucidarium* (Elucidation), demonstrates d'Ailly's further reading in both astrology and eschatology. The main purpose of the work was to correct and complete the *Vigintiloquium* and the *Concordantia* based on d'Ailly's increased knowledge of astrology. There are indications, however, that he was still pursuing eschatological themes as well. The old question of the age of the world comes up again, and d'Ailly admits that there is no sure answer.[31] He hazards a number of guesses, all giving the world some 5,300 or 5,400 years' duration before the birth of Christ.[32] D'Ailly also continues his investigation into Saturn-Jupiter conjunctions. Forecasting their occurrences, he locates conjunctions as far in the future as A.D. 1915, 3072, and 3469.[33] D'Ailly apparently did not see any astrological evidence for the imminent end of the world.

D'Ailly still evinced interest in an astrological interpretation of the Schism, eager for any clues about its future course. He noted that there had been a conjunction of Saturn, Jupiter, and the sun in 1365. This conjunction should have occurred in Libra, but "on account of their direct motion" (as opposed to retrograde) the planets were conjoined in Scorpio. Now Scorpio was an "enemy of religion," and therefore some say that this conjunction signified the Great Schism.[34] There was also a conjunction in 1405, on January 12 in Aquarius. Since this conjunction marked a return to the airy triplicity (the signs Gemini, Libra, and Aquarius), from which the planets had departed in 1365, some said that it would bring an end to the Schism, begun under the influence of Scorpio.[35] D'Ailly did not pass judgment on this opinion. Perhaps he was waiting to see if the conjunction's effects would appear belatedly in the actions of the Council of Constance.

Finally, d'Ailly reported a curious theory about the ages of the world. Its unnamed author drew a parallel between the four cardinal points of the zodiac, the four ages of man, and the four ages of the world.[36] The world's first age was that of Adam, and it corresponded to the quadrant of the zodiac beginning with the sign Aries. Its second age, youth, corresponded to the portion of the zodiac beginning with Cancer, and this age's dominant figure was Moses. The third age of the world began with Christ's birth, paralleling the quarter beginning with Libra. The world's decline would come in the age of Antichrist, and it would represent the winter quarter of the zodiac, beginning with Capricorn. According to this theory, there should elapse as many years from Adam to the time of Christ as there would be from the Incarnation to the end of the world.[37] (Thus, the End would be somewhere between A.D. 5300 and 5400, according to d'Ailly's figures for the age of the world.) D'Ailly was careful to state that this opinion was not certain, but only a probable conjecture. Nonetheless, it suited his present anti-apocalyptic sentiments.

D'Ailly did not name his source for this theory; perhaps its author was unknown to him. The only other redaction of this work known to me is as an anonymous treatise De antichristo, which formed part of a prophetic anthology compiled in the late thirteenth century by Pierre de Limoges (d. 1306).[38] It would seem most likely that d'Ailly copied the work from this or another such anthology and not from any astrological miscellany. The treatise is far too unsophisticated astrologically to have warranted the attention of a serious astrologer. D'Ailly's version follows the thirteenth-century De antichristo almost verbatim, with one significant alteration. Following the passage cited earlier, in which the world is predicted to endure as many years after Christ as it had before, the thirteenth-century version continues: "But the years from the beginning of the world up to the time of the Lord have been noted; therefore also how many there are from the Incarnation of the Lord up to the end of the world."[39] D'Ailly omits these words.

If this passage also appeared in d'Ailly's exemplar, it is curious that he chose not to reproduce it in his Elucidarium. He may have sensed that, given the work's various inquiries into the dating of Creation, to point out that these numbers were known would simply be too blatant a violation of the "Non est vestrum" commandment. Or perhaps d'Ailly realized the inconsistency of saying that world's age was known, since he himself had posed no less than four different figures for its duration. Whatever the case, the omitted sentence was really not necessary. Any reader wishing to calculate the End had all the information he needed in d'Ailly's treatise. It is revealing that d'Ailly incorporated any of the *De antichristo* into his *Elucidarium*. The *Elucidarium* was essentially a catchall for gleanings from the reading that d'Ailly must have pursued in the summer of 1414. The treatise was obviously compiled as d'Ailly's deeper knowledge in astrology pointed to errors in his earlier works. His excerpting of the *De antichristo* demonstrates that, in the months before the Council of Constance, his reading included eschatology as well.

At last the council was convened. D'Ailly arrived in Constance on November 17, 1414, apparently in high spirits.[40] In the sermon he preached for Advent (December 2) before the council, d'Ailly expressed both his hope for the church and his sense of the importance of the council's actions.[41] The text for his sermon was, significantly, an apocalyptic one, and its words had astrological resonance as well: "There shall be signs in the sun, and in the moon, and in the stars" (Luke 21:25). As he had in his 1385 Advent sermon, d'Ailly interpreted this text as referring to the pope, the emperor, and the ecclesiastical powers, seeing in the Schism an obscuration of their light. Now, he urged the council, let us see in these luminaries a sign of hope. "Therefore this is a marvelous and propitious change on high," he declared, "which we now see is beginning in the sun, the moon, and the stars, when we see this Sacred Council congregated for this purpose: that our holy Mother the Church, once miserably deformed by schismatic division, now joyfully will be reformed through peaceful union."[42] With this end in mind, d'Ailly became practically rhapsodic: "O! blessed eyes which should have deserved to see this. O! happy place, this city of Constance."[43] This emotional outburst is almost without precedent in his writings. Notwithstanding his hopes that the church might at last be healed, d'Ailly made clear to his audience that the crisis had not yet passed. Citing Joachim and Hildegard, he warned, "if this Holy Synod does not provide a fitting remedy for this evil [Schism], I boldly affirm that after such horrendous lightnings as we have seen, more horrible thunders will follow."[44]

In this sermon, as in his writings immediately preceding the Council of Constance, d'Ailly expressed the view that the church stood at a decisive crossroads. Increasingly hopeful that the Schism would be healed, d'Ailly no longer preached the imminence of the End. Astrological calculations and new eschatological sources backed up his hopes. Should the council fail in its mission, however, the outlook was grim. The Schism would surely be the *discessio* of 2 Thessalonians, and Antichrist would soon arrive. D'Ailly's view of the situation gave his words an urgency and a fervor that must have been a factor in his leadership at the Council of Constance.[45] He eloquently and emotionally expressed the need for decisive action. As he informed John XXIII, "from all of these preambles to Antichrist, we must likely believe and vehemently fear that if we are not converted to God by true penitence, and he to us by holy mercy, there will be popular sedition against prelates and the clergy, and such a great tribulation that the mind dreads to think of it and the tongue to tell of it."[46]

D'AILLY'S MATURE ESCHATOLOGY: THE *DE PERSECUTIONIBUS ECCLESIE*

After the successful conclusion of the Council of Constance, d'Ailly retired to Avignon to the house given him by John XXIII. There, in 1418, he composed a final eschatological work, the *De persecutionibus ecclesie* (On the persecutions of the church).[47] In this treatise, all of the various elements of d'Ailly's eschatological speculations converged: Scripture, history, prophecy, and astrology. D'Ailly now set out his mature interpretation of the book of Revelation and its relation to church history and the events of his own lifetime. The Schism underwent yet another interpretation. No longer the preamble to Antichrist of d'Ailly's early years or the uncertain crisis point of the year 1414, in his final view, the Schism was only one of a series of persecutions foretold by Revelation. After its conclusion there would be a reformation of the church and then the final onslaught of Antichrist, predicted once again for the astrologically significant year of 1789.

In the *De persecutionibus ecclesie*, d'Ailly followed the expositions of Revelation composed by Nicholas of Lyra (ca. 1270–1340) and Peter Auriol (1280–1322).[48] According to Auriol's interpretation, the visions of Revelation corresponded to six periods in the history of the church. The first vision of the seven seals (Rev. 5–8:1) represented the period from the founding of the church up to the time of Julian the Apostate (361–63). The church's persecution by heretics lasting to the reign of the emperor Maurice (582–602) was foretold by the second vision of the seven angels with trumpets (Rev. 8:2–11). The third vision (Rev. 12) represented the church's sufferings at the hands of Islam. The fourth vision, of seven angels and seven plagues (Rev. 15–16), depicted the numerous evils befalling the church in the years from Charlemagne to the emperor Henry IV (1056–1106). So far d'Ailly was in accord.[49]

The fifth vision of Revelation tells of the whore of Babylon, the beast with seven heads, the fall of Babylon, and the binding of Satan for a thousand years (Rev. 17–20:7). The sixth and final vision describes Satan's final, brief persecution, the Last Judgment, and the New Jerusalem (Rev. 20:7–22). It was clear to all three interpreters that the sixth vision referred to the three and one-half year reign of Antichrist and the end of the world. Both d'Ailly and Nicholas of Lyra had difficulty, however, in accepting Peter Auriol's explication of the fifth vision. According to Auriol, the fifth vision represented the period of time from

Henry IV to the advent of Antichrist. The damnation of the harlot represented the church's victory over "Muhammad's sect" under Godfrey and Baldwin, first kings of Jerusalem.[50] D'Ailly worried that Auriol's interpretation left no room for the description of subsequent events. Under Auriol's exegesis, Revelation foretold nothing beyond the First Crusade save Antichrist's persecution.

In a revealing passage, d'Ailly outlined his objections to Auriol's view. According to Auriol, he protested, everything prophesied in chapters 17 through 20 had already been fulfilled for some years, whereas the following text was universally interpreted to refer to Antichrist, *"whose advent even now does not appear near."*[51] It did not seem fitting, he continued, that John would have foretold so many notable events in the church's history, while omitting many which had occurred between the time of the First Crusade and the present. In particular, he felt it would be odd for Revelation to have omitted mention of the Great Schism, which Joachim, Cyril, and Hildegard had foreseen.[52] Hence he proposed an alternative explanation of this vision. D'Ailly now felt confident in saying Antichrist's advent did not even *appear* to be near. In the 1380s his reign had seemed imminent. In 1414, d'Ailly saw signs of Antichrist's approach, but hoped it could be averted. Now with the Schism ended, the threat was removed, and d'Ailly could at last understand the Schism's place in the scheme of Revelation.

The fifth vision, he proposed, corresponded to the Great Schism, its ending, and the subsequent history of the church up to the time of Antichrist.[53] The whore astride the beast represented the schismatic church, while the city of Babylon referred to Rome and the entire schismatic obedience. The beast of seven heads and ten horns stood for the temporal powers that supported the schismatic church.[54] Just as the harlot and the beast were overcome in the text, so too Revelation described the Schism's end. In Revelation 20, with the binding of Satan, d'Ailly foresaw a period of reformation and peace for the church, which he equated with the age of peace foretold by pseudo-Methodius. "After the victory of the king of the Greeks or the Romans, about which [pseudo-] Methodius speaks," he wrote, "there will follow peace and tranquil concord for the church. . . . then the ecclesiastical and temporal powers will be in agreement, and they will procure the perfect reformation of the church and, thereby, its true peace and union."[55] Following this peace, Satan would be loosed, signaling the onslaught of Antichrist and the impending Last Judgment.

D'Ailly's final pronouncement on the church's future was, in fact, optimistic. Following its "schismatic persecution," the church could look forward to a time of "victory and consolation" before the final torments of Antichrist.[56] When d'Ailly listed, as he had in the 1414 *Concordantia*, pseudo-Methodius's eight preambles to Antichrist, only the first of these, the Schism, was described in the present tense. All the others he set in the future.[57] On the other hand, d'Ailly clearly harbored no millenarian beliefs. He made it certain that the thousand-year binding of Satan was figurative and did not represent a millennium of peace on earth. D'Ailly preferred the interpretation of Nicholas of Lyra, who saw in the one thousand years a symbol for the entire period of history extending from Christ to Antichrist.[58] All the same, d'Ailly did not wish his readers to think that Antichrist's advent was too close. Hence he devoted the third section of his treatise to an attempt to forecast the coming of Antichrist.

D'Ailly offered a number of ways of making such a prediction, most of them using astrology. He noted that, although there could be no human certitude about the time of Antichrist's advent, one could use astrology to obtain a plausible, but indeterminate, conjecture of when he might come.[59] It would seem that the most likely astrological prediction in d'Ailly's eyes was his own prognostication from 1414, in which he slated Antichrist's arrival for 1789. This is the first astrological argument that d'Ailly presents in his inquiry, and he reproduces it almost verbatim from the *Concordantia astronomie cum hystorica narratione*, and without comment. Rather he repeats his wonderfully hedged prediction from 1414: "from these [astrological observations] it plausibly can be concluded that perhaps around that time [1789] Antichrist will come with his damnable law or sect."[60]

By 1418, d'Ailly knew that this prediction was faulty. In the *Elucidarium*, which he completed only a few months after the *Concordantia*, d'Ailly acknowledged that he had assigned Saturn–Jupiter conjunctions to the wrong years in his previous works. One such conjunction in 1692 or 1693 was involved in the 1789 prediction; in the *Elucidarium*, d'Ailly had redated it variously for 1915 and 1761.[61] It is not clear why he repeated this prediction, which rested on at least one erroneous calculation. Perhaps it simply was more convenient for him to copy from his previous work. Perhaps in Avignon he no longer had access to the materials required to make a new calculation. Perhaps he felt that "speaking indeterminately" ("indeterminate loquendo"), the error was not significant. Clearly d'Ailly was not afraid to criticize his earlier work. He had done that amply in the *Elucidarium*. In all likelihood, the date of the conjunction was of secondary importance for the prediction, and it was the contemporaneous completion of Saturn's tenth orbit and the eighth sphere's "standing still" that attracted d'Ailly to the date 1789. Knowing well the limitations of his prediction, d'Ailly included it here without a single critical comment. This fact can only indicate his continued faith in this prognostication.

After his own conjecture, d'Ailly brought forth a number of other attempts to predict Antichrist's advent, perhaps arranged in order of decreasing probability. First, he reproduced the concluding passage from the *De antichristo* of John of Paris, a new author in d'Ailly's eschatological repertory, yet an important source for his political philosophy.[62] John had predicted the appearance of the son of perdition within two hundred years of the time of his writing in 1300.[63] Another group of calculations aimed to determine the duration of the Islamic religion, for pseudo-Methodius had predicted that its total destruction would occur before Antichrist's advent.[64] Following Scripture and Albumasar, d'Ailly advanced a number of figures for the religion's duration: 666, 693, 584, and, later, 1,141 years. He noted that, even if one began to count from the death of Muhammad, the first three figures had long been surpassed without the sect's having disappeared.[65] He suggested that one might perhaps number the religion's years from the time of the expansion of the Arab kingdom in the eleventh century. Even so, d'Ailly concluded, this theory was greatly in doubt.

By the same reasoning, d'Ailly protested, Albumasar had estimated that the Roman law (i.e., Christianity), would endure 1,460 years, the "maximum years" of the sun.[66] Hence, "the Christian law should not endure more than forty-two years from the present year, which is 1418."[67] Further, the Arab law was to endure 1,141 years, the maximum years of Venus, its planet.[68] D'Ailly did not add the obvious conclusion that Islam would thereby outlast Christianity by several centuries, even if one computed its beginnings with Muhammad's birth in circa 570. This so completely violated Christian eschatology as to render Albumasar's theory useless. D'Ailly instead referred the reader back to Albumasar's interpretation of the conjunction signifying Islam. Therein might lie the reason behind this "diversity," he suggested.[69]

Given the foregoing criticisms of Albumasar, and d'Ailly's rather hopeful interpretation of Revelation, the treatise's final paragraph is rather puzzling. It does not seem possible to conclude anything certain from these astronomical conjectures, he states,

> but nonetheless, from these and the other [theories] presented above, one can have a plausible indication that within one hundred years from the present there will be a great change in laws and sects, and especially as regards the Christian law and church. And it is expedient to foreknow this somehow, so that Christians may constantly prepare themselves to suffer tribulations.[70]

D'Ailly may have had in mind Albumasar's prediction of Christianity's 1,460-year duration when he predicted a "great change" within the next century. On the other hand, his preceding arguments would seem to discredit this theory. He may also have wished to remind his readers of John of Paris's prognostication of the world's end around 1500. How these final words are to be reconciled with the prediction for 1789 d'Ailly nowhere addresses. His final warning makes no mention of Antichrist. Perhaps d'Ailly envisioned the next century as bringing only the torments that would precede Antichrist's advent. In this manner, both predictions could be salvaged.

Were he forced to choose only one of these prognostications, d'Ailly would surely have had to come down on the side of his own calculation. John of Paris's argument, although d'Ailly did not mention the fact, was based on the notion that the world would endure only 6,000 (or, speaking roundly, 6,500) years. D'Ailly had dismissed this theory as "superstitious" in a number of writings.

Similarly, he had presented Albumasar's theory as erroneous. D'Ailly knew, however, that any attempt to predict the End was a conjecture at best. Even a flawed prognostication might contain some grain of truth. Hence d'Ailly warned his readers of possible tribulations in the century to come. But he must have placed the most credence in his twice-given prediction for 1789.

D'AILLY'S ANTI-APOCALYPTIC
MESSAGE

In 1414 d'Ailly warned all who would listen that the church was at a decisive crossroads. Should the Council of Constance end the Schism, the church would be saved. Should the council fail, Antichrist would soon come. Although d'Ailly stressed the critical need for proper action at Constance, his astrological calculations reveal that already in May 1414 he dared to hope and to assume that the apocalypse was not at hand. As early as 1403, in the *De reformatione ecclesiae*, d'Ailly was contemplating a reinterpretation of his eschatological views. And in 1410 he put forth the framework for examining this question using astrology. It seems a curious paradox that the longer the Schism endured, the less convinced d'Ailly became of his early apocalypticism, for with the worsening of the church's division, the greater must have appeared its identification with the *discessio* of 2 Thessalonians. Although the prospect of the Council of Constance gave d'Ailly cause for hope in 1414, his rejection of his earlier interpretation of the Schism also would seem to indicate a disillusionment with apocalyptic beliefs in general. D'Ailly probably became dissatisfied with such teachings for two reasons. First, he came to associate apocalyptic notions with disorderly and disobedient behavior. Second, and most important, d'Ailly realized that the belief that the End was an immutable and imminent happening worked against any attempts to settle the Schism.

Norman Cohn has stressed the link between millenarian beliefs and political resistance in the Middle Ages. Although scholars have tempered Cohn's interpretation in recent years, there is a sense in d'Ailly's lifetime in which disobedience to church authority could be related to apocalyptic sentiments.[71] A Frederick of Brunswick, for example, was convicted at Speyer in 1392 for his dangerous preaching. Central to Frederick's message was the notion that Antichrist was to come within four and a half years and that he, Frederick, was the herald of a great Franciscan "repairer," who would slay the fiend and reign as pope and emperor for a thousand years.[72] D'Ailly himself was strongly concerned with maintaining order among the faithful. When he attempted to paint the horrors of Antichrist's time for John XXIII, it was in terms of popular uprisings against the clergy.[73] In his *De falsis prophetis I*, d'Ailly stressed the necessity of properly examining the credentials of those itinerant preachers who claimed to have been sent by church authority.[74]

In his own diocese of Cambrai, d'Ailly had direct experience with a group that denied the authority of the church and whose beliefs, at least in the minds of the clergy, were linked to a Joachimist apocalypticism. On June 12, 1411, d'Ailly questioned Wilhelmus de Hildernissem, one of the leaders of a sect known as the *homines intelligentiae* (men of intelligence).[75] Wilhelmus and his coleader, "a certain illiterate layman of some sixty years, called Aegidius Cantoris," were charged with disseminating a number of errors and practicing something akin to free love.[76] Among the most disturbing of the sect's teachings was its disrespect for the clergy. Aegidius had taught that in the end all men and the Devil would be saved. His followers thus saw little need for the statutes and precepts of the church, or for prayers, confession, or penitence.[77] Although Wilhelmus denied ever having heard such a notion, d'Ailly and his inquisitors attributed to the sect yet another error, that of believing that they were living in the Joachimite age of the Holy Spirit. According to the charges, the members of Wilhelmus's sect held that all Scriptures and church teachings belonged to the two prior ages and were now invalid.[78] Whether or not Wilhelmus's protestation of innocence was sincere, it is noteworthy that d'Ailly associated an anticlerical sect with certain apocalyptic beliefs.

At the Council of Constance, d'Ailly showed increasing concern with threats to the authority of the church. In the letter written to Pope John XXIII before the opening of the council, d'Ailly remarked on the number of errors, heresies, and new sects he had witnessed as papal legate in Germany.[79] In 1415 he presided over the council's trial and execution for heresy of the Bohemian dissenter John Hus. Among Hus's more egregious errors were the teaching that a sinner could not hold a position of authority over others and his equation of the pope with Antichrist.[80] In the eyes of d'Ailly and the other leaders at Constance, Hus posed a clear threat to church order. Given such disturbances of ecclesiastical authority, it is perhaps not surprising that d'Ailly began to promulgate his antiapocalyptic notions at this period.

Although there is no definite link drawn between apocalypticism and the teachings of John Hus or the other sects mentioned by d'Ailly, a further episode at the Council of Constance makes clear the relationship between apocalyptic preaching and disobedience to the clergy. On June 21, 1416, Jean Gerson wrote a letter to the itinerant preacher Vincent Ferrer, rebuking him for the group of flagellants who accompanied him on his journeys. D'Ailly, Gerson's friend and former master, appended a postscript, greeting Vincent, recalling their meeting in Italy, and urging his attendance at Constance.[81] He had heard many rumors, Gerson wrote to the preacher, about "that sect of people whipping themselves." Such behavior had been condemned in the past, Gerson noted. And while, as Vincent's friends testified, "you do not approve [of that sect]," Gerson complained, "nor do you reprove [it] effectively."[82] Gerson's stern admonition apparently did not have the desired effect, for on July 18, 1417, he brought up the matter of flagellation before the council at large.[83]

Gerson reminded the assembly that the church had always reproved flagellants. Nonetheless such sects had arisen "within the memory of many living men" in Lotharingia, Germany, and many parts of France. He made clear their threat to stability. "The hierarchic order would be confounded," he warned darkly, ". . . if anyone, at his own will, could institute or encourage a new rite, without leader, without stable law, without order, where there are youths and virgins, elders with younger persons, and rich and poor together as one."[84]

Gerson offered a number of suggestions for dealing with such sects, such as teachings and sermons aimed at curtailing their practices. A final proposal was apparently intended to extirpate flagellation by striking at its source. If there should be any sermon preached about the Last Judgment or Antichrist, Gerson urged, "let it be made in general terms, concluding that any person will have in death his own near and uncertain Judgment."[85] In other words, Gerson condemned the practice of preaching that Antichrist's reign was at hand. It should carefully be observed, he continued, that such preachers in no way should act in a contentious manner, nor should they show any signs of contempt for prelates or the clergy in general, "neither in making their sermons, nor in hearing confessions."[86] Concluding, Gerson suggested that Vincent Ferrer either appear before Constance or segregate himself from the society of flagellants.[87]

Gerson's words before the council drew a distinct connection between apocalyptic preaching, flagellation, and disobedience to the church's authority. One can well imagine that d'Ailly approved of his sentiments. D'Ailly may even have had knowledge of a band of flagellants beyond those who traveled with Saint Vincent Ferrer. In 1414–16 in Thuringia authorities uncovered a sect that combined flagellation, anticlericalism, and apocalypticism. The group was inspired by the teachings of Conrad Schmidt, who had prophesied the end of the world for 1369, and who was probably burned in the same year. His followers believed that Schmidt, by transmigration of the soul, was now Enoch, while a companion had become Elias. This pair, according to medieval lore, would return to preach against Antichrist in the world's last days. The members of this sect further taught that "baptism by blood" (i.e., flagellation) had replaced penance and all other sacraments and that it was absolutely necessary for salvation. Some four hundred flagellants were burned in Thuringia in the years 1414 to 1416.[88] Whether or not d'Ailly, as legate to Germany, knew of the Thuringian flagellants, he had abundant evidence by the time of the Council of Constance that apocalyptic preaching could have dangerous consequences.

If the threat to ecclesiastical authority was one of the reasons behind d'Ailly's anti-apocalyptic sentiments in his final years, his concern with ending the Schism must have given him further support in his views. D'Ailly's position becomes apparent in the letter he wrote to John XXIII outlining his belief that the church was at a crisis point in 1414. D'Ailly acknowledged in that letter that the present misfortunes of the church resembled the preambles of Antichrist. Yet he reproved the error of those who greeted the present with despair, noting that the remedy for sin was difficult but not impossible.[89] Although d'Ailly did not explicitly draw the inference, it is easily made: those who view the Schism with despair, that is, as a sure sign of the End, pose a distinct impediment to those who would find a means to bring peace to the church. To work to end the Schism, one must move beyond apocalypticism.

There is evidence to suggest that another of the leading conciliarists, Henry of Langenstein (or Henry of Hesse, 1325–97), underwent a similar transformation in his views of the Schism and the end of the world. Although d'Ailly would attack Henry's views on astrology in 1414, the two were in firm concord on the necessity of ending the Schism. Like Pierre d'Ailly, Henry seems originally to have interpreted the Schism as a preamble to Antichrist. In 1383, he was busily engaged in gathering as much information as he could about this advent, reading the Sibyls, Hildegard of Bingen, and a number of pseudo-Joachimist works. In 1388 Henry composed a treatise on the Hebrew alphabet based upon the pseudo-Joachimist *De semine scripturarum*. A sermon from 1390 similarly contained high praise for Joachim, while condemning the use of astrology to predict the Last Things. In 1392, however, Henry completed a treatise attacking the prophecies of Telesphorus of Cosenza. He now denounced all attempts to foretell the End.[90]

Henry's treatise had as its main target the prophecy put forth by Telesphorus in which he predicted a time of turmoil under the Schism, then peace, to be followed by the appearance of Antichrist. Henry ridiculed Telesphorus's claim that he had been visited by an angel, who urged him to read the works of Joachim and Cyril.[91] Similarly, he turned on Joachim, who, he said, had not prophesied with God's inspiration, but rather had divined by means of human industry.[92] In addition, Henry roundly denounced astrologers' attempts to predict the end of the Schism or the coming of Antichrist.[93] Worse still to Henry, Telesphorus had claimed that publishing his vision would help to restore peace to the church. Henry protested, however, that "such prophecies, made with the intervention of diabolical craft, do not work toward peace, but rather, delaying peace, prolong the controversy and further strengthen the discord in people's hearts."[94] Refusing to hazard any opinion on the time of the apocalypse, Henry suggested that there were always signs of its approach, so that men would constantly live in fear of the Last Judgment.[95]

Beatrice Hirsch-Reich, who pointed out the change in Henry's attitudes about prophecy, suggested that it might be explained by the opposition of fellow academics as well as the pro-French bias of Telesphorus's treatise. Marjorie Reeves has proposed that his attack on Telesphorus represents a disillusionment with prophecy and a profound pessimism about the church's future.[96] Henry had declared firmly in 1392 that he foresaw no great reformation of the church prior to the death of Antichrist but anticipated a continued decline.[97] It would seem, however, that Henry's treatise also expressed the sense that an apocalyptic interpretation of present events served only to hinder efforts to end the Schism.

Indeed, by 1394, Henry had abandoned his grim view of the post-Schism church. In that year he wrote a letter to Pierre d'Ailly, as chancellor of the University of Paris. If we hold that the church will only decline after the end of the Schism, or will be restored to less than its former virtue, he declared, people may strive less fervently for its union and the restoration of peace. "It should be held to be more likely," he continued, "that God, . . . when he decrees the church is to be liberated from the evil of schismatic persecution with which he has by measure disturbed it, certainly shall reform it for the better."[98] Henry, like Pierre d'Ailly, had come to recognize the need for calm, clear thinking to bring an end to the Schism. Apocalyptic predictions of future gloom for the church were simply of no help.

As did most of Christendom, d'Ailly viewed the outbreak of the Great Schism as a sign of Antichrist's approach and desperately searched eschatological writers for clues to when the fiend would arrive. "Others during this Schism," Henry of Langenstein noted, "have begun to speak as if prophets or soothsayers, or more truly, diviners. To their predictions, or more truly, empty talk, all have given their ears."[99] D'Ailly found in authors like Hildegard of Bingen and Joachim of Fiore confirmation of his increasingly apocalyptic sentiments during those early years of the Schism. He scrutinized such writings and preached the imminence of Antichrist's reign.

In writings dating from the years 1400 to 1414, d'Ailly reexamined his earlier apocalypticism. In the months preceding the opening of the Council of Constance, he dared to propose that the End was not at hand. D'Ailly now lay down the elements of a system by which to predict Antichrist's advent using astrology, an endeavor he had roundly rejected in his early years. "Although human certitude cannot be had about his advent," d'Ailly now wrote, "nonetheless, by indeterminately saying he will come about that time we can have plausible conjecture and likely indication by astrological judgments."[100] D'Ailly's astrological conclusions agreed with his new interpretation of the current situation. On the eve of the Council of Constance, he sensed the church was at a crucial point. Only if the council could find a remedy to the Schism could the persecution of Antichrist be forestalled. Astrology confirmed d'Ailly in his belief that the council would restore peace in the church, for the stars told him that Antichrist would not arrive until 1789. Secure in his new eschatology, backed up by his astrological calculations, d'Ailly preached in his opening sermon at Constance a message of joy and hope.

Astrology and eschatological speculations continued to be linked in d'Ailly's mind in his treatise on the persecutions of the church completed in 1418. In this final study of the Last Things, d'Ailly was at last able to understand the place of the Schism in the program foretold by Revelation. In an interpretation that attempted to blend Scriptural exegesis with astrological calculations, he described the Schism in terms of Revelation's whore of Babylon. Following its end, he predicted, the church would experience the consolation of a period of peace and reformation. D'Ailly repeated his astrological prediction that Antichrist would arrive in the year 1789, but he darkly warned of changes to come in the next one hundred years.

In the years after 1400, then, d'Ailly began to object to the notion that the world was near its End. He had encountered disturbing behavior associated, at least in his mind, with apocalyptic beliefs: the sect of the *homines intelligentiae* in his own diocese and the flagellants who accompanied Vincent Ferrer in his travels. At the same time, d'Ailly had come to realize that an emphasis on the imminence of Antichrist's reign stood in the way of his attempts to end the Schism. D'Ailly now stressed that it was an error to view the future as immutable and urged that a proper remedy for the Schism could save the church.

Believing that the apocalypse was at hand called for a collective despair for which the only remedy was an individual response, to pray and repent for one's own sins. If the Schism were indeed the *discessio* of Scripture, then the schismatic church (the Roman obedience in d'Ailly's mind) would represent the false prophets heralding Antichrist. To hold council with them would be unthinkable and, given an immutably impending apocalypse, of little use. By contrast, if a reformation of the church could avert the disaster prepared for men, then the remedy for the crisis would be some sort of collective penance. Together, Christians must heal the Schism and reform the church. The idea that the apocalypse might be postponed removed the despair that had prevented some from working to end the Schism. Further, by this interpretation the members of the schismatic church were merely sinful humans and not the minions of Antichrist. D'Ailly's revised view of the Schism pointed toward the council as solution.

In 1394 the faculty of the University of Paris outlined three possible remedies to the Schism: the *via cessionis*, whereby the two popes would both abdicate; the *via compromissi*, whereby an impartial group of arbiters would judge the matter; and the *via concilii*, whereby a general council of the church would adjudicate. The third of these was seen as the most extreme, and least desirable, solution. Gerson advocated the first two modes of dealing with the Schism until as late as 1408. Interestingly, he termed them the *via poenitentiae* (the way of personal penance).[101]

The shift that the Parisian theologians underwent during the years of Schism, from hesitation about to embrace of the conciliar solution, in some way parallels d'Ailly's eventual dissatisfaction with apocalypticism. The apocalyptic view of the Schism indicated personal penance as the only possible response. In calling upon his audiences to repent because the Judgment was at hand, might d'Ailly not also have hoped to persuade the rival pontiffs to follow the *via poenitentiae* and resign their offices or submit to a board of judges? And in months before the Council of Constance, when he urged that proper action could postpone the torments of the apocalypse, he surely hoped to persuade his contemporaries to attend the council and to see it through its task.

In revising his view of the Schism, d'Ailly found in astrology a tool with which to refute predictions of the imminence of the End. Therein must lie the reason he chose to promulgate researches that, by his own admission, were somewhat eyebrow-raising. "Now we know," he wrote, "that certain persons have objected that it would be more suitable for one of our profession, and also our age, to be occupied with theological rather than with these mathematical studies."[102] To his critics d'Ailly protested that his writings "pertained to the utility of the church."[103] D'Ailly's astrological researches strengthened his conviction, in 1414, that the church would survive the Schism to endure many years more. He persevered in his astrology because it was fascinating to him and because it gave him hope about the future. He publicized his interests, we must believe, because astrology could combat those apocalyptic beliefs which, as d'Ailly had come to see, were impediments to ending the Schism.

CHAPTER SEVEN

The Concordance of Astrology and Theology

PIERRE D'AILLY lived in an age engulfed by crisis. Plague, famine, rebellion, political instability, war, economic depression, and Schism all left their mark on the Europe of his lifetime. The momentous occurrences of his times forced d'Ailly and his contemporaries to look for new interpretations of the disasters and for innovative methods of solving them. For d'Ailly, caught up in the political storms of the Great Schism and the competing factions in France, the course of events provoked a reevaluation of his thinking on two fundamental issues. First, he moved from an interpretation of the Schism in apocalyptic terms to one in which healing the division would postpone the End. Second, he dropped his early hesitations in favor of a wholehearted embrace of astrology. Indeed, a key analytical tool in d'Ailly's reexamination of the Schism was his knowledge of the stars. Astrology emboldened him to claim, on the eve of the Council of Constance, that the Schism did not signal Antichrist's advent, but rather that the apocalypse lay embedded in a series of planetary events some centuries in the future.

Why astrology? Why, in particular, was scrutiny of the heavens so crucial to this churchman's view of the Schism? The answer lies, it seems, in d'Ailly's concordance of astrology and theology—that is, first, in his insistence that astrology be considered a "natural theology" and, second, in his implication, by the use he made of the stars, that astrology was also a valid science, useful because it lay outside of the realm of prophecy and revelation. That is, he established astral causality to be an essential component of the divine plan, one entirely in keeping with the central feature of his theology, the dialectic of God's absolute and ordained power. And yet, he relied upon astrology to interpret the apocalypse precisely because it was nontheological. It offered him evidence drawn from sources other than prophecy and revelation, which, as he argued, could be contradictory, problematic, and even deceptive.

Thus, d'Ailly's attempts to foresee the future, using both prophecy and astrology, hearkened back to his earliest theological speculations about God's foreknowledge and his omnipotence. God's omniscience, d'Ailly had argued, did not rule out the contingency of the future. To preserve this latter point, he insisted that even divine revelations about the future could not be considered to be infallible predictions. Given the uncertainty of even revealed statements, he may have hoped to find independent confirmation of prophecies by using astrology. By arguing that one could apply astrological judgments to prophetic statements, d'Ailly made strong claims for the science of the stars and also under-scored its connections with Christian theology.

Was d'Ailly's obsession with astrology also motivated by some prospect of economic or political gain? Bernard Guenée, d'Ailly's most recent biographer, has stressed the man's skill and ambition as a politician, citing opportune moments of sickness, well-timed changes of opinion, and friendships in high places.[1] And yet there is little indication that d'Ailly stood to gain anything material from his use of astrology; if anything, his interest in the stars would seem to have been a mark against him.[2] Indeed, throughout the period of d'Ailly's greatest involvement with astrology (1414–15), the only honors remaining for him to win were the papacy and posthumous canonization, neither of which was a likely possibility. It is difficult to interpret d'Ailly's defense of stellar causality as a ploy for self-aggrandizement.

Further, there is little evidence to indicate that d'Ailly *could* have used astrology in such manner even if he had desired to do so. Granted, he certainly knew how to erect an astrological figure. But it is not clear that d'Ailly ever studied the rules for interpreting one. When he included horoscopes in his writings, he most often left interpretation of the chart "to more experienced astrologers."[3] If d'Ailly did offer a prediction based on a figure of the heavens, it was invariably taken verbatim from some other source.[4] Looking at the pattern of borrowings from other astrologers in d'Ailly's treatises, one might reasonably infer that he did not even read the sections of astrological primers dealing with rules for interpretation of charts. Indeed the only real astrological predictions that d'Ailly himself ever made were for the appearance of Antichrist in 1789, the birth of a new religion around the year 1600, and a vague admonition about great changes in the Christian religion within the next one hundred years.[5]

D'Ailly's astrological predictions, exceptional as they were, are nonetheless revealing. They confirm what he expressed to his readers: that he viewed astrology as a science useful to students of theology, precisely because it could reveal the broad patterns of the past, present, and future. His study of the stars was a rational attempt to understand the troubles of his age. It was an ambitious and optimistic undertaking. By searching astrology for the overarching structures of earthly events, d'Ailly was able to discern the (to him) true implications of the Great Schism and to offer his contemporaries comfort and encouragement. Human reason had deciphered the meaning of the crisis, and human efforts would bring it to an end.

ASTROLOGY AS "NATURAL THEOLOGY"

Although belief in the power of the stars was near universal in the later Middle Ages, the acceptance of both astrology and the Christian religion inevitably entailed friction. D'Ailly acknowledged these conflicts and attempted to resolve them both in his astrological writings and, early in his career and in a more general way, in his commentary on Peter Lombard's *Sentences*. The most obvious problem was that of reconciling astral determinism and human free will, for if the stars caused all that happened on earth, man was unaccountable for his sins. D'Ailly also dealt with the threat to God's omnipotence implied by the existence of regular and predictable laws of nature. His resolution of such conflicts was nowhere entirely successful.[6] Given his inability fully to answer these issues, one might reasonably ask if d'Ailly put more faith in the stars than he did in Christian revelation and prophecy.

D'Ailly's astrological writings, as we have seen, dealt broadly with the tensions between Christianity and the science of the stars. Following Thomas Aquinas and others, d'Ailly protected human free will by exempting it from the control of the heavenly bodies. Astrological theories, on the other hand, still held true because men often acted under the influence of their bodily passions. To safeguard God's omnipotence, d'Ailly insisted on the operation of both natural and supernatural causality in the present world. God always has the ability, in this schema, to bypass the laws of nature and to act in a supernatural or miraculous manner. In the intervening moments, the laws of nature hold force, for which astrology remains a valid means of interpretation.

If d'Ailly set out in general a concordance of astrology and theology, in a number of key passages his words betray the unsteadiness of his compromise. In particular he seems to have worried that his acceptance of astral causality might appear to contain a belief that everything happens of necessity. In one treatise he warned the readers that he ascribed nothing to "fatal necessity," but rather "to natural causality and inclination, which the free will can resist in those matters which are subject to its faculty (with God's concurrent aid); divine omnipotence also can efficaciously obviate the same by a sole command of the will."[7] In this one sentence d'Ailly both acknowledged that astrology, carried to its logical extreme, would deny free action to both man and God, and he limited the science to avoid this heresy. In other works d'Ailly stressed the contingency of astrological predictions, as for example in the language with which he qualified his forecast of Antichrist's advent. "If the world shall have lasted to that time, which God alone knows," he wrote, ". . . it plausibly can be concluded that perhaps around that time Antichrist will come."[8]

D'Ailly's balancing of astral and divine causality parallels an important aspect of his theology. As we have noted, theologians of the later Middle Ages had long accommodated the existence of a predictable world order with an omnipotent deity by distinguishing God's *potentia absoluta* (his absolute power) from his *potentia ordinata* (ordained power).[9] It was a shorthand way of expressing the belief that God *can* do many things he does not *will* to do. *Potentia absoluta* guaranteed that the Creator was in fact omnipotent, whereas *potentia ordinata* assured that God would act in certain agreed-upon ways. The covenant by which God freely bound himself to act according to his ordained power allowed for predictable laws of nature and, more important, of salvation. For proponents of this type of theology, such as d'Ailly, William of Ockham, and Gabriel Biel, only God's free choice established what were the meritorious deeds that would earn salvation, but the concept of *potentia ordinata* insured that the chosen path to grace would always be the same one.

D'Ailly employed such terms in his commentary on Peter Lombard's *Sentences* (required of all who would attain the rank of master in theology) and in other writings. In the *Sentences* commentary of 1376–77, he defined the dialectic of absolute and ordained power in two ways. *De potentia absoluta*, God could do anything that did not involve a contradiction. God's ordained power, on the other hand, encompassed that which God had ordained to be done in this world order; and anything that did not contradict a truth in Holy Scripture.[10] In other words, God has agreed to act within an established natural and spiritual order, an order that exists only because God has freely chosen it. The *absoluta/ordinata* distinction thus preserved both God's omnipotence and predictable rules of nature and for salvation. Through such terminology d'Ailly addressed on a broad scale the questions that would be implicit in his later acceptance of astrology, which, too, opposed the course of nature to an omnipotent God.

One might reasonably expect, therefore, to find mention of God's absolute and ordained power in d'Ailly's defense of God's omnipotence vis-à-vis astral determinism. Curiously, however, he nowhere uses these terms in his discussions of astrology. Rather he opposes the natural causality of the stars with the supernatural causality at work in events such as the birth of Christ. Francis Oakley has suggested that d'Ailly used the word "supernaturally" as a synonym for "by the absolute power."[11] From this inference Oakley concludes that d'Ailly intended that God's absolute power was at work in supernatural occurrences such as miracles, while the usual course of nature (and, presumably, astrology) fell within the purview of *potentia ordinata*. Had this been d'Ailly's view, one would expect to find reference to God's absolute power in those passages of the *De legibus et sectis* in which d'Ailly exempted Christianity from the control of the stars.

That he does not use such language suggests that he understood the distinction in a much more traditional theological manner. *Potentia absoluta*, in this view, encompasses all of the logically possible choices originally open to God. *Potentia ordinata* represents God's free choice to carry out only some of these logically possible acts. Within God's ordained plan, however, some occurrences are natural, that is, involving natural causes such as astral influence, and some are supernatural or miraculous. Miracles, then, represent a subset of God's actions within his ordained power and not an incursion of *potentia absoluta*.[12] This is, I believe, the sense behind d'Ailly's twofold definition of *potentia ordinata* cited earlier; in its two parts are embodied the natural and supernatural causes of his defense of astrology.

Although this distinction regarding God's use of his absolute power may appear to be little more than dry theological debate, it has broad consequences for one's view of the world. Modern scholars have, therefore, quite rightly devoted much attention to the question of whether late medieval theologians believed God would actually use his absolute power to override the present order of things. In his theological writings d'Ailly, as did his contemporaries, speculated on what things God could do by his absolute power. God has the ability, *de potentia absoluta*, for example, to deceive men, to save a person who lacks grace, and to damn one who has it.[13] The prospect of a world in which God capriciously exercises his absolute power can thus be quite a frightening one. The older view of the later Middle Ages stressed just this possibility in its theology, charging men such as d'Ailly with skepticism, fideism, and undermining the scholastic achievement.[14] More recent scholars have emphasized the dialectic of *potentia ordinata* and *potentia absoluta* in this theology and have largely doubted that medieval theologians believed God would act *de potentia absoluta* in this world, and certainly not in the arbitrary manner implied in their speculations.[15]

Leonard Kennedy recently has suggested that in the *Sentences* commentary, d'Ailly did believe that God would exercise his absolute power in this world and that to hold otherwise would unduly constrain the Creator. According to this author, d'Ailly's view of God led to a thorough skepticism.[16] D'Ailly's later discussions of astrology, however, suggest a different interpretation of his thought. If he believed that the world and its Creator were completely unpredictable, there would have been little reason for him to look to the stars for guidance. Further, d'Ailly's discussions of supernatural and natural causality with respect to astrology imply his faith that *potentia ordinata* always ruled the world. In the *De legibus et sectis*, he cited the virgin birth of Christ as an example of a supernatural event.[17] Surely this central point of Christian history fit within both of his definitions of God's ordained power. God had decreed it as part of the present world order, and it did not contradict a Scriptural truth. Although he did not refer to God's *potentia ordinata* in his defense of astrology, d'Ailly implied that God's covenant both established and guaranteed the present order of things, including astral causality.

The theological distinction between God's absolute and ordained power contained a dialectical tension between two views of God. Excessive stress on God's absolute power could lead to skepticism and the view that the world was in no way predictable or consistent. Such a God could seem an arbitrary, capricious, and frightening being. Overemphasizing the ordained power, however, could in effect take away God's omnipotence by binding him to a certain, established rule. D'Ailly, in my view, tended toward the second of these errors, sensing that this was a more understandable and reassuring world view. He was certainly aware that astrological causality could threaten divine freedom. This realization was the reason, I believe, behind statements such as d'Ailly's curious remark that "God is the true sage who alone rules the stars."[18] At crucial points in his discussions of astrology, d'Ailly felt compelled to nod to the omnipotence of God before proceeding as if the laws of nature would always obtain.

ASTROLOGY, PROPHECY, AND THE
CONTINGENCY OF THE FUTURE

D'Ailly's description of God's absolute and ordained power both underscored the tensions involved in his dubbing astrology a "natural theology" and paved the way for his future defense of the science. Similarly, his early investigation of the problem of God's foreknowledge provided a partial explanation of his later use of astrology. In his *Sentences* commentary, d'Ailly dwelled upon the tensions between God's foreknowledge and human free will. If God has certain knowledge of the future, how is it that one's future actions are not necessary? Anxious to preserve the contingency of the future, d'Ailly in effect limited the value of prophecy by insisting that even revealed statements about the future were contingent and fallible. He thus may have come to embrace astrology precisely because it lay outside of the realm of prophecy and theology. Indeed, the resolution of the true character of the future also had immediate implications for study of the apocalypse. Theologians who discussed the problem of so-called future contingent statements frequently used as an example the proposition "Antichrist will exist."

In his commentary on Lombard's *Sentences*, d'Ailly insisted on both God's infallible knowledge of the future and the complete contingency of future events. He explained how the two were not incompatible by describing God's existence outside of time. Just as our eyes give us knowledge about the existence of present things, he asserted, God's "eye" is eternal, and hence has an immediate intuition of the existence of things past, present, and future.[19] That the future was not predetermined, however, led to some rather interesting conclusions. Revelation, for example, is fallible. God has the ability *de potentia ordinata* to deceive man in revealing the future.[20] Likewise a supposed prophecy can turn out never to have been a prophecy in that it is possible that the event it predicted will not happen. By the same principle, d'Ailly adduces that it will be within Antichrist's power to make it that Christ was not a prophet in the things he predicted about Antichrist.[21] Was prophecy to be trusted? Calvin Normore has suggested that d'Ailly had no intention of ruling out the possibility of certain revelation. "We can know," he explains, "that God *will not* deceive us; what we cannot know (because it is false) is that God *cannot* deceive us."[22]

If "Antichrist will exist" was merely an example from a scholastic exercise when d'Ailly commented on the *Sentences* in 1376–77, it was soon to take on added significance. The division in the church from 1378 on, as we have seen, served as a focus for apocalyptic expectations. Christians in all periods have been able to point to the sorry state of human morals as an indication that the End is at hand. But the Great Schism added further and, to many, convincing proof by appearing to fulfill Paul's prediction of a *discessio* in 2 Thessalonians 2:3. D'Ailly, as we have seen, investigated the possible connections between the Schism and

Antichrist's advent using Scripture, prophecies, and, later, astrology. Although originally sure that the Schism signaled the imminence of the apocalypse, he came to believe that proper action by Christendom could avert this disaster.

The view of the Schism as a preamble to Armageddon depended upon a common interpretation of Scripture. As early as 1403 d'Ailly suggested, however, that the Schism be viewed as an opportunity for reform and not a necessary prelude to the End.[23] Did d'Ailly's shift in his view of the Schism mean that he thought the passage in 2 Thessalonians was an example of the deceptive revelation he discussed in his *Sentences* commentary? He perhaps now viewed that passage in the light of William of Ockham's view of revelation (an interpretation he had explicitly rejected in the commentary on the *Sentences*), to wit, that any revelation contained a disguised conditional.[24] Hence, in this instance, the prediction that Antichrist would come following a Schism depended on the implied clause, "unless the Schism is healed." Or perhaps he saw in the apocalyptic view of the Schism the sort of indirect and accidental deception mentioned in the *Sentences* commentary, which occurs when men misunderstand or misinterpret revelation.[25] In this case, the prediction of 2 Thessalonians could still be true, but it did not apply to the current Schism. D'Ailly would not have had to look to philosophical debates to explain away the vast bulk of apocalyptic utterances; they were simply examples of the false prophecy about which Scripture had warned.

In short, an intellectual of d'Ailly's time might, with good reason, be disappointed with traditional religious ways of knowing about the future. At best, revelation could be ambiguous, and one could not be certain of having discovered its proper interpretation. Further, predictions not contained in Scripture could very well turn out not to have been prophecy. Hence they too were not reliable. Finally, there was the philosophical conclusion debated in the universities that God could immediately *de potentia absoluta* or mediately *de potentia ordinata* promulgate deceptive revelation. Anyone trained in that tradition could certainly question how accurate any statement about the future might be.

In a time of crisis, and particularly one with the eschatological overtones of the Great Schism, forecasting the future turn of events took on particular importance. In such a circumstance the limitations of visions, prophecies, and revelation would doubtless become troublesome. In d'Ailly's case astrology offered a way of complementing and compensating for the shortcomings of other ways of looking at the future. He turned to astrology as he was changing his interpretation of the apocalyptic implications of the Great Schism. D'Ailly in the years after 1400 countered the ready equation of the Schism with the *discessio* of 2 Thessalonians and doubted those would-be prophets who said Antichrist had already been born. Yet merely to substitute his view that a reform of the church would ward off the apocalypse was to speak within the same limited system of revelation and prophecy. In other words, d'Ailly could be misinterpreting revelation just as those who predicted the imminent End were. Astrology provided a glimpse of the future that was external to the traditional religious ways of prediction. It, too, could err; God was omnipotent, the future was contingent, and the science was complicated. Yet astrology held out a special appeal for d'Ailly because it was so different from the other means of prediction. Because astrology did not rely on revelation, it could confirm d'Ailly's new interpretation of events without reference to the troublesome types of prognostication upon which the old view had depended.[26]

One might with reason ask whether d'Ailly in these years gave more credence to astrology than he did to prophecy. The answer is no and yes. D'Ailly himself certainly would not have thought he placed astrology over religion. He behaved as if both astrology and Scripture were valid indicators of the future, as for example in his *De persecutionibus ecclesie* (On the persecutions of the church), in which he used both means to investigate the end of the world. Yet it is doubtful that d'Ailly would have or could have read in the stars a future that contradicted or was not contained in Scripture. On the other hand, if revelation was superior in predicting *what* was to pass in the future, astrology, d'Ailly would seem to have thought, could give a better indication of *when*. Again, d'Ailly's astrological prediction of Antichrist is instructive. The stars indicated merely that there was to be a great change in religion in 1789. Only Scripture told him that this change would be led by Antichrist. But for those who interpreted Scripture to say that the apocalypse was imminent, d'Ailly could point to compelling astrological evidence that this event was still some time off.

The philosophical and theological debates of the fourteenth-century universities underscored the difficulties involved in predicting the future. To preserve the contingency of the future and the omniscience and omnipotence of God, it was necessary to stress the limits on man's knowledge of what was to come. It is difficult to know just to what extent his early discussions of such issues penetrated into the later thought of a man like d'Ailly. Yet the problems with revelation and prophecy can go a long way toward explaining the appeal of astrology to such a mind. Astrology served to indicate to d'Ailly the broad pattern of the future at a time when Europe was in the midst of a crisis that largely was understood in apocalyptic terms. The science of the stars filled in some of the gaps in human knowledge left by the philosophical determination about future contingents.

Pierre d'Ailly, then, lived in a world filled with crisis, disaster, and dissolution, but from which there arose many rational, innovative responses. He was a man of action, a significant force in the politics of both the University of Paris and the church. Yet he was also a man of thought. D'Ailly attempted to understand the upheavals of his times and to interpret them within the context of God's overall plan for the world. To that end he used all of the means at his disposal: Scripture, prophecy, history, and, eventually, astrology. He saw in the existence of the Great Schism a frightening sign that Antichrist's advent was near. By using astrology to investigate the apocalypse, he was able to locate that event in the safely distant future.

D'Ailly treated astrology as if it were a science, one that could be of great use to theology. A synthesizer rather than an innovator, he contributed little new to the development of astrology, and yet his work was cited and respected by future readers. He believed that he lived in a world that was wholly contingent upon God's will, but that God had agreed that the world would adhere to a certain order. He came to hold that a human science like astrology could help us to understand our world. Astrology confirmed d'Ailly's rejection of his early apocalypticism. Astrological calculations buoyed his spirits in the months prior to the opening of the Council of Constance and gave him hope that he and his colleagues would bring the Schism to an end. Through his study of the stars, d'Ailly discerned the Schism's true meaning. In so doing, he affirmed the power of human reason to approach the secrets of God.

A Note on the Availability of
d'Ailly's Writings on Astrology

BECAUSE OF d'Ailly's great reputation both before and after his death, many of his works were edited in the early years of printing and in subsequent centuries as well. Although the demand was greatest for d'Ailly's philosophical, theological, and ecclesiological writings, his major astrological treatises survive in at least two incunables. (Salembier alludes to a third edition, but it is not clear that any copies of this printing remain.) His astrological works also circulated widely in manuscript form, generally all grouped together with other works on like subjects.[1] For most of d'Ailly's scientific treatises, one has a wide choice of where to read them.

What follows is a highly selective inventory of my own sources for d'Ailly's writings on the stars. I have grouped this listing into three categories: modern editions, early printed editions, and manuscripts. Particularly in the latter two categories, I make no claims to completeness, for the catalogue would be enormous. In Ernst Zinner's compilation of astronomical manuscripts in Germanic lands, for example, there are eighty entries devoted to Pierre d'Ailly alone. For fuller details, one should turn to Salembier's bibliographies of 1886 and 1932, Palémon Glorieux's bibliographical note of 1965, and Thorndike and Kibre's catalogue of Latin scientific manuscripts.[2]

Modern Editions

1706. Jean Gerson. *Opera omnia*. Edited by Louis Ellies Dupin. 5 vols. Antwerp: Sumptibus societatis.

Dupin edited a number of d'Ailly's treatises, letters, and sermons. There are many errors in his edition, however, and when I have used his text, I have checked it against manuscript sources.

Of d'Ailly's writings on the stars, Dupin's edition contains:

De legibus et sectis contra superstitiosos astronomos
De falsis prophetis tractatus duo
Principium in cursum Bibliae
Sermo de adventu (December 1414)
Epistola ad Joannem Gersonium (*Incipit*: "Postquam scripseram . . .")
Epistola ad Papam Joannem XXIII (*Incipit*: "Dudum . . .")

1728–36. Charles du Plessis d'Argentré, ed. *Collectio judiciorum de novis erroribus*. 3 vols. Paris: Lambert Coffin.

Contains d'Ailly's *Errores sectae hominum intelligentiae* and records of the trial of the astrologer Symon de Phares.

1761. Etienne Baluze. *Miscellanea novo ordine digesta: Tomus secundus—Monumenta Sacra*. Lucca: Vincentius Junctinius.

Another edition of d'Ailly's *Errores sectae hominum intelligentiae*.

1877. Paul Tschackert. *Peter von Ailli: Zur Geschichte des grossen abendlandischen Schisma und der Reformconcilien von Pisa und Constanz*. Gotha: Friedrich Andreas Perthes. Reprint. Amsterdam: Rodopi, 1968.

Tschackert prints excerpts from a number of d'Ailly's works in his appendix, including:

Sermo in die omnium sanctorum (November 1, 1416)
Sermo de beato Bernardo

1889. Paul Fredericq, ed. *Corpus documentorum inquisitionis haereticae pravitatis Neerlandicae*. 5 vols. Ghent: J. Vuylsteke.

Also prints d'Ailly's *Errores sectae hominum intelligentiae*.

1904. Noël Valois. "Un ouvrage inédit de Pierre d'Ailly, le *De persecutionibus ecclesiae*." *Bibliothèque de l'Ecole des Chartes* 65: 557–74.

Prints extensive excerpts.

1930. Edmond Buron, ed. Imago mundi *de Pierre d'Ailly: Texte latin et traduction française des quatre textes cosmographiques de d'Ailly et des notes marginales de Cristophe Colomb. Etude sur les sources de l'auteur*. 3 vols. Paris: Maisonneuve Frères.

1948. E. F. Keeber, trans. *Imago mundi [of Pierre d'Ailly]*. Wilmington, N.C.: Linprint.

1960–73. Jean Gerson. *Oeuvres complètes*. Edited by Palémon Glorieux. 10 vols. Paris: Desclée & Cie.

Glorieux edited letters between Gerson and d'Ailly, including d'Ailly's *Apologia astrologiae defensiva ad Joannem Gersonium* and *Epistola ad Joannem Gersonium* (*Incipit:* "Postquam scripseram . . .").

1972. Palémon Glorieux. "Deux élogues de la saint écriture par Pierre d'Ailly." *Mélanges de science religieuse* 29: 113–29.

French translation of d'Ailly's *Principium in cursum Bibliae*.

1986. Olaf Pluta. *Die philosophische Psychologie des Peter von Ailly: Ein Beitrag zur Geschichte der Philosophie des späten Mittelalters*. Bochumer Studien zur Philosophie, 6. Amsterdam: B. R. Grüner.

Contains an edition of d'Ailly's *De anima*.

———. Marguerite Chappuis. *Le Traité de Pierre d'Ailly sur la Consolation de Boèce*. (In preparation).

Chappuis discussed her planned edition of d'Ailly's *Tractatus utilis super Boecii de consolatione philosophie* in an article in *Freiburger Zeitschrift für Philosophie und Theologie* 31 (1984): 89–107.

Early Printed Editions

Many of the works I discuss are found in three incunabular collections of d'Ailly's writings.

Tractatus de imagine mundi et varia ejusdem auctoris et Joannis Gersonis opuscula. Louvain: Johann de Westphalia, ca. 1483.

This is the volume of d'Ailly's writings read and annotated by Columbus. For that reason, many copies of this book are in American library collections. Furthermore, a limited number of facsimiles of Columbus's copy were made in 1927: *Imago mundi by Petrus de Aliaco (Pierre d'Ailly) with annotations by Christopher Columbus* (Boston: Massachusetts Historical Society, 1927). The volume contains the following works by d'Ailly:

> *Imago mundi*
> *Epilogus mappe mundi*
> *De legibus et sectis contra superstitiosos astronomos*
> *De correctione kalendarii*
> *De vero ciclo lunari*
> *Cosmographie tractatus duo*
> *Vigintiloquium de concordantia astronomice veritatis cum theologia*
> *De concordia astronomice veritatis et narrationis hystorice*
> *Elucidarium astronomice corcordie cum theologia et cum hystorica narratione*
> *Apologetica defensio astronomice veritatis*
> *Alia secunda apologetica defensio eiusdem*
> *Tractatus de concordia discordantium astronomorum*

Concordantia astronomie cum theologia. Concordantia astronomie cum hystorica narratione. Et elucidarium duorum precedentium. Augsburg: Erhard Ratdolt, 1490.

Ratdolt specialized in printing astrological works. For the three treatises that it contains, his edition is vastly preferable to the 1483 volume above, whose text is faulty at times.

Tractatus et sermones. Strasbourg: [Printer of Jordanus de Quedlinburg], 1490.

This is a large collection of writings by d'Ailly. An earlier edition of this work was printed in Brussels in 1484. Among its contents are:

> *De anima*
> *Sermo de beato Dominico confessore*
> *Sermo de beato Francisco confessore*
> *Sermo primus de adventu domini* (On: "Ecce salvator tuus venit")
> *Sermo secundus de adventu* (December 1414)
> *Sermo (tertius) de quadruplici adventu domini et specialiter de adventu ad iudicium* (December 1385)

The following works exist in many early printings. I note here the editions I have used.

Pierre d'Ailly. *Quaestiones super libros Sententiarum cum quibusdam in fine adjunctis.* Strasbourg: n.p., 1490. Reprint. Frankfurt: Minerva, 1968.

Pierre d'Ailly. *Questiones Magistri Petri de Aylliaco cardinalis cameracensis super primum, tertium et quartum libros sententiarum.* Paris: Jean Petit, 1505(?).

Both volumes contain, inter alia, d'Ailly's *Sentences* commentary and his *Principium in cursum Bibliae.*

Pierre d'Ailly. [*Questiones* on Sacrobosco's *Sphera*]. In Joannes de Sacrobosco, *Uberrimum sphere mundi commentum intersertis etiam questionibus domini Petri de Aliaco.* Paris: Jean Petit, 1498.

Pierre d'Ailly. *Tractatus Petri de Eliaco episcopi Cameracensis super libros metheororum de impressionibus aeris. Ac de hijs quae in prima, secunda, atque tertia regionibus aeris fiunt.* Strasbourg: Johannes Prüs, 1504.

Pierre d'Ailly. *Tractatus de anima.* Paris: Jean Petit, 1505.

Manuscript Sources

I have also consulted the following manuscripts, both to check the accuracy of the various editions of d'Ailly's writings and to read works that have never been (or have been only partially) edited.

Cracow. Biblioteka Jagiellonska. MS 575, fols. 108r–111r; MS 584, fols. 56r–59r; and MS 586, fols. 66v–70r.

These are all fifteenth-century manuscripts containing scientific writings by d'Ailly. At the folios indicated each contains a work on conjunctions by d'Ailly, beginning "Pro declaratione decem dictarum figurarum . . ." These are the only known copies of this work.

Marseilles. Bibliothèque Municipale. MS 1156, fols. 1–8; 11–30.

This is the sole manuscript copy of d'Ailly's *De persecutionibus ecclesie,* from which Valois published his partial edition in 1904. Valois omits much of the astrological section of the treatise.

Paris. Bibliothèque Nationale. MS Lat. 3122.

This is an important collection of writings from the early years of d'Ailly's career, assembled for his own use. The treatises therein date from 1372 to 1388 or 1394. D'Ailly willed this manuscript to the College of Navarre at his death. Among its contents are:

Sermo de adventu (On: "Scitote quoniam regnum dei . . .")
Sermo de beato Bernardo (of which Tschackert prints excerpts)
De falsis prophetis tractatus duo (this is the manuscript upon which Dupin bases his edition of these treatises)
Tractatus utilis super Boecii de consolatione philosophie (this is the best and fullest version of the text, according to Chappuis)

Vatican City. Biblioteca Apostolica Vaticana. MS Reg. lat. 689A.

Published in facsimile. Gilbert Ouy. *Le recueil épistolaire autographe de Pierre d'Ailly et les notes d'Italie de Jean de Montreuil.* Umbrae Codicum Occidentalium, 9. Amsterdam: North-Holland, 1966. This manuscript contains a copy, in d'Ailly's own hand, of his *Epistola ad Papam Joannem XXIII (Incipit:* "Dudum . . .").

Vienna. Österreichische Nationalbibliothek. MS 5138.

The manuscript dates from 1474. It contains a number of astrological works, including the following by d'Ailly:

De concordantia theologiae et astronomiae (= *Vigintiloquium*)

De concordantia astronomiae et historiae

Elucidarium astronomicae concordiae cum theologica et historica veritate

Apologeticae defensiones duae astronomicae veritatis

Opusculum de themate coelesti tempore creationis et de conjunctionibus insequentibus
   (= *De figura inceptionis mundi* . . . This work has never been printed.)

De concordantia discordantium astronomorum

Vienna. Österreichische Nationalbibliothek. MS 5266.

This manuscript dates from the fifteenth century, after 1434. It contains a large number of astrological and astronomical works, most of which are by d'Ailly. Of d'Ailly's works, there are:

De concordantia astronomicae veritatis cum theologia (= *Vigintiloquium*)

De concordia astronomiae cum historia

Elucidarium astronomicae concordiae cum theologica et historica veritate

Apologetica defensio prima astronomicae veritatis

Apologetica defensio altera astronomicae veritatis

De figura inceptionis mundi et conjunctionibus sequentibus (never printed)

De concordia discordantium astronomorum

De legibus et sectis contra superstitiosos astronomos

Exhortatio ad concilium generale Constantiense super kalendarii correctione

A Chronology of d'Ailly's Works
Dealing with Astrology

IN CASES where d'Ailly himself put a date on the work, I simply give the date without comment. For the other works, where the time of composition is only approximate, see the notes. Lost are a sermon on "*Vidimus stellam eius in oriente*," mentioned in d'Ailly's *Apologetica defensio astronomice veritatis (I)*; *De astronomia*, mentioned in *De anima*; and *Elucidationes in astronomicon Manilii* (Venice, 1490). The two latter are indicated in Paul Tschackert, *Peter von Ailli: Zur Geschichte des grossen abendländischen Schisma und der Reformconcilien von Pisa und Constanz* (Gotha: Friedrich Andreas Perthes, 1877; repr., Amsterdam: Rodopi, 1968), pp. 358, 360; and Salembier 1886, pp. xxiii, xxv. The *Elucidationes in astronomicon Manilii* indicated by Salembier and Tschackert does not appear in the British Library's Incunable Short Title Catalogue (the BLAISE-LINE ISTC). Given that Manilius was rediscovered by Poggio Bracciolini only in 1416, it is possible that d'Ailly read and commented on his work before his death in 1420, but equally likely that such a commentary never existed or was merely attributed to d'Ailly.

Other significant events in d'Ailly's career appear in brackets.

[1367]	[Licentiate in arts]
Fall 1375	*Principium in cursum Biblie*[1]
Between 1372 and 1396	*Tractatus super libros metheororum de impressionibus aeris*[2]
Between 1372 and 1395?	*Questiones* on Joannes de Sacrobosco's *Sphera*[3]
Between 1372 and 1395	*Sermo de adventu domini* (On: "Ecce salvator tuus venit")[4]
[1376–77]	[Commentary on Peter Lombard's *Sentences*]
Between 1377 and 1381	*De anima*[5]
Between 1377 and 1381	*Tractatus utilis super Boecii de consolatione philosophie*[6]
[April 11, 1381]	[Licentiate in theology]
Between 1378 and 1388	*De falsis prophetis II*[7]
[1384]	[Rector of the College of Navarre]
December 3, 1385	*Sermo (tertius) de quadruplici adventu domini et specialiter de adventu ad iudicium*[8]
[1389]	[Chancellor of the University of Paris]
[1395]	[Bishop of Le Puy]
[1396]	[Bishop of Noyon]
[1397]	[Bishop of Cambrai]
[1403]	[*De materia concilii generalis*]
[1409]	[Council of Pisa]
August 12, 1410	*Imago mundi*[9]

1410	*Epilogus mappe mundi*
December 24, 1410	*De legibus et sectis contra superstitiosos astronomos*[10]
[June 6, 1411]	[Cardinal]
[January 10, 1412]	[*Apologia concilii Pisani*]
1414	*Vigintiloquium (= Concordantia astronomie cum theologia)*[11]
May 10, 1414	*Concordantia astronomie cum hystorica narratione*[12]
June 18, 1414	*Epistola ad Papam Joannem XXIII (Incipit:* "Dudum . . .")
September 24, 1414	*Elucidarium*[13]
September 26, 1414	*Apologetica defensio astronomice veritatis*
October 3, 1414	*Alia secunda apologetica defensio eiusdem*
1414	*Pro declaratione decem dictarum figurarum*[14]
1414	*De figura inceptionis mundi et coniunctionibus mediis sequentibus*[15]
[November 1414]	[Opening of the Council of Constance]
December 2, 1414	*Sermo de adventu (On:* "Erunt signa in sole . . .")[16]
January 5, 1415	*De concordia discordantium astronomorum*[17]
November 1, 1416	*Sermo in die omnium sanctorum*[18]
[May 30, 1417]	[*Modus seu forma elegendi Summum Pontificem*]
1418	*De persecutionibus ecclesie*[19]
December 1419	*Apologia astrologiae defensiva*[20]
December 1419	*Epistola ad Joannem Gersonium (Incipit:* "Postquam scripseram . . .")[21]

d'Ailly 1490	*Concordantia astronomie cum theologia. Concordantia astronomie cum hystorica narratione. Et elucidarium duorum precedentium.* Augsburg: Erhard Ratdolt, 1490.
PL	J.-P. Migne, ed. *Patrologiae cursus completus. Series latina.* Paris: Garnier Fratres, 1844–64.
Salembier 1886	Louis Salembier. *Petrus de Alliaco.* Lille: J. Lefort, 1886.
Thorndike, *HMES*	Lynn Thorndike. *A History of Magic and Experimental Science.* 8 vols. New York: Columbia University Press, 1923–58.

a.	*articulus*
diff.	*differentia*
q.	*quaestio*
tr.	*tractatus*
v.	*verbum; verbi* (in reference to Pierre d'Ailly's *Vigintiloquium* and pseudo-Ptolemy's *Centiloquium*, both of which are divided into *verbi* and not chapters)

In the early printed editions of Pierre d'Ailly's works, the pages are not numbered. I have cited these works by signature and folio numbers. When these indications are lacking, I have assigned numbers in brackets, following the sequence in the volume; for example, fol. [a5r] would follow immediately after fol. a4v.

Chapter One
Introduction

1. See esp. Samuel Eliot Morison, *Admiral of the Ocean Sea: A Life of Christopher Colum-bus* (Boston: Little, Brown, 1942), e.g., p. 6, where he describes "the Columbus of action, the Discoverer who held the key to the future in his hand, and knew in exactly which of a million possible keyholes it would turn the lock."

2. E.g., Kirkpatrick Sale, *The Conquest of Paradise: Christopher Columbus and the Colum-bian Legacy* (New York: Knopf, 1990), pp. 17–18: "It is hard to know from his [Colum-bus's] later writings exactly what impelled those dreams, but the standard historians' line that the European conquest of America was driven by 'God, gold, and glory' would seem to be pretty much on the mark in this case. . . . His dedication to gold . . . would even determine the course of every one of his explorations in the Caribbean."

3. Pauline Moffit Watts, "Prophecy and Discovery: On the Spiritual Origins of Chris-topher Columbus's 'Enterprise of the Indies,'" *American Historical Review* 90 (1985): 93–99, citing Columbus's *Libro de las Profecias* (Book of prophecies) and the admiral's prefa-tory letter to Ferdinand and Isabella.

4. Watts's article is the first description of Columbus's career fully to take into account the apocalyptic strain in his thought and d'Ailly's influence in that matter. Columbus's marginal annotations on d'Ailly survive; they are available in facsimile in a number of American libraries in *Imago mundi by Petrus de Aliaco (Pierre d'Ailly) with annotations by Christopher Columbus* (Boston: Massachusetts Historical Society, 1927). For the influence of d'Ailly's geographical writings on Columbus, see *Imago mundi de Pierre d'Ailly: Texte latin et traduction française des quatre textes cosmographiques de d'Ailly et des notes marginales de Cristophe Colomb. Etude sur les sources de l'auteur par Edmond Buron*, 3 vols. (Paris: Maison-neuve Frères, 1930).

5. This latter is the picture of those believing in astrology presented in T. M. Luhrmann, *Persuasions of the Witch's Craft: Ritual Magic in Contemporary England* (Cam-bridge, Mass.: Harvard University Press, 1989).

6. Thorndike, *HMES*. Thorndike's work remains a fundamental starting point be-cause of its enormous scope. Sarton is quoted in Otto Neugebauer, "The Study of Wretched Subjects," *Isis* 42 (1951), reprinted in Neugebauer, *Astronomy and History: Selected Essays* (New York: Springer-Verlag, 1983), p. 3. Several of the essays in this collection bear testimony to Neugebauer's conviction that a knowledge of astrological concepts is fundamental to an understanding of the history of astronomy, as evidenced in his collection (with H. B. Van Hoesen), *Greek Horoscopes*, Memoirs of the American Philosophical Society, 48 (Philadelphia: American Philosophical Society, 1959).

7. Edward S. Kennedy, "Ramifications of the World-Year Concept in Islamic Astrol-ogy," *Proceedings of the Tenth International Congress of the History of Science* (1962): 23–43; Edward S. Kennedy and David Pingree, *The Astrological History of Masha'allah*, Harvard Monographs in the History of Science (Cambridge, Mass.: Harvard University Press, 1971); David Pingree, *The Thousands of Abu Ma'shar*, Studies of the Warburg Institute, 30

(London: Warburg Institute, 1968). Richard Lemay has emphasized the importance of Abu Ma'shar's astrological texts as vehicles for the transmission of Aristotelian physics to the West in his *Abu Ma'shar and Latin Aristotelianism in the Twelfth Century: The Recovery of Aristotle's Natural Philosophy through Arabic Astrology* (Beirut: American University, 1962). Also see Bernard R. Goldstein and David Pingree, "Levi ben Gerson's Prognostication for the Conjunction of 1345," *Transactions of the American Philosophical Society* 80, pt. 6 (1990).

8. John D. North, "Astrology and the Fortunes of Churches," *Centaurus* 24 (1980): 181–211; "Celestial Influence—The Major Premiss of Astrology," in Paola Zambelli, ed., *"Astrologi hallucinati": Stars and the End of the World in Luther's Time* (Berlin and New York: Walter de Gruyter, 1986), pp. 45–100; *Chaucer's Universe* (Oxford: Clarendon Press, 1988); "Chronology and the Age of the World," in Wolfgang Yourgrau and Allen D. Beck, eds., *Cosmology, History, and Theology* (New York: Plenum Press, 1977), pp. 307–33; *Horoscopes and History* (London: Warburg Institute, 1986); "Medieval Concepts of Celestial Influence: A Survey," in Patrick Curry, ed., *Astrology, Science and Society: Historical Essays* (Woodbridge, Suffolk: Boydell Press, 1987), pp. 5–18; and *Richard of Wallingford: An Edition of His Writings with Introductions, English Translation and Commentary by John D. North*, 3 vols. (Oxford: Clarendon Press, 1976).

9. E.g., M.L.W. Laistner, "The Western Church and Astrology during the Early Middle Ages," *Harvard Theological Review* 34 (1941): 251–75, reprinted in his *The Intellectual Heritage of the Early Middle Ages: Selected Essays*, ed. Chester G. Starr (Ithaca, N.Y.: Cornell University Press, 1957); and Theodore Otto Wedel, *The Medieval Attitude towards Astrology Particularly in England* (New Haven: Yale University Press, 1920). For more recent studies of theologians' responses to astrology, see also Marie-Thérèse d'Alverny, "Astrologues et théologiens au XIIe siècle," in *Mélanges offerts à M.-D. Chenu*, Bibliothèque thomiste, 37 (Paris: J. Vrin, 1967), pp. 31–50; and Thomas Litt, *Les corps célestes dans l'univers de Saint Thomas d'Aquin* (Louvain: B. Nauwelaerts and Paris: Publications Universitaires, 1963). There are some notable exceptions. Besides the work of Thorndike mentioned in n. 6, one might cite Franz Johannes Boll, Carl Bezold, and Wilhelm Gundel, *Sternglaube und Sterndeutung: Die Geschichte und das Wesen der Astrologie* (Darmstadt: Wissenschaftliche Buchgesellschaft, 1974); Wilhelm Gundel, *Sternglaube, Sternreligion und Sternorakel* (Leipzig: Verlag Quelle & Meyer, 1933); Friedrich von Bezold, "Astrologische Geschichtsconstruction im Mittelalter," *Deutsche Zeitschrift für Geschichtswissenschaft* (1892): 29–72; and, for ancient Greek and Roman astrology, Auguste Bouché-Leclercq, *L'astrologie grecque* (Brussels: Leroux, 1899; repr., Aalen: Scientia Verlag, 1979).

10. Herbert Pruckner, *Studien zu den astrologischen Schriften des Heinrich von Langenstein*, Studien der Bibliothek Warburg, 14 (Leipzig: Teubner, 1933); G. W. Coopland, *Nicole Oresme and the Astrologers: A Study of His Livre de Divinacions* (Cambridge, Mass.: Harvard University Press, 1952).

11. Symon de Phares, *Recueil des plus célèbres astrologues et quelques hommes docts, faict par Symon de Phares du temps de Charles VIIIe*, ed. Ernest Wickersheimer (Paris: Champion, 1929). Symon de Phares's practice of astrology was condemned by the University of Paris in 1494; his catalogue of famous astrologers was composed as a partial defense. Wickersheimer's edition forms the basis for the account of astrology given in Alexander Murray, *Reason and Society in the Middle Ages* (Oxford: Clarendon Press, 1978), esp. pp. 207–9; and

Maxime Préaud, *Les astrologues à la fin du Moyen Age*, Collection Lattès/Histoire, Groupes et sociétés (Paris: J. C. Lattès, 1984).

12. S. J. Tester, *A History of Western Astrology* (Woodbridge, Suffolk: Boydell Press, 1987).

13. Valerie I. J. Flint, *The Rise of Magic in Early Medieval Europe* (Princeton, N.J.: Princeton University Press, 1991).

14. Préaud, *Les astrologues*.

15. Hilary M. Carey, "Astrology at the English Court in the Later Middle Ages," in Patrick Curry, ed., *Astrology, Science and Society: Historical Essays* (Woodbridge, Suffolk: Boydell Press, 1987), pp. 41–56; and *Courting Disaster: Astrology at the English Court and University in the Later Middle Ages* (New York: St. Martin's Press, 1992).

16. The sermon for All Saints' Day 1416, *Sermo in die omnium sanctorum*. Extensive excerpts from this sermon are printed in Paul Tschackert, *Peter von Ailli: Zur Geschichte des grossen abendländischen Schisma und der Reformconcilien von Pisa und Constanz* (Gotha: Friedrich Andreas Perthes, 1877; repr., Amsterdam: Rodopi, 1968), pp. [41]–[50].

17. Tschackert, *Peter von Ailli*; Salembier 1886 (the author's Latin dissertation); and *Le cardinal Pierre d'Ailly, chancelier de l'Université de Paris, évêque du Puy et de Cambrai 1350–1420* (Tourcoing: Imprimerie Georges Frère, 1932) (published posthumously). Salembier terms d'Ailly's astrology "a very grave matter" ("materiam gravissimam"); Salembier 1886, p. 177. Lynn Thorndike offers an informative and sympathetic view of d'Ailly's views on the stars, but makes no effort to relate d'Ailly's interest in the stars to other aspects of his life or thought; Thorndike, *HMES*, 4: chap. 42. Stefano Caroti has also examined d'Ailly's astrological writings, mainly as regards the influence on d'Ailly of Nicole Oresme's polemics against astrology; Caroti, *La critica contro l'astrologia di Nicole Oresme e la sua influenza nel medioevo e nel Rinascimento*, Atti dell'Accademia nazionale dei Lincei, Memorie, Classe di scienze morali, storiche e filologiche, ser. 8, vol. 23, fasc. 6 (1979), pp. 545–684.

18. Bernard Guenée, *Entre l'Eglise et l'Etat: Quatre vies de prélats français à la fin du Moyen Age (XIIIe–XVe siècle)* (Paris: Gallimard, 1987), pp. 269 (d'Ailly's study of astrology related to anxiety) and 169 (d'Ailly's ideas do not change over time in most subjects). Guenée's book has recently appeared in English translation as *Between Church and State: The Lives of Four French Prelates in the Late Middle Ages*, trans. Arthur Goldhammer (Chicago: University of Chicago Press, 1991); my citations throughout are to the French edition.

19. E.g., Carey, *Courting Disaster*, p. 110 (d'Ailly "wrote extensively against the practice of divination"); and José Maria da Cruz Pontes, "Astrologie et apologétique au moyen âge," in Christian Wenin, ed., *L'homme et son univers au Moyen Age*, 2 vols., Philosophes médiévaux, 26–27 (Louvain: Institut Supérieur de Philosophie, 1986). He states, "Les humanistes, depuis Pierre d'Ailly jusqu'à Pic de la Mirandole, réfuteront la validité de l'astrologie" (2:637). The treatise in question was edited in Dupin's edition of Gerson's works (Jean Gerson, *Opera omnia*, ed. Louis Ellies Dupin, 5 vols. [Antwerp: Sumptibus societatis, 1706], 1: cols. 778ff.) and is, therefore, the most widely available of d'Ailly's astrological treatises. The same edition (1: cols. 511–603) contains d'Ailly's *De falsis prophetis II*, which in Dupin's edition bears the descriptive note "in quo adversus vanam astrologiae superstitionem fuse disserit." (This work is an early attack on the use of astrology, not from d'Ailly's later period of interest in the stars.)

20. On d'Ailly's philosophy, see Marguerite Chappuis, Ludger Kaczmarek, and Olaf Pluta, "Die philosophischen Schriften des Peter von Ailly: Authentizität und Chronologie," *Freiburger Zeitschrift für Philosophie und Theologie* 33 (1986): 593–615; Richard Desharnais, "Reassessing Nominalism: A Note on the Epistemology and Metaphysics of Pierre d'Ailly," *Franciscan Studies* 34 (1974): 296–305; Maurice P. de Gandillac, "De l'usage et de la valeur des arguments probables dans les questions du Cardinal Pierre d'Ailly sur le 'Livre des Sentences,'" *Archives d'histoire doctrinale et littéraire du moyen âge* 8 (1933): 43–91; Leonard A. Kennedy, *Peter of Ailly and the Harvest of Fourteenth-Century Philosophy*, Studies in the History of Philosophy, 2 (Lewiston, N.Y.: Edwin Mellen Press, 1986); Bernhard Meller, *Studien zur Erkenntnislehre des Peter von Ailly* (Freiburg: Verlag Herder, 1954); Olaf Pluta, *Die philosophische Psychologie des Peter von Ailly: Ein Beitrag zur Geschichte der Philosophie des späten Mittelalters*, Bochumer Studien zur Philosophie, 7 (Amsterdam: B. R. Grüner, 1986); Salembier 1886; *Conceptus et Insolubiles [of Pierre d'Ailly]: An Annotated Translation*, trans. and ed. Paul Vincent Spade (Dordrecht and Boston: D. Reidel, 1980); and Tschackert, *Peter von Ailli*.

On his theology, see William Courtenay, "Covenant and Causality in Pierre d'Ailly," *Speculum* 46 (1971): 94–119; Kennedy, *Peter of Ailly*; George Lindbeck, "Nominalism and the Problem of Meaning as Illustrated by Pierre d'Ailly on Predestination and Justification," *Harvard Theological Review* 52 (1959): 43–60; Alfonso Maierù, "Logique et théologie trinitaire: Pierre d'Ailly," in Zénon Kaluza and Paul Vignaux, eds., *Preuve et Raisons à l'Université de Paris* (Paris: J. Vrin, 1984), pp. 253–68; Francis Oakley, "Pierre d'Ailly and the Absolute Power of God," *Harvard Theological Review* 56 (1963): 59–73; Louis B. Pascoe, "Theological Dimensions of Pierre d'Ailly's Teaching on the Papal Plenitude of Power," *Annuarium historiae conciliorum* 11 (1979): 357–66; Salembier 1886; and Tschackert, *Peter von Ailli*.

On d'Ailly's ecclesiological and political thought and activities, see Alan E. Bernstein, *Pierre d'Ailly and the Blanchard Affair: University and Chancellor of Paris at the Beginning of the Great Schism* (Leiden: E. J. Brill, 1978); John Patrick McGowan, *Pierre d'Ailly and the Council of Constance* (Washington, D.C.: Catholic University of America Press, 1936); Francis Oakley, *The Political Thought of Pierre d'Ailly: The Voluntarist Tradition* (New Haven: Yale University Press, 1964); and Tschackert, *Peter von Ailli*.

For d'Ailly's geographical writings, see *Imago mundi* (ed. Buron); Charles Guignebert, *De imagine mundi ceterisque Petri de Alliaco geographicis opusculis* (Paris: E. Leroux, 1902); and Louis Salembier, *Pierre d'Ailly et la découverte de l'Amérique* (Paris: Letouzey et Ané, 1912).

21. D'Ailly wrote, "Ego siquidem per 5 circiter annos post dictam coniunctionem natus sum" ("I myself was born around five years after that conjunction [of March 14, 1345]"). Pierre d'Ailly, *Concordantia astronomie cum hystorica narratione*, in d'Ailly 1490, chap. 57, fol. [d7r]. In the old style of dating the year did not begin until March 25, hence the conjunction was really in 1346 by the modern dating style. If d'Ailly was born exactly five years later or a little more, his birth year would be 1351; if less, 1350. His year of birth has traditionally been given as 1350. Guenée and Max Lieberman both give 1351. Guenée, *Entre l'Eglise et l'Etat*, p. 125 (based on the conjunction's date); Max Lieberman, "Pierre d'Ailly, Jean Gerson et le culte de saint Joseph, II," *Cahiers de Joséphologie* 14 (1966): 289 (based on the testimony of Jean Campani that d'Ailly was approaching his sixty-ninth year in January 1419 [old style; 1420 according to the new style]).

In addition to the biographies by Guenée, Salembier, and Tschackert mentioned earlier, brief accounts of d'Ailly's life appear in A. Coville, "Ailly, Pierre de," in *Dictionnaire de biographie française* (Paris: Letouzey et Ané, 1933–); Francis Oakley, "Pierre d'Ailly," in B. A. Gerrish, ed., *Reformers in Profile* (Philadelphia: Fortress Press, 1967), pp. 40–57; and Gilbert Ouy, "Ailly, Pierre d'," in *Lexikon des Mittelalters* (Munich: Artemis, 1980–).

22. On the affair of Jean Blanchard, see Bernstein, *Pierre d'Ailly and the Blanchard Affair*; on Monzon, see the article of Palémon Glorieux, "Pierre d'Ailly et Saint Thomas," in *Littérature et religion: Mélanges offerts à Joseph Coppin* (Lille: Facultés Catholiques, 1966), pp. 45–54.

23. Because of their rejection of the "realist" position on universals, d'Ailly and other late medieval thinkers have frequently been termed "nominalists." This term is not entirely an accurate representation of the Ockhamist position on universals. Furthermore, it carries with it a negative connotation, based upon the pejorative view of late medieval philosophy common earlier in this century. For these reasons, I deliberately avoid the term here.

24. See, e.g., Desharnais: "Ailly's epistemology agrees with the general doctrine of the major schoolmen. . . . ecumenism is . . . part and parcel of Ailly's philosophical synthesis. . . . he put together harmoniously the spirit which animated both the older (*via antiqua*) and the newer (*via moderna*) ways of reasoning." "Reassessing Nominalism," pp. 304–5.

25. The most accessible summary of d'Ailly's philosophy and theology is that in Oakley, *Political Thought*, chap. 1. Helpful surveys also appear in Salembier 1886, pp. 141–76 and 195–303; and Tschackert, *Peter von Ailli*, chap. 6 ("Ailli's wissenschaftlicher Standpunkt").

26. Oakley, *Political Thought*.

27. Many authors have addressed the subject of d'Ailly's borrowings. See esp. Zénon Kaluza, "Le Traité de Pierre d'Ailly sur l'Oraison dominicale," *Freiburger Zeitschrift für Philosophie und Theologie* 32 (1985): 286–93; Lieberman, "Pierre d'Ailly, Jean Gerson et le culte de saint Joseph, II," pp. 312–14; and Olaf Pluta, "Albert von Köln und Peter von Ailly," *Freiburger Zeitschrift für Philosophie und Theologie* 32 (1985): 261–71 (the dependence of d'Ailly's *De anima* on Albertus Magnus).

28. E.g., Jean de Bruges, in 1444. See Jean-Patrice Boudet, "Simon de Phares et les rapports entre astrologie et prophétie à la fin du Moyen Age," *Mélanges de l'Ecole Française de Rome. Moyen Age* 102 (1990): 639. The early sixteenth-century German astrologer Virdung von Hassfurt copied d'Ailly's astrological prediction for 1789. Max Steinmetz, "Johann Virdung von Hassfurt, sein Leben und seine astrologischen Flugschriften," in Hans-Joachim Köhler, ed., *Flugschriften als Massenmedium der Reformationszeit*, Spätmittelalter und Frühe Neuzeit, 13 (Stuttgart: Klett-Cotta, 1981), pp. 365–66.

29. Bernd Moeller, "Piety in Germany around 1500," in Steven Ozment, ed., *The Reformation in Medieval Perspective* (Chicago: Quadrangle Books, 1971), pp. 57–58.

30. His *Sentences* commentary, for example, was printed seven times in the fifteenth century (Kennedy, *Peter of Ailly*, p. 3). There were ten editions of the *De anima* (Treatise on the soul) between 1490 and 1520 (Pluta, *Die philosophische Psychologie*, pp. vi, xxii–xxix). And there exist some two hundred manuscripts of the *Imago mundi* (Colette Beaune, "La notion d'Europe dans les livres d'astrologie du XV^c siècle," in *La conscience Européene au XVe et au XVIe siècle*, Collection de l'Ecole Normale Supérieure de Jeunes Filles, 22 [Paris: Ecole Normale Supérieure de Jeunes Filles, 1982], p. 3).

31. E.g., Zénon Kaluza: "La pratique instrumentale des connaissances dans tous les domaines l'a obligé à chercher toujours un guide sûr. Telle est probablement la raison du

haut niveau intellectual de ses écrits et de leur manque d'originalité" ("Le traité de Pierre d'Ailly," p. 293). He offers the best discussion of d'Ailly's plagiarisms and the reasons to read him in any case. Max Lieberman states, "il faut convenir que le cardinal de Cambrai est un écrivain remarquable par son défaut d'originalité" ("Pierre d'Ailly, Jean Gerson et le culte de saint Joseph, II," p. 312). As Alan Bernstein notes, we can nonetheless appreciate his role in using, adapting, and transmitting the ideas of his sources. Bernstein, *Pierre d'Ailly and the Blanchard Affair*, p. 183.

32. See Richard Lemay, "The Teaching of Astrology in Medieval Universities, Principally at Paris in the Fourteenth Century," *Manuscripta* 20 (1976): 199–202 (for the foundation of a college of medicine and astrology at Paris in the 1360s and the granting of permission to lecture on astrology in the faculty of arts at Paris on feast days). For an edition and translation of Sacrobosco's text and of a number of medieval commentaries, see Lynn Thorndike, *The Sphere of Sacrobosco and Its Commentators* (Chicago: University of Chicago Press, 1949).

33. See Francis S. Benjamin, Jr., and G. J. Toomer, *Campanus of Novara and Medieval Planetary Theory. Theorica planetarum, edited with an introduction, English translation, and commentary* (Madison: University of Wisconsin Press, 1971). Sacrobosco devotes approximately one page (in Thorndike's edition) to a general description of planetary movements, offering little more information than the fact that each planet's motion (save the sun's) requires the actions of an epicycle, a deferent, and an equant. (These terms will be explained later.) Campanus, by way of contrast, devotes approximately twenty-five pages (in Benjamin and Toomer's edition) to the theory for Venus, Mars, Jupiter, and Saturn, while Mercury, the planet with the most complex motion, receives more than forty pages of commentary.

34. Emmanuel Poulle, "The Alfonsine Tables and Alfonso X of Castille," *Journal for the History of Astronomy* 19 (1988): 99–105. Poulle demonstrates that the Latin "Alfonsine tables" in fact originated in Paris in the 1320s. They differ in several important respects from the Castillian versions produced for or by Alfonso. Also see Poulle, ed. and trans., *Les tables alphonsines avec les canons de Jean de Saxe: Edition, traduction et commentaire* (Paris: Editions du C.N.R.S., 1984). An excellent introduction to the various medieval astronomical tables and their use appears in North, *Chaucer's Universe*, pp. 147–53.

35. North summarizes the contents of Alchabitius's work (ibid., chap. 5).

36. *Epilogus mappe mundi*, in Pierre d'Ailly, *Tractatus de imagine mundi et varia ejusdem auctoris et Joannis Gersonis opuscula* (Louvain: Johann de Westphalia, ca. 1483), fol. e2r: "In primis supponendum est quod celum est figure sperice seu rotunde."

37. D'Ailly noted that theologians recognized in addition a tenth and an eleventh sphere, the empyrean and the crystalline heavens. Ibid., fol. e2v.

38. A good introduction to the history of astronomical theory and the efforts to "save the phenomena" can be found in E. J. Dijksterhuis, *The Mechanization of the World Picture*, trans. C. Dikshoorn (Oxford: Oxford University Press, 1961), esp. pp. 15–67. For the elements of the Ptolemaic system, in addition to Dijksterhuis's treatment, there are excellent summaries in North, *Chaucer's Universe*, pp. 22–26, 134–69; North, *Richard of Wallingford*, 3: app. 29; Benjamin and Toomer, *Campanus*, pp. 39–56; and Geoffrey Chaucer, *The Equatorie of the Planetis, Edited from Peterhouse MS. 75.I*, ed. Derek J. Price (Cambridge: Cambridge University Press, 1955), pp. 93–118.

39. This point is fixed; it does not rotate as the small circle in figure 1 might seem to imply.

40. The illustrator of figure 1 has confused the deferent center and the equant point and, thereby, the circles labeled "deferens" and "equans." The center of the planet's epicycle moves on the deferent (the circle incorrectly labeled "equans" in figure 1); its angular motion is uniform with respect to the equant. The artist's confusion should serve to remind us that the details of astronomy were far from common knowledge.

41. The moon and Mercury required more complicated models to explain their motions; this explanation basically holds for Venus, Mars, Saturn, and Jupiter, although the three outer planets maintain a different relationship to the sun than do Mercury and Venus. For further details and distinctions, see Benjamin and Toomer, *Campanus*, pp. 47–56; North, *Richard of Wallingford*, 3: app. 29; and *Equatorie* (ed. Price), pp. 99–104.

42. North, *Chaucer's Universe*, pp. 139, 143.

43. The treatise is edited and analyzed by Francis J. Carmody in *The Astronomical Works of Thabit B. Qurra* (Berkeley: University of California Press, 1960), pp. 84–113; and appears in translation with commentary by O. Neugebauer in "Thâbit ben Qurra 'On the Solar Year' and 'On the Motion of the Eighth Sphere,'" *Proceedings of the American Philosophical Society* 106 (1962): 290–99. See also North, *Richard of Wallingford*, 3: app. 25. On the Alfonsine tables, see Poulle, "The Alfonsine Tables," p. 100.

44. The basics of medieval astrology are outlined in North, *Chaucer's Universe*, pp. 194–234; and J. C. Eade, *The Forgotten Sky: A Guide to Astrology in English Literature* (Oxford: Clarendon Press, 1984), pp. 39–103. Also useful is Bouché-Leclercq, *L'astrologie grecque*. One might also consult Willy Hartner, "The Mercury Horoscope of Marcantonio Michiel of Venice: A Study in the History of Renaissance Astrology and Astronomy," in Hartner, *Oriens-occidens: Ausgewählte Schriften zur Wissenschafts- und Kulturgeschichte. Festschrift zum 60. Geburtstag* (Hildesheim: Georg Olms, 1968), pp. [91]–[105]. A good overview of the history and techniques of astrology is found in David Pingree, "Astrology," in *Dictionary of the History of Ideas* (New York: Charles Scribner's Sons, 1968, 1973).

45. For astrology in literature see, for example, Chaucer's Wife of Bath's Tale, lines 609–13: "For certes, I am al Venerien / In feelynge, and myn herte is Marcien. / Venus me yaf my lust, my likerousnesse, / And Mars yaf me my sturdy hardynesse / Myn ascendent was Taur, and Mars therinne." Chaucer, *Canterbury Tales*, in *The Complete Works of Geoffrey Chaucer*, ed. F. N. Robinson (Boston: Houghton Mifflin, 1933). For a popular prophecy that presupposed knowledge of planetary characters, see Robert E. Lerner, *The Powers of Prophecy: The Cedar of Lebanon Vision from the Mongol Onslaught to the Dawn of the Enlightenment* (Berkeley: University of California Press, 1983).

46. North notes that the calculations involved in determining a single planetary longitude would fill a small sheet of paper, while "a practised astronomer could probably have computed a full set of longitudes . . . in a couple of hours or less." Using an *equatorium*, one could perform the same task in a matter of minutes. *Chaucer's Universe*, pp. 153, 156. On such instruments, also see North, *Richard of Wallingford* (he designed an instrument called the *albion*); Benjamin and Toomer, *Campanus* (he includes instructions for making an *equatorium*); Emmanuel Poulle, *Les instruments de la théorie des planètes selon Ptolémée: Equatoires et horlogerie planétaire du XIIIᵉ au XVIᵉ siècle*, 2 vols. (Geneva: Droz, and Paris: Champion, 1980); *Equatorie* (ed. Price) (Chaucer's *equatorium*); and Lynn White, Jr., "Medical Astrologers and Late Medieval Technology," *Viator* 6 (1975): 295–308.

47. There are useful descriptions of the development of astrology in Pingree's article in the *Dictionary of the History of Ideas*; Tester, *History of Western Astrology*, pp. 11–29; and

O. Neugebauer, *The Exact Sciences in Antiquity*, 2d ed. (New York: Dover, 1969), esp. pp. 170–77.

48. I use "temperament" in the sense in which it is defined in the theory of the four humors. A melancholic temperament is characterized by an excess of black bile, which is cold and dry and causes sullenness, depression, and anger. One can find such a characterization of the planets, for example, in Alchabitius, *Libellus isagogicus abdilazi id est servi gloriosi dei qui dicitur Alchabitius ad magisterium iudiciorum astrorum interpretatus a Joanne hispalensi . . .*, in *Alchabitius cum commento. Noviter impresso* (Venice: Melchior Sessa, 1512), diff. 2.

49. Pierre d'Ailly, [*Questiones* on Sacrobosco's *Sphera*], in Joannes de Sacrobosco, *Uberrimum sphere mundi commentum intersertis etiam questionibus domini Petri de Aliaco* (Paris: Jean Petit, 1498), chap. 2, q. 2 (of d'Ailly's questions), fol. [g ii v]: "cum sit questio de qua oporteat discernere sexum sicut fit in pluribus videlicet ut de puero in ventre matris utrum sit vir vel mulier vel de utroque sexu tunc si fuerit significator inventus in aliquo gradu masculino illud atestatur masculinitati et si in gradu feminino illud atestatur femininitati." Note that in this case it is a question of individual degrees within signs and not the signs themselves bearing significance. A list of masculine and feminine degrees can be found in Alchabitius, diff. 1, fol. 5v.

50. D'Ailly, *Questiones* on Sacrobosco, chap. 2, q. 2, fol. [g ii v]: "Et est advertendum quod ad inceptionem alicuius operis bene respiciendum est ipsam lunam si fuerit in signo fixo vel mobili vel communi quia si fuerit in mobili tunc opus inceptum non haberet durationem, si in fixo debet diu durare, si in communi tunc debet durare communiter."

51. Alchabitius, diff. 1, fols. 4v–5v.

52. For different methods for erecting the houses, see esp. North, *Horoscopes and History*, pp. 1–69. Also useful is Eade, *Forgotten Sky*, pp. 42–50. I describe subsequently what North terms the "standard method."

53. On the use of the astrolabe, see North, *Horoscopes and History*, pp. 56–69; and *Chaucer's Universe*, pp. 64–65, 84. North notes that the fourteenth-century calendar of Nicholas of Lynn, for example, has columns listing the cusps of houses. *Chaucer's Universe*, p. 104.

54. The houses in figure 2, allowing for scribal error or slight miscalculation, have been erected according to this "standard method."

55. Alchabitius, diff. 1, fols. 7r–8r. The significations of the twelve houses are fairly standard.

56. Alchabitius, diff. 1, fol. 8v.

57. These more complex cases are treated in Alchabitius, diff. 4–5. North offers an excellent discussion of such considerations in *Chaucer's Universe*, pp. 213–52.

58. See d'Ailly's statements in the *Epilogus mappe mundi*, fol. e3; and the *Elucidarium*, in d'Ailly 1490, chap. 24. This sentiment was also expressed by Messahalla (Masha'allah) in *Epistola Messahalae de rebus eclipsium, et de coniunctionibus Planetarum, in revolutionibus annorum mundi, breviter elucidata*, trans. Johannes Hispalensis, chap. 9, printed in *Messahalae, antiquissimi ac laudatissimi inter arabes astrologi, Libri tres* (Nuremberg: Ioannes Montanus, 1549); and by Avenesra (Abraham ibn Ezra), in *Liber coniunctionum*, chap. 1, printed in *Abrahe Avenaris Judei astrologi peritissimi in re iudiciali opera: ab excellentissimo philosopho Petro de Abano post accuratam castigationem in latinum traducta* (Venice: Petrus Liechtenstein, 1507).

59. Albumasar, *De magnis coniunctionibus, annorum revolutionibus, ac eorum profectionibus, octo continens tractatus* (Augsburg: Erhard Ratdolt, 1489), tr. 1, diff. 1, fol. a3v.

60. D'Ailly, *Concordantia astronomie cum hystorica narratione*, chap. 1, fol. b7v: "Et significat super mutationes imperiorum et regnorum et super ignitas impressiones in aere, super diluvium et super terre motum et super gravitatem annone."

61. The four types of conjunctions are defined in the *Concordantia astronomie cum hystorica narratione*, chap. 1, but d'Ailly is not entirely clear here. I follow the extremely lucid explanation in North, *Chaucer's Universe*, pp. 370–74. Also see North, "Astrology and the Fortunes of Churches," pp. 181–211; and Eugenio Garin, *Astrology in the Renaissance: The Zodiac of Life*, trans. Carolyn Jackson and June Allen (London and Boston: Routledge and Kegan Paul, 1983), pp. 15–29.

62. These correspondences were generally expressed pictorially in a figure termed the "zodiac man." See Charles Clark, "The Zodiac Man in Medieval Medical Astrology," *Journal of the Rocky Mountain Medieval and Renaissance Association* 3 (1982): 13–38.

63. In his *Super sapientiam Salomonis*, cited in North, *Chaucer's Universe*, p. 131.

64. For a discussion of astrological forecasts of the weather, see Stuart Jenks, "Astrometeorology in the Middle Ages," *Isis* 74 (1983): 185–210. Jenks notes that such treatises were aimed either at beginners in the field or at professional astrologers and not at the group most obviously in need of information about the weather, farmers. Weather forecasts also appeared in the more popular annual predictions that circulated in the later Middle Ages. See Philippe Contamine, "Les prédictions annuelles astrologiques à la fin du Moyen Age: Genre littéraire et témoin de leur temps," in *Histoire sociale, sensibilités collectives et mentalités: Mélanges Robert Mandrou* (Paris: Presses Universitaires de France, 1985), p. 198.

65. Watts, "Prophecy and Discovery," pp. 73–74: "Morison's image of Columbus [as innovator and explorer] coincides with but one of two images that Columbus had of himself. The second image tended, particularly in his later years, to take precedence over the first and seemed to consume him. . . . This second self-image is epitomized in the signature that Columbus adopted: Christoferens."

Chapter Two
The Medieval Debate about Astrology

1. Theodore Otto Wedel, *The Medieval Attitude towards Astrology Particularly in England* (New Haven: Yale University Press, 1920), pp. 16–17; Wedel cites Gregory, *Homily XX on Epiphany*; Cassiodorus (who says Basil and Augustine oppose astrology); Lactantius, *Liber Divinarum Institutionum*; Tatian, *Oratio ad Graecos*; and Origen, *Comment. in Matth.* See also M.L.W. Laistner, "The Western Church and Astrology during the Early Middle Ages," *Harvard Theological Review* 34 (1941): 251–75, reprinted in Laistner, *The Intellectual Heritage of the Early Middle Ages: Selected Essays*, ed. Chester G. Starr (Ithaca, N.Y.: Cornell University Press, 1957), chap. 4, pp. 57–82; Valerie I. J. Flint, *The Rise of Magic in Early Medieval Europe* (Princeton, N.J.: Princeton University Press, 1991), pp. 87–101; and her "The Transmission of Astrology in the Early Middle Ages," *Viator* 21 (1990): 1–27. The attack on astrological fatalism did not begin with the church fathers. Their arguments against the stars' power over humans stemmed from centuries of debate about fatalism and free will. The objections which Augustine and others raise against astrologi-

cal necessity have been traced back as far as the teachings of Carneades in the second century B.C. For the history of these arguments in antiquity, see David Amand, *Fatalisme et liberté dans l'antiquité grecque: Recherches sur la survivance de l'argumentation morale antifataliste de Carnéade chez les philosophes grecs et les théologiens chrétiens des quatre premiers siècles* (Louvain: Bibliothèque de l'Université, 1945).

2. Augustine, [*De civitate dei*] *The City of God against the Pagans*, with English translation by William M. Green, 7 vols., Loeb Classical Library (Cambridge, Mass.: Harvard University Press, 1963), book 5, chaps. 1–7. All the translations in the text are Green's. Augustine also discusses astrology in the *Confessions*, book 7, chap. 6, and book 4, chap. 3; the letter to Lampadius; *De doctrina christianorum* 2.21; *Ad faustum* 2.5 (the star of the Magi); *De diversis quaestionibus* 83; and *De genesi ad litteram* 2.17.

3. This was a favorite argument of earlier authors as well (e.g., Cicero, Favorinus, and Sextus Empiricus) and is traceable to Carneades. Wedel, *Medieval Attitude towards Astrology*, pp. 11–12, n. 3; Amand, *Fatalisme et liberté*, pp. 51–53. Augustine dealt with many of these issues relating to astrology in a briefer fashion in the *Confessions*, book 7.

4. *City of God*, book 5, chap. 1 (2:134–36 of Loeb ed.): attacking "illi . . . qui positionem stellarum quodam modo decernentium qualis quisque sit et quid ei proveniat boni quidve mali accidat ex Dei voluntate suspendunt."

5. Ibid. (2:136 of Loeb ed.): "Quale inde iudicium de hominum factis Deo relinquitur, quibus caelestis necessitas adhibetur, cum dominus ille sit et siderum et hominum?"

6. Note that Augustine is not concerned with any threat to God's omnipotence posed by astrology (such as the argument that stellar control takes events out of the hands of God) as d'Ailly will be later in dealing with the issue of astrology and Christianity. Rather he wishes to stress that man, and not God or the stars, bears the ultimate responsibility for sin.

7. *City of God*, book 5, chap. 1 (2:136 of Loeb ed): " . . . dicuntur stellae significare potius ista quam facere . . . (non enim mediocriter doctorum hominum fuit ista sententia)." This is probably a reference to Plotinus's opinion about the stars. See Plotinus, *The Enneads*, trans. Stephen McKenna, 2d ed. (London: Faber and Faber, 1956), 2.3, pp. 96–98: "We may think of the stars as letters perpetually being inscribed on the heavens or inscribed once for all and yet moving as they pursue the other tasks allotted to them: upon these main tasks will follow the quality of signifying. . . . If all this be true, we must at once admit signification, though, neither singly nor collectively, can we ascribe to the stars any efficacy except in what concerns the [material] All and in what is of their own residuary function."

8. *City of God*, book 5, chap. 1 (2:136–38 of Loeb ed.): "quid fit quod nihil umquam dicere potuerunt cur in vita geminorum . . . sit plerumque tanta diversitas . . .?"

9. *City of God*, book 5, chap. 2 (2:138 of Loeb ed.): "Cicero dicit Hippocratem, nobilissimum medicum, scriptum reliquisse quosdam fratres, cum simul aegrotare coepissent et eodem levaretur, geminos suspicatum." Augustine's source may have been Cicero's *De fato*.

10. Ibid. In the *Confessions* (book 7, chap. 6) Augustine gave an example of a slave's child and a patrician's son born at exactly the same hour to vastly different fates.

11. *City of God*, book 5, chaps. 2–3. In chap. 4, Augustine discusses the example of Jacob and Esau.

12. Ibid., book 5, chap. 3 (2:144 of Loeb ed.): "Nam si tam multum in caelo interest quod constellationibus conprehendi non potest ut alteri geminorum hereditas obveniat,

alteri non obveniat, cur audent ceteris qui gemini non sunt, cum inspexerint eorum constellationes, talia pronuntiare quae ad illud secretum pertinent quod nemo potest conprehendere et momentis adnotare nascentium?"

13. Ibid., chap. 6 (2:154–59 of Loeb ed.): "igitur non usquequaque absurde dici potest ad solas corporum differentias adflatus quosdam valere sidereos, . . . non autem et animi voluntates positionibus siderum subdi . . . Quid enim tam ad corpus per[t]inens quam corporis sexus? Et tamen sub eadem positione siderum diversi sexus gemini concipi potuerunt. Unde quid insipientius dici aut credi potest quam sideram positionem, quae ad horam conceptionis eadem ambobus fuit, facere non potuisse ut, cum quo habebat eandem constellationem, sexum diversum a fratre non haberet; et positionem siderum, quae fuit ad horam nascentium, facere potuisse ut ab eo tam multum virginali sanctitate distaret?" (Augustine is referring to a pair of opposite-sex twins known to him; she is a holy virgin.) Wedel notes that Augustine's suggestion that the heavens might affect bodies but not souls contains the seeds of Aquinas's later solution to the problem. Wedel, *The Medieval Attitude towards Astrology*, pp. 23–24.

14. *City of God*, book 5, chap. 6 (2:162–63 in Loeb ed.): "His omnibus consideratis non inmerito creditur, cum astrologi mirabiliter multa vera respondent, occulto instinctu fieri spirituum non bonorum."

15. Isidorus Hispalensis Episcopus, *Etymologiarum sive originum libri XX*, ed. W. M. Lindsay, 2 vols. (Oxford: Clarendon Press, 1910), book 3, chap. 27, and book 8, chap. 9.

16. Ibid., book 3, chap. 27.

17. Ibid. "Astrologia vero partim naturalis, partim superstitiosa est. Naturalis, dum exequitur solis et lunae cursus, vel stellarum certas temporum stationes. Superstitiosa vero est illa quam mathematici sequuntur, qui in stellis auguriantur, quique etiam duodecim caeli signa per singula animae vel corporis membra disponunt, siderumque cursu nativitates hominum et mores praedicere conantur." In book 4, chap. 13, of the *Etymologiae*, however, Isidore voices support for astrological medicine: "Postremo et astronomiam notam habebit, per quam contempletur rationem astrorum, et mutationem temporum. Nam sicut ait quidam medicorum, cum ipsorum qualitatibus et nostra corpora commutantur."

18. Ibid., book 8, chap. 9: "Primum autem idem stellarum interpretes magi nuncupabantur, sicut de his legitur qui in Evangelio natum Christum adnuntiaverunt; postea hoc nomine soli Mathematici. Cuius artis scientia usque ad Evangelium fuit concessa, ut Christo edito nemo exinde nativitatem alicuius de caelo interpretaretur." Isidore is following Tertullian, *De Idolatria*, book 9, chap. 4 (*PL*, 1:672).

19. Laistner, "The Western Church and Astrology," pp. 264, 275.

20. E.g., S. J. Tester, *A History of Western Astrology* (Woodbridge, Suffolk: Boydell Press, 1987), pp. 132–42. Tester notes that the oldest surviving manuscripts of Julius Firmicus Maternus's *Mathesis* date from the eleventh century.

21. Flint, *The Rise of Magic*, pp. 93, 97. She defines "magic" as "the exercise of a preternatural control over nature by human beings, with the assistance of forces more powerful than they" (p. 3). She includes under this rubric a number of practices not explicitly termed *magia* in the written sources, among the most "testing and informative" of which she deems astrology.

22. Ibid., p. 99.

23. Ibid., pp. 141–45.

24. Ibid., p. 93. As Hilary M. Carey has noted, without a set of tables drawn up for

one's own meridian, practical astrology is almost impossible. Carey, *Courting Disaster: Astrology at the English Court and University in the Later Middle Ages* (New York: St. Martin's Press, 1992), p. 27.

25. Flint, *The Rise of Magic*, p. 142. She suggests that the Carolingians' preoccupation with the paganism of the Saxons and Slavs led them to put forth such "controlled compromises" as a rehabilitation of astrology.

26. Gratian, *Decretum*, pars 2, causa 26. Citing Isidore, Augustine, and Jerome, the great legal scholar pronounced that divination by the stars was illicit: "Non liceat Christianos tenere traditiones Gentilium, et observare, vel colere elementa, aut lunae aut stellarum cursus, aut inanem signorum fallaciam pro domo facienda, aut propter segetes vel arbores plantandas, vel conjugia socianda" (q. 5, chap. 3; *PL*, 187:1346).

27. See Charles Burnett, ed., *Adelard of Bath: An English Scientist and Arabist of the Early Twelfth Century* (London: Warburg Institute, 1987).

28. Bernard Silvestris, *Cosmographia*, ed. Peter Dronke (Leiden: E. J. Brill, 1978), Megacosmos, chap. 3, lines 33–36: "Scribit enim celum stellis, totumque figurat / Quod de fatali lege venire potest. / Presignat qualique modo qualique tempore / Omnia sidereus secula motus agat" ("The heavens write by means of the stars and shape all things that can come to pass by the law of fate. They presignify in what manner or what time the sidereal motions impel all things in this world"). Even Christ's birth is prefigured in the heavens: "Exemplar speciemque dei virguncula Christum / Parturit, et verum secula numen habent"; Megacosmos, chap. 3, lines 53–54. On Bernard Silvestris, see Brian Stock, *Myth and Science in the Twelfth Century: A Study of Bernard Silvestris* (Princeton, N.J.: Princeton University Press, 1972). Stock notes (p. 132) that in Bernard's conception, history is to be understood as the unfolding of God's predetermined order. God's ministers, angels and lesser deities, are the agents of this unfolding; the heavens reveal the pattern in their changes and motions.

29. See Marie-Dominique Chenu, "Nature and Man—The Renaissance of the Twelfth Century," in Chenu, *Nature, Man, and Society in the Twelfth Century: Essays on New Theological Perspectives in the Latin West*, ed. and trans. J. Taylor and Lester Little (Chicago: University of Chicago Press, 1968); and Tullio Gregory, "La nouvelle idée de nature et de savoir scientifique au XIIe siècle," in John E. Murdoch and Edith Sylla, eds., *The Cultural Context of Medieval Learning*, Boston Studies in the Philosophy of Science, 26 (Dordrecht, Holland, and Boston: D. Reidel, 1975), p. 203.

30. On translations of the important ninth-century Arab astrologer Abu Ma'shar (Albumasar), see Richard Lemay, *Abu Ma'shar and Latin Aristotelianism in the Twelfth Century: The Recovery of Aristotle's Natural Philosophy through Arabic Astrology* (Beirut: American University, 1962), esp. introd. and chap. 1; and Lemay, "The True Place of Astrology in Medieval Science and Philosophy: Towards a Definition," in Patrick Curry, ed., *Astrology, Science and Society: Historical Essays* (Woodbridge, Suffolk: Boydell Press, 1987), pp. 65–68. Tester lists the more important translators and their works in his *History of Western Astrology*, p. 152.

31. *De generatione et corruptione* 2.10, cited by John D. North, "Medieval Concepts of Celestial Influence: A Survey," in Patrick Curry, ed., *Astrology, Science and Society: Historical Essays* (Woodbridge, Suffolk: Boydell Press, 1987), p. 5. See also his "Celestial Influence—The Major Premiss of Astrology," in Paola Zambelli, ed., *"Astrologi hallucinati": Stars and the End of the World in Luther's Time* (Berlin and New York: Walter de Gruyter, 1986), pp. 45–100.

32. *Introductorium maius*, tr. 1; described in Lemay, *Abu Ma'shar*, pp. 42–85.

33. See John D. North, "Some Norman Horoscopes," in Burnett, *Adelard of Bath: An English Scientist and Arabist of the Early Twelfth Century*, pp. 151, 155, 160; and North, *Horoscopes and History* (London: Warburg Institute, 1986), pp. 96–107.

34. Carey, *Courting Disaster*, p. 31. The two were Michael Scot and a Master Theodore.

35. This statement was wrongly attributed to Ptolemy. For a discussion of its use in the Middle Ages, see G. W. Coopland, *Nicole Oresme and the Astrologers: A Study of His Livre de Divinacions* (Cambridge, Mass.: Harvard University Press, 1952), app. 4, pp. 175–77. Indications of the twelfth-century struggle to reconcile astrology and theology (and of the debate about astrology's use and validity) are given in Marie-Thérèse d'Alverny, "Astrologues et théologiens au XIIe siècle," in *Mélanges offerts à M.-D. Chenu*, Bibliothèque thomiste, 37 (Paris: Vrin, 1967), pp. 31–50.

36. On Albertus Magnus's astrology, see Thorndike, *HMES*, 2:577–88; and Paola Zambelli, "Albert le Grand et l'astrologie," *Recherches de théologie ancienne et médiévale* 49 (1982): 141–58.

37. Thorndike, *HMES*, 2:582.

38. Ibid., p. 584.

39. Ibid., pp. 587–88.

40. Albertus Magnus, *Speculum astronomiae*, Edizione a cura di Stefano Caroti, Michela Pereira, Stefano Zamponi, sotto la direzione di Paola Zambelli (Pisa: Domus Galilaeana, 1977), chap. 15, p. 45: "De electionibus vero est quaestio minus difficilis, non enim libertas arbitrii ex electione horae laudabilis coercetur, quin potius in magnarum rerum inceptionibus electionem horae contemnere est arbitrii praecipitatio, non libertas." See also Thorndike, *HMES*, 2:699–700.

41. On Bacon's astrology, see Thorndike, *HMES*, 2:660–74. The bulk of his writing on the stars can be found in *The "Opus maius" of Roger Bacon*, ed. John Henry Bridges, 2 vols. (Oxford: Clarendon Press, 1897), vol. 1, pars 4, esp. pp. 238–69.

42. See John D. North, "Astrology and the Fortunes of Churches," *Centaurus* 24 (1980): 190–91 (on Bacon and the theory of conjunctions); and pp. 184–89 (on Albumasar's version of the theory).

43. *Opus maius* (ed. Bridges), vol. 1, pars. 4, pp. 253–54: "Magnum enim solatium fidei nostrae possumus habere, postquam philosophi qui ducti sunt solo motu rationis nobis consentiunt, et sectam seu professionem fidei Christianae confirmant et nobiscum concordant in stabilitate hujus sectae."

44. North, "Astrology and the Fortunes of Churches," p. 191.

45. E.g., for Pierre d'Ailly as well as for a number of fifteenth-century Cracow authors cited by Mieczyslaw Markowski, including a Thomas von Strzempino who follows Aquinas word for word (a fact that Markowski does not mention). Markowski, "Der Standpunkt der Gelehrten des späten Mittelalters und der Renaissance dem astrologischen Determinismus gegenüber," *Studia Mediewistyczne* 23 (1984): 25–27.

Thomas Aquinas's views on the stars appear in the *Summa theologiae*, Ia q. 115, a. 4; and IIa IIae, q. 95, a. 5; the *Summa contra gentiles*, III, chap. 84; and elsewhere. See esp. Thomas Litt, *Les corps célestes dans l'univers de Saint Thomas d'Aquin* (Louvain: B. Nauwelaerts, and Paris: Publications Universitaires, 1963); also Thorndike, *HMES*, 2:603–15.

46. Thomas Aquinas, *Summa theologiae*, Latin text and English translation (Cambridge: Blackfriars, 1964), IIa IIae q. 95, a. 5: "Nullum autem corpus potest imprimere in rem

incorpoream. Unde impossibile est quod corpora caelestia directe imprimant in intellectum et voluntatem." (All translations in the text are from this edition.)

47. Ibid., Ia q. 115, a. 3: "motus horum inferiorum corporum, qui sunt varii et multiformes, reducuntur in motum corporis caelestis sicut in causam."

48. Ibid., Ia q. 115, a. 4: "*Sciendum est tamen quod indirecte et per accidens impressiones* corporum caelestium ad intellectum et voluntatem pertinere possunt, inquantum scilicet tam intellectus quam voluntas aliquo modo ab inferioribus viribus accipiunt, quae organis corporeis alligantur. . . . Nam intellectus ex necessitate accipit ab inferioribus viribus apprehensivis."

49. Ibid.: "Unde et ipsi astrologi dicunt quod sapiens homo dominatur astris, inquantum scilicet dominatur suis passionibus."

50. Ibid.: "plures hominum sequuntur passiones. . . . Et ideo astrologi ut in pluribus vera possunt praedicare, et maxime in commune."

51. Ibid., IIa IIae q. 95, a. 5, "utrum divinatio quae fit per astra sit illicita."

52. Ibid., IIa IIae q. 95, a. 5, ad 2. Aquinas says there are two reasons that the astrologers' prognostications are sometimes true—sometimes because of men not resisting passions (especially for predictions about the masses), and sometimes because of the intervention of demons: "plures hominum passiones corporales sequuntur, et ideo actus eorum disponuntur, ut in pluribus, secundum inclinationem caelestium corporum . . . praecipue in communibus eventibus, qui dependent ex multitudine. Alio modo, propter daemones se immiscentes."

53. See Bernard Capp, *English Almanacs, 1500–1800: Astrology and the Popular Press* (Ithaca, N.Y.: Cornell University Press, 1979); and Philippe Contamine, "Les prédictions annuelles astrologiques à la fin du Moyen Age: Genre littéraire et témoin de leur temps," in *Histoire sociale, sensibilités collectives et mentalités, Mélanges Robert Mandrou* (Paris: Presses Universitaires de France, 1985), pp. 191–204.

54. Lemay discusses the revulsion for the dependence on Arabic learning; "The True Place of Astrology," pp. 71–72.

55. See Anna Campbell, *The Black Death and Men of Learning* (New York: Columbia University Press, 1931), pp. 14–17, 39–43.

56. Richard Lemay, "The Teaching of Astronomy in Medieval Universities, Principally at Paris in the Fourteenth Century," *Manuscripta* 20 (1976): 200, citing the astrologer Symon de Phares.

57. See Hilary M. Carey, "Astrology at the English Court in the Later Middle Ages," in Patrick Curry, ed., *Astrology, Science and Society: Historical Essays* (Woodbridge, Suffolk: Boydell Press, 1987), p. 46. According to Symon de Phares, Charles had a number of astrological works translated into French. See *Recueil des plus célèbres astrologues et quelques hommes docts, faict par Symon de Phares du temps de Charles VIIIe*, ed. Ernest Wickersheimer, (Paris: Champion, 1929), p. 228. These translations are discussed in Lys Ann Shore, "A Case Study in Medieval Nonliterary Translation: Scientific Texts from Latin to French," in Jeanette Beer, ed. *Medieval Translators and Their Craft* (Kalamazoo: Western Michigan University Press, 1989), pp. 307–10. Hilary Carey notes a numbers of factors contributing to the rise of astrology in late medieval England; by the time of Richard II (1377–99), nobles at court regularly consulted astrologers. Carey, *Courting Disaster*, pp. 17–20, 22.

58. The Condemnations appear in Heinrich Denifle and Emile Chatelain, eds., *Chartularium Universitatis Parisiensis*, 4 vols. (Paris: Delalain, 1889–97), 1:543ff.; they are edited and arranged topically in Pierre Mandonnet, *Siger de Brabant et l'averroïsme latin au XIIIme*

siècle, 2d ed., 2 vols. in 1 (Louvain: Institut Supérieur de Philosophie, 1908 and 1911), pt. 2, pp. 175–91. See also the commentary of Roland Hissette in his *Enquête sur les 219 articles condamnés à Paris le 7 mars 1277*, Philosophes médiévaux, 22 (Louvain: Publications Universitaires, 1977). The attack on astrology is discussed in Tester, *History of Western Astrology*, p. 177 (he gives six propositions clearly concerning astrology); and Philippe Contamine, "Les prédictions annuelles astrologiques à la fin du Moyen Age: Genre littéraire et témoin de leur temps," p. 192 (he lists twenty-seven dealing with astrology, plus one from Condemnations that Tempier issued on December 10, 1270: "Quod omnia quae hic in inferioribus aguntur, subsunt necessitati corporum coelestium," again the concern being with necessity).

59. No. 104 (no. 143, *Chartularium*): "Quod ex diversis signis coeli signantur diversae conditiones in hominibus tam donorum spiritualium, quam rerum temporalium." Mandonnet, *Siger de Brabant*, p. 183.

60. No. 154 (no. 162, *Chartularium*): "Quod voluntas nostra subiacet potestati corporum coelestium." Mandonnet, *Siger de Brabant*, p. 187. Hissette concludes it unlikely that such extreme determinism had any real adherents in Paris, but rather that most believed in the sort of indirect influence described by Aquinas. Hissette, *Enquête*, p. 237.

61. No. 92 (no. 6, *Chartularium*): "Quod redeuntibus corporibus coelestibus omnibus in idem punctum, quod fit in XXX sex milibus annorum, redibunt idem effectus, qui sunt modo." Mandonnet, *Siger de Brabant*, p. 183. There were, of course, other competing figures for the length of the Great Year, given different rates of precession. Hissette traces the source of Tempier's condemned thesis to Boethius of Dacia, but Boethius's text gives no figure for the length of time required for the return of the heavenly bodies to their original posts. Hissette finds no Parisians supporting this thesis (Hissette, *Enquête*, p. 158).

62. No. 105 (no. 207, *Chartularium*): "Quod in hora generationis hominis in corpore suo et per consequens in anima, quae sequitur corpus, ex ordine causarum superiorum et inferiorum inest homini dispositio inclinans ad tales actiones vel eventus.—Error, nisi intelligatur de eventibus naturalibus, et per viam dispositionis." Mandonnet, *Siger de Brabant*, p. 184.

63. The literature on Oresme is immense. For a recent bibliography see Bert Hansen, *Nicole Oresme and the Marvels of Nature: A Study of His* De causis mirabilium *with Critical Edition, Translation, and Commentary*, Studies and Texts, 68 (Toronto: Pontifical Institute of Medieval Studies, 1985). For Oresme's views on astrology, see Hansen, pp. 17–25; Thorndike, *HMES*, 3:398–423; Coopland, *Nicole Oresme and the Astrologers* (texts of *Livre de divinacions* and *Tractatus contra iudiciarios astronomos*); Pierre Maurice Marie Duhem, *Le système du monde: Histoire des doctrines cosmologiques de Platon à Copernic*, 10 vols. (Paris: Hermann, 1913–59), 8:462–83; Edward Grant, ed., *Nicole Oresme and the Kinematics of Circular Motion: Tractatus de commensurabilitate vel incommensurabilitate motuum coeli, edited with an Introduction, English Translation, and Commentary* (Madison: University of Wisconsin Press, 1971); Max Lejbowicz, "Chronologie des écrits anti-astrologiques de Nicole Oresme. Etude sur un cas de scepticisme dans la deuxième moitié du XIV^c siècle," in Jeannine Quillet, ed., *Autour de Nicole Oresme*, Actes du Colloque Oresme organisé a l'Université de Paris, 12 (Paris: J. Vrin, 1990), pp. 119–76; and esp. Stefano Caroti, *La critica contro l'astrologia di Nicole Oresme e la sua influenza nel medioevo e nel Rinascimento*, Atti dell'Accademia Nazionale dei Lincei, Memorie, Classe di scienze morali, storiche e filologiche, ser. 8, vol. 23, fasc. 6 (1979): 545–684; Caroti, "Nicole Oresme: *Quaestio contra*

divinatores horoscopios," *Archives d'histoire doctrinale et littéraire du moyen âge* 43 (1976): 201–310; and Caroti, "Nicole Oresme's Polemic against Astrology in His 'Quodlibeta,'" in Patrick Curry, ed., *Astrology, Science and Society: Historical Essays* (Woodbridge, Suffolk: Boydell Press, 1987), pp. 75–93.

64. Of all Oresme's anti-astrological writings, these are the mildest and the most traditional, perhaps, Caroti suggests, because of his fear of the astrologers at the court of Charles V. See Caroti, *La critica contro l'astrologia di Nicole Oresme*, p. 561.

65. Both the *Tractatus* and the *Livre* most likely date from the period 1348–65. See Coopland, *Nicole Oresme and the Astrologers*, pp. 1, 3; Caroti, *La critica contro l'astrologia di Nicole Oresme*, p. 546.

66. See Coopland, *Nicole Oresme and the Astrologers*, p. 23.

67. The questions asked of astrologers all involved events dependent upon the exercise of free will, he maintained. Caroti, "Nicole Oresme's Polemic against Astrology in his 'Quodlibeta,'" p. 87, citing Oresme's *Quodlibeta*.

68. *Livre*, chap. 9; *Tractatus*, chap. 4 (both of these works appear in Coopland, *Nicole Oresme and the Astrologers*).

69. *Tractatus*, chap. 4; *Livre*, chap. 3; *Quaestio*, pt. 3, cited in Caroti, *La critica contro l'astrologia di Nicole Oresme*, p. 602.

70. *Tractatus*, chap. 4.

71. *Livre*, chap. 12; *Tractatus*, chap. 6.

72. *Tractatus*, chap. 4.

73. *Tractatus*, chap. 4; *Quaestio*, pt. 2, cited in Caroti, *La critica contro l'astrologia di Nicole Oresme*, p. 595.

74. *Tractatus*, chap. 2; *Livre*, chap. 8.

75. See Thorndike, *HMES*, 3:407; Caroti, *La critica contro l'astrologia di Nicole Oresme*, p. 81.

76. Caroti, *La critica contro l'astrologia di Nicole Oresme*, p. 592.

77. Ibid., p. 591. Oresme says that he himself had studied astrology as a youth.

78. The pertinent works are his *De proportionibus proportionum, Ad pauca respicientes, De commensurabilitate vel incommensurabilitate motuum coeli*. See Grant, *Nicole Oresme and the Kinematics of Circular Motion*; Caroti, *La critica contro l'astrologia di Nicole Oresme*, pp. 583–87, and Thorndike, *HMES* 3:405–6.

79. Caroti, *La critica contro l'astrologia di Nicole Oresme*, p. 585.

80. Ibid., p. 586, citing Oresme's *De proportionibus*.

81. Oresme laid out the theory of "configurations" in the treatise *De configurationibus qualitatum*. I follow Caroti's summary (ibid., pp. 571–75).

82. Ibid., p. 582.

83. Ibid., pp. 598–99, citing the *Quaestio*. The "lumen, motus, et influentia" ascribed to the heavens by Oresme were the standard instrumentalities of heavenly influence according to medieval theorists. See Edward Grant, "Medieval and Renaissance Scholastic Conceptions of the Influence of the Celestial Region on the Terrestrial," *Journal of Medieval and Renaissance Studies* 17 (1987): 10–11.

84. "Quamvis enim complexionum diversitas aliqualiter sit a celo quod animas hominum ad varios mores, sine tamen fati necessitate, inclinat; tamen hoc astrologi nequeunt prescire tum quia motuum celi proporciones sunt inscibiles ut alibi demonstravi tum quia vires astrorum ignote sunt, tum quia eorum indicia [or *iudicia* = judgments?] frivolis

persuasionibus vallata sunt." *Tractatus*, chap. 4, ed. in Coopland, *Nicole Oresme and the Astrologers*, p. 131.

85. His work shows a deep influence from Oresme. He became a bachelor of arts at Paris in 1363 and was active in Paris until his move to the new University of Vienna in 1381 or 1382. His astrological writings all belong to the Paris years. For recent studies of Henry's life and thought, see Nicholas Steneck, *Science and Creation in the Middle Ages: Henry of Langenstein (d. 1397) on Genesis* (Notre Dame, Ind.: University of Notre Dame Press, 1976); and Michael H. Shank, *"Unless You Believe, You Shall Not Understand." Logic, University, and Society in Late Medieval Vienna* (Princeton, N.J.: Princeton University Press, 1988). For Henry's attitude about astrology and an edition of his major anti-astrological works, see Herbert Pruckner, *Studien zu den astrologischen Schriften des Heinrich von Langenstein*, Studien der Bibliothek Warburg, 14 (Leipzig: Teubner, 1933); Caroti, *La critica contro l'astrologia di Nicole Oresme*, pp. 613–29; and Thorndike, *HMES*, 3:472–510.

86. In the treatise *De reductione effectuum in virtutes communes*. See Caroti, *La critica contro l'astrologia di Nicole Oresme*, pp. 620, 623; and Thorndike, *HMES*, 3:481.

87. In the treatise *De habitudine causarum*. See Caroti, *La critica contro l'astrologia di Nicole Oresme*, pp. 615–16; and Thorndike, *HMES*, 3:477.

88. *De reductione effectuum*; Caroti, *La critica contro l'astrologia di Nicole Oresme*, p. 623; and Thorndike, *HMES*, 3:486.

89. Caroti, *La critica contro l'astrologia di Nicole Oresme*, p. 622; Thorndike, *HMES*, 3:489.

90. Both printed by Pruckner, *Studien*, pp. 89–138, 139–206.

91. *Tractatus contra coniunctionistas*, pt. I, chap. 8, ed. Pruckner, *Studien*, pp. 151–52: "Stupenda absurditas in philosophia, quod planeta naturaliter agens et non libere existens in ange [auge] epicycli forcius influit omnibus ceteris paribus, ex quo ex parte situs in ange [auge] nulla virtus sibi imprimitur nec acquiritur . . . quoniam eque naturaliter situatur in quolibet puncto sui orbis." This passage is cited by Caroti, *La critica contro l'astrologia di Nicole Oresme*, p. 627; and Thorndike, *HMES*, 3:497.

92. *Tractatus contra coniunctionistas*, pt. I, chap. 7, p. 147: "Quid enim facit ad vigorem raro et non nisi post magnum tempus ibi semel coniungi; ymmo ad debilitatem facit, qui interim influencia precedentis coniunctionis magis totaliter evanuit, quam si statim post ibi iterum coniungerentur."

93. *Quaestio de cometa*, ed. Pruckner, *Studien*, chap. 9, p. 114: "Et non minus irrationabile est, quod significationem comete durare dicunt per annos, menses, vel dies . . ."

94. *Tractatus contra coniunctionistas*, pt. II, chap. 5, pp. 184–85. Cited by Caroti, *La critica contro l'astrologia di Nicole Oresme*, p. 628.

95. See Coopland, *Nicole Oresme and the Astrologers*, p. 8.

96. For the small circulation of Oresme's works and for the literary borrowings from Oresme, see Coopland, *Nicole Oresme and the Astrologers*, pp. 8–12. For the *Songe du vieil pelerin*, see Philippe de Mézières, Chancellor of Cyprus, *Le songe du vieil pelerin*, ed. G. W. Coopland, 2 vols. (Cambridge: Cambridge University Press, 1969). For Oresme's influence on other writers on astrology see Caroti, *La critica contro l'astrologia di Nicole Oresme*, pp. 613–84; and Thorndike, *HMES*, 3:423. According to Caroti's thorough study of Oresme's influence, he was known almost exclusively by the *Tractatus* and the *Livre* and not by the harsher anti-astrological writings.

97. Thorndike, *HMES*, 3:508. Symon de Phares listed him as one expert in the art of astrology: "il estoit erudit en la science des etoiles . . . il predist plusiers beaulx jugements." *Recueil des plus célèbres astrologues . . . par Symon de Phares* (ed. Wickersheimer), p. 223.

98. See Tester, *History of Western Astrology*, pp. 240–43; and Patrick Curry, *Prophecy and Power: Astrology in Early Modern England* (Cambridge: Polity Press, 1989), pp. 114–19.

99. For a convenient summary of aspects of the debate in and after d'Ailly's lifetime, see Markowski, "Der Standpunkt der Gelehrten . . . dem astrologischen Determinismus gegenüber."

100. *De falsis prophetis II*, printed in Jean Gerson, *Opera omnia*, ed. Louis Elhes Dupin, 5 vols. (Antwerp: Sumptibus societatis, 1706), 1: cols. 511–603. D'Ailly wrote two treatises entitled *De falsis prophetis*; this treatise, although traditionally numbered second may, in fact, have been first in order of composition. See Max Lieberman, "Chronologie Gersonienne. VIII. Gerson et D'Ailly (III)," *Romania* 81 (1960): 82–84. For the date, see appendix 2.

101. He quotes large sections of Augustine, *De civitate dei*, book 5; Thomas Aquinas, *Summa theologiae*, IIa IIae, q. 95, a. 5, and *Summa contra gentiles*, III, chap. 84; and Nicole Oresme, *Tractatus*. He does not quote Oresme's more scientific works, such as the treatise on commensurability, or his more thoroughgoing *Quaestio contra judiciarios astronomos*. On d'Ailly's use of Oresme, see Caroti, *La critica contro l'astrologia di Nicole Oresme*, pp. 629–43; and Coopland, *Nicole Oresme and the Astrologers*, p. 11. D'Ailly also cites (via Oresme) Gratian, Isidore of Seville, John of Salisbury, and Hugh of St. Victor, among others. Oresme was grand master of d'Ailly's own College of Navarre in Paris. It is not inconceivable that d'Ailly would have known Oresme.

102. Pierre d'Ailly, *Apologetica defensio astronomice veritatis (I)*, in Pierre d'Ailly, *Tractatus de imagine mundi et varia ejusdem auctoris et Joannis Gersonis opuscula* (Louvain: Johann de Westphalia, ca. 1483), fol. [gg6r]: "medium tenere docui inter extremas duas opiniones quarum una astronomicam potestatem nimis extollit alia nimis deprimit."

103. The distinction was the same as that drawn by Isidore of Seville, *Etymologiae*, book 3, chap. 27. D'Ailly may have taken this distinction from Hugh of St. Victor (via Oresme). See *De falsis prophetis II*, col. 594, where d'Ailly quotes Hugh of St. Victor, following Oresme's *Tractatus*, chap. 5.

104. Pierre d'Ailly, *Concordantia astronomie cum theologia* (most frequently called the *Vigintiloquium*), in d'Ailly 1490, v. 2, fol. a3r, listing three errors that have rendered astrology suspect: "Primus est eorum qui ex astris omnia futura necessitate fatali evenire senserunt. Secundus est eorum qui astronomicis libris plures supersticiones execrabiles artis magice miscuerunt. Tercius est eorum qui terminos astronomice potestatis respectu liberi arbitrii et quarumdam rerum que solum subsunt divine ac supernaturali potestati superbe et supersticiose excesserunt."

105. Gerson's *Contra doctrinam medici cuiusdam in monte pessulano sculpentis in numismate figuram leonis*.

106. *De falsis prophetis II*, cols. 590–93, following Aquinas's *Summa theologiae*, IIa IIae, q. 95, and the *Summa contra gentiles*, III, chap. 84.

107. Pierre d'Ailly, *De legibus et sectis contra superstitiosos astronomos*, in Pierre d'Ailly, *Tractatus de imagine mundi et varia ejusdem auctoris et Joannis Gersonis opuscula* (Louvain: Johann de Westphalia, ca. 1483), chap. 6, fol. F4v: "Et tunc anima corpori unita excitatur

fortiter et inducitur efficaciter, licet non cogatur," following Bacon's *Opus maius* (ed. Bridges), vol. 1, pars 4, p. 267.

108. Ibid.: "ut sic opiniones et secte ac mutationes consuetudinum inducantur per aliquem famosum in populo et potentem," following the *Opus maius* (ed. Bridges), vol. 1, pars 4, p. 267.

109. Esp. the *De legibus et sectis*, the *Vigintiloquium* (or *Concordantia astronomie cum theologia*), and the *Elucidarium*.

110. J. D. North says that he is nowhere consistent in his definitions of what is involved in these two. North, "Astrology and the Fortunes of Churches," p. 201. What North sees as inconsistency is more likely a development in d'Ailly's views on astrology; with the passage of time he grants more and more power to the stars and hence attributes more and more events to natural causality.

111. *De legibus et sectis*, chap. 7, fol. [F5r]: "Verbi gratia Christus legislator noster a sua nativitate valde bonam complexionem naturalem dicitur habuisse. Non est ergo fidei dissonum et est rationi naturale consonum quod sub bona celi dispositione seu constellatione natus fuerit a qua complectionis bonitas naturaleter [naturaliter] in eo dependere potuit."

112. *Concordantia astronomie cum hystorica narratione*, in d'Ailly 1490, fol. [b8r]: "Tamen [Deus] disposuit cum causis naturaliter operari nisi ubi miraculosa operatio intervenerit."

113. And, of course, the workings of human free will, as d'Ailly made clear: "In his tamen lectorem cupimus esse premonitum quod nihil fatali necessitati ascribimus sed naturali causalitati et inclinationi cui liberum arbitrium in his que sue subiciantur facultati potest dei concurrente auxilio resistere." Ibid.

114. *Apologetica defensio astronomice veritatis (I)*, in Pierre d'Ailly, *Tractatus de imagine mundi*, fol. [gg6v]: "fides non cogit dicere quod huius sancte prolis nativitas omnem huiusmodi astrorum influxum excluserit sicut nec cogit dicere quod sol eam non calefecit."

115. Ibid., fol. [gg8v]: "ut excludam in his ea que divinitus speciali privilegio non nature sed gratie facta esse Christiana tradidit auctoritas."

116. He called it "natural theology" in the *Vigintiloquium*, v. 1, fol. a2v: "Et ideo astronomia non inconvenienter naturalis theologia nominatur . . ."

117. On d'Ailly's use of Oresme, see Caroti, *La critica contro l'astrologia di Nicole Oresme*, pp. 629–43. He finds d'Ailly to be fairly positive on astrology in the sections of the *De falsis prophetis* other than the long final *quaestio* on divination by the stars, in which d'Ailly condemns astrology.

118. Edward Grant insists that this is plagiarism; d'Ailly does not even acknowledge that his source is "*quidam doctor*" or "*aliquis*," as he usually does when quoting an unnamed author. See Grant, *Nicole Oresme and the Kinematics of Circular Motion*, pp. 130–32, esp. n. 121.

119. *Apologetica defensio astronomice veritatis (I)*, fol. [gg8v]: "rationes quantum ad hoc magis probare videntur sciendi difficultatem quam scientie impossibilitatem." Also see *Secunda apologetica defensio astronomice veritatis*, in Pierre d'Ailly, *Tractatus de imagine mundi*, fol. hh2: "Omnia alia et illa ibidem obiecta probant huius scientie maximam esse difficultatem et maxime incertitudinaliter iudicando de nativitatibus . . . sed non probant huius scientie falsitatem aut impossibilitatem."

120. *De legibus et sectis*, chaps. 8 and 9.

121. Ibid., chap. 10. D'Ailly was an expert on this matter and addressed a treatise on the calendar to Pope John XXIII.

122. Ibid., fol. g4r–g4v: "Nam cum in astrologia de motibus tot et tante difficultates sint et incertitudines ut predictum est, necessario oportet quod in astronomia de judiciis plures sint et maiores cum hec ex illa dependeat et ultra eam multa incerta presupponat. . . . Ideo videtur esse humana superbia ex huiusmodi coniecturis velle pertingere ad talem prescientiam contingentium futurorum, unde talem prescientiam deus sibi soli creditur reservasse, saltem respectu aliquorum futurorum." D'Ailly leaves open a loophole in the last clause, noting that God has reserved for himself the foreknowledge of the future, "at least with respect to *some* future contingents." Presumably, then, the foreknowledge of the rest is open to man.

123. *Secunda apologetica defensio astronomice veritatis*, fol. hh2v: "de his talis haberi potest per hanc scientiam cognitio que debet reputari ad aliqua iudicia sufficiens."

124. *Vigintiloquium*, v. 4, fol. a3v: "Constellatio aliqualiter disponit et inclinat humanam voluntatem. Celi maxime obediunt homines brutales qui malis inclinationibus non resistunt per virtutes."

125. *Apologetica defensio astronomice veritatis (I)*, fol. [gg8v]: "dico quod iudicia astronomica de nativitatibus . . . pre ceteris aliis iudiciis astrorum valde difficilia sunt et incerta et hoc satis confitentur peritiores astronomi."

126. *Secunda apologetica defensio astronomice veritatis*, fol. hh2v: "Et si quandoque etiam aliqui forte minus periti astronomi errent in iudicando vel etiam si inter peritos sit quandoque in iudiciis diversitas aut non plena concordia non est tamen propter hoc ista scientia contemnanda nec eius studia contemnendum."

127. *Apologetica defensio astronomice veritatis (I)*, fols. [gg6v]–[gg7r].

128. Ibid., fol. [gg7v]: "quid mirum si eis [planetis] specialiter virtutem dederit [deus] ut secundum eorum ordinem rebus aliis hic inferius producendis ordo daretur sicut embrioni et rebus ceteris suo modo."

129. Ibid.: "Et ponam in hoc exemplum notabile in quo concordant non solum astronomi sed philosophi et medici. Nec discordant theologi videlicet quod sicut dicit Campanus [of Novarra], ex ordinatione creatoris gloriosi dispositum est ut singuli planete successive in quolibet orizonte dominentur singulis horis successive. . . . Patet propositum quia ex hac consideratione philosophi denominaverunt quemlibet diem ab illo planeta qui in prima eius hora dominium optinet."

130. *Secunda apologetica defensio astronomice veritatis*, fol. hhr: "Secundo dicit [Oresme] quod . . . propositiones que in libris iudicialibus posite sunt ut plurimum sunt ficte et voluntarie divinate. . . . Qua enim ratione potest dici de grada zodiaci quia unus est lucidus et unus masculus . . .?"

131. Ibid., fols. hh2v–hh3r: "[the ancients set forth rules for making astrological judgments] partim per experie[n]tiam partim per rationem naturalem. . . . Nam aliqui astronomi 7 planetas applicant ad climata. . . . Et hoc concluserunt antiqui ex conditionibus terrarum et moribus hominum eas inhabitantium. . . . Nam preter hos duos modos acquirendi scientiam alios duos posuit Alkindus in libro suo de radicibus stellarum videlicet quod aliqua cognita fuerunt a causa aliqua per revelationem . . . . Notandum est quod Alkindus non loquitur de illa revelatione que immediate fit dei inspiratione sed de illa que fit mediante celi impressione." Al-Kindi was a ninth-century Arab philosopher. He wrote extensively on many subjects, including astrology, and his most important work on the stars is the *De radiis* cited by d'Ailly here. It has been edited by M.-T. d'Alverny and

F. Hudry in *Archives d'histoire doctrinale et littéraire du moyen âge* 41 (1974): 139–260. I have been unable, however, to locate d'Ailly's reference in their edition of the *De radiis*.

132. *Apologetica defensio astronomice veritatis (I)*, fol. hh3r: "non est tamen incredibile et quod alique huiusmodi veritates etiam per divinam inspirationem quibusdam revelate sunt. Et hoc maxime sanctis patriarchis et prophetis quorum aliqui in scientia astronomie leguntur fuisse periti sicut de Abraham et Moyse narravit Josephus et Christiane recitant hystorie. Et hoc facit utique ad laudem et exaltationem huius veritatis scientie astronomice et ad honorem et gloriam almi conditoris siderum qui est deus benedictus in secula seculorum."

133. Nicole Oresme, *Tractatus contra astronomos*, in Coopland, *Nicole Oresme and the Astrologers*, chap. 1, p. 124 (astrology revealed to the patriarchs is put forth as a defense of the science); and chap. 7, pp. 139–40 (this argument is answered by the assertion that astrologers had falsely attributed these books to the patriarchs).

134. See Edward Grant, "The Condemnations of 1277, God's Absolute Power, and Physical Thought in the Late Middle Ages," *Viator* 10 (1979): 211–17, reprinted in Edward Grant, *Studies in Medieval Science and Natural Philosophy* (London: Variorum Reprints, 1981). Grant argues that in the years after 1277, God's absolute power to do anything short of a logical contradiction became a powerful analytical tool in natural philosophy, sparking the contemplation of all sorts of hypothetical possibilities.

135. "Plurimi cogitaverunt astrologi Deum ociosam vitam ducere cunctaque ministrari per celum." Salutati, *De fato*, 3.1, cited by Charles Trinkaus, "Coluccio Salutati's Critique of Astrology in the Context of His Natural Philosophy," *Speculum* 64 (1989): 68.

136. T. M. Luhrmann, *Persuasions of the Witch's Craft: Ritual Magic in Contemporary England* (Cambridge, Mass.: Harvard University Press, 1989), pp. 7–8. In a striking parallel to d'Ailly's dismissal of charges that astrology was disproved by the contradictions and inaccuracies of its predictions, Luhrmann's magicians attributed failure of their spells "to faulty technique, not to fragile theory" (p. 137).

137. Ibid., pp. 271–72, citing Leon Festinger, H. W. Riecken, and S. Schachter, *When Prophecy Fails* (New York: Harper and Row, 1956); and Festinger, *A Theory of Cognitive Dissonance* (Stanford: Stanford University Press, 1957). G.E.R. Lloyd, *Magic, Reason and Experience: Studies in the Origin and Development of Greek Science* (Cambridge: Cambridge University Press, 1979), e.g., pp. 5, 29, 33, 40, 45. Lloyd rejects the view that rational science came to replace magic in ancient Greece, but rather argues that the development of magical thought may indeed have depended on and followed models drawn from physical science.

Chapter Three
The Making of an Astrologer

1. *Le recueil épistolaire autographe de Pierre d'Ailly et les notes d'Italie de Jean de Montreuil*, ed. Gilbert Ouy, Umbrae Codicum Occidentalium, 9 (Amsterdam: North-Holland, 1966), p. xvi. This information is based on an analysis of autograph manuscripts. D'Ailly changed his writing style in the years 1414–18; Ouy found marginalia in d'Ailly's new hand in manuscripts of his early works.

2. For details of the ambivalent attitude about the stars, see chapter 2.

3. Heinrich Denifle and Emile Chatelain, eds., *Chartularium Universitatis Parisiensis*, 4 vols. (Paris: Delalain, 1889–97), 4:35 ("Conclusio facultatis theologie super materia fidei

nunc agitata novissime determinata," September 19, 1398): "XXVIIus articulus, quod cogitationes nostre intellectuales et volitiones interiores immediate causentur a celo, et quod per aliquam traditionem magicam possunt sciri, et quod per illam de eis certitudinaliter judicare sit licitum. Error."

4. *Principium in cursum Biblie,* in Pierre d'Ailly, *Questiones Magistri Petri de Aylliaco cardinalis cameracensis super primum, tertium et quartum libros sententiarum* (Paris: Jean Petit, 1505?), fol. 274v: "quosdam ego reperio sibi nomen mathesis usurpantes indebite, ex quorum infamia tota fere Mathematicorum scola infamis redditur et suspecta."

5. Ibid., fol. 275r: "Sed ad veros mathematicos transeo qui . . . in astrologia celorum influentias scientifice perscrutantur."

6. Ibid.: "ita desipiunt ut Christi doctrinam, Christi vitam, Christi opera, Christi miracula causis naturalibus a celestibus influentiis audeant deputare." Of course the unorthodox consequence of such teaching is that Christ's passion was undertaken involuntarily, being determined by the stars.

7. Ibid.

8. Ibid. D'Ailly attributes this statement to Augustine, *De natura demonum*; this book was in fact the work of Rhabanus Maurus. As mentioned in chapter 2, this statement goes back to Tertullian, *De idolatria*, and was repeated by Isidore of Seville and Gratian, among others.

9. This brief description of mathematical knowledge was, however, hardly the place to go into any details of the science.

10. [*Questiones* on Sacrobosco's *Sphera*] in Joannes de Sacrobosco, *Uberrimum sphere mundi commentum intersertis etiam questionibus domini Petri de Aliaco* (Paris: Jean Petit, 1498), fol. b4r. "Alia pars [astrologie] considerat in generali motus, situs et figuras corporum celestium per rationes mathematicas. Et ista habetur ab actore in isto tractatu."

11. Ibid., fol. [d5r]: "Sol per aspectos suos fortunatus est, malus tamen est per coniunctionem eius cum aliquibus planetis in aliquibus signis; masculinus, diurnus et operatur calorem et siccitatem."

12. Ibid., fol. g2v: "Sed iterum huiusmodi signorum quidam gradus dicitur lucidi, alii tenebrosi, alii fumosi; et causa est quia si ascendens in nativitate alicuius pueri fuerit in gradu lucido et etiam luna, tunc talis puer debet esse pulcher et lucidus; si in gradu tenebroso, minus pulcher et turpis; si in fumosa, tunc debet tenere medium." Since each sign of the zodiac contains thirty degrees, this division of the sign indicates the extreme precision necessary to make an accurate astrological judgment.

13. Ibid., fol. n1r: ". . . illa eclipsis que fuit tempore passionis Christi non fuit naturalis immo miraculosa quia fiebat versus oppositionem. Et illud patet per auctorem in fine huius tractatus ubi allegat sanctum Dyonisium quia illo tunc erat paganus et magnus astrologus. Qui dixit sic aut deus nature patitur aut totalis mundi machina destruitur [either the God of nature suffers or the whole machinery of the world is being destroyed]." A similar statement appears in the *Historia scholastica* of Petrus Comestor (Peter the Eater). "Dixit Dionysius Areopagita quod Deus naturae patiebatur." *Historia scholastica*, chap. 175 (*PL*, 198: 1651). A brief discussion of the date of the Crucifixion and possible nearly contemporary solar eclipses can be found in D. Justin Schove, *Chronology of Eclipses and Comets, AD 1–1000* (Woodbridge, Suffolk: Boydell Press, 1984), pp. 6–7.

14. *Tractatus Petri de Eliaco episcopi Cameracensis super libros Metheororum de impressionibus aeris. Ac de his quae in prima, secunda, atque tertia regionibus aeris fiunt* (Strasbourg: Johannes

Prüs, 1504), fols. 4v–5r: "Signant rixas: quia siccitas est principium caloris ignei, qui est principium iracundiae (secundum Galienum). Ira est causa rixarum. Rixae vero preliorum. Prelia vero sunt causae mortis principum et aliorum pugnantium."

15. *Tractatus de anima* (Paris: Jean Petit, 1505), chap. 15, pt. 3, fol. f4v: "Habens autem habitum et non cogitans actualiter de usu illius vel obiecto potest moveri ad cogitandum de illo non solum a voluntate sed aliunde multipliciter, scilicet vel a motore extrinseco ut ab influentia celi, vel a motore intrinseco ut a dispositione corporali, vel alia causa latente." For a recent study of this work, with critical edition, see Olaf Pluta, *Die philosophische Psychologie des Peter von Ailly: Ein Beitrag zur Geschichte der Philosophie des späten Mittelalters*, Bochumer Studien zur Philosophie, 6 (Amsterdam: B. R. Grüner, 1987).

16. *Tractatus utilis super Boecii de consolatione philosophie* (Paris, Bibliothèque Nationale, MS Lat. 3122), q. 2, a. 2: "Quantum ad secundum articulum principalem videndum est utrum omnia que fiunt disponantur secundum ordinem fatalis necessitatis." Fols. 158v–161r.

17. Ibid., fol. 159r: "Sed tale fatum reprobat fides et omnis doctrina sanctorum."

18. Ibid., fol. 159v: ". . . capiendo fatum . . . pro fato significante vel inclinante . . . non est negandum ymmo simpliciter et catholice concedendum fatum esse."

19. Ibid., fol. 159v: "Volo dicere quod non est negandum quin ex dispositione celestium possibilis sit nobis per astronomiam aliqualis notitia de eventibus futurorum [*sic*]."

20. Ibid.: "Non dico tamen nec dicunt etiam sapientes astronomi quod talis constellatio vel influentia necessitet voluntatem."

21. Ibid., fol. 160r: "Ita astrologus ex dispositione corporum celestium sicut ex causa remota et per hunc modum potest [judicare]," quoting Thomas Aquinas, *Summa contra gentiles*, III, chap. 84.

22. *Tractatus utilis*, fol. 159v: "Apparet ex predictis quod . . . querere etiam scienciam stellarum utilis est et licita etiam Christianis."

23. Ibid., fol. 160r: "Nolunt ergo sancti quod relicta morali philosophia vel sciencia que pertinet ad anime salutem astrologia studeatur."

24. Ibid.: "Unde sic Tullius ait: studium conservare in res obscuras difficiles et non necessarias vicium est quod curiositas vocatur." D'Ailly's great pupil Gerson would echo such sentiments in a 1402 discourse entitled *Contra curiositatem studentium*: "Curiositas est vitium quo dimissis utilioribus homo convertit studium suum ad minus utilia vel inattingibilia sibi vel noxia" ("Curiosity is a vice by which a person turns his attention to things less useful, unattainable, or harmful to him"). Jean Gerson, *Oeuvres complètes*, ed. Palémon Glorieux, 10 vols. (Paris: Desclée & Cie, 1960–73), 3:230. Gerson blamed *curiositas* for leading scholars into pride, error, contentious debates, and obscure doctrines.

25. *Tractatus utilis*, fol. 160r: "De quibus loquitur Policraticus ubi super. Philosophi inquit dum nature nimium auctoritatis attribuunt in auctorem nature adversando fidei plerumque impingunt. Et ideo sancti huius scientie studium non propter sciencias sed propter studentes dampnaverunt. Et maxime eo tempore quo homines huic studio nimis tenaciter adherebant." John of Salisbury expresses a similar sentiment, although not in this language, in the *Policraticus*, book 2, chap. 19. In d'Ailly's *Sermo [primus] de adventu*, he would attribute this passage to Thomas Bradwardine. Pierre d'Ailly, *Sermo primus de adventu domini*, in Pierre d'Ailly, *Tractatus et sermones* (Strasbourg: [Printer of Jordanus de Quedlinburg], 1490; repr., Frankfurt am Main: Minerva, 1971), fol. s3v.

26. The firmest dates we have for the pair of treatises entitled *De falsis prophetis* put them between 1378 and 1388; the Boethius commentary dates from 1377–81. A number

of considerations have led me to consider the *De falsis prophetis II* to postdate the commentary: first, d'Ailly's greater knowledge about astrology in the *De falsis prophetis II*; second, the fact that d'Ailly apparently lifts one paragraph from the Boethius commentary into the work on prophecy; and finally, a consideration of the subject matter. In the Boethius commentary, d'Ailly states that he is not allowed to deal with certain theological topics, being merely in the faculty of arts (quoted in Marguerite Chappuis, Ludger Kaczmarek, and Olaf Pluta, "Die philosophischen Schriften des Peter von Ailly: Authentizität und Chronologie," *Freiburger Zeitschrift für Philosophie und Theologie* 33 [1986]: 602). He expresses no such reluctance in the *De falsis prophetis II*.

27. *De falsis prophetis II*, in Jean Gerson, *Opera omnia*, ed. Louis Ellies Dupin, 5 vols. (Antwerp: Sumptibus societatis, 1706), 1: col. 594: "Quaedam autem sunt futura contingentia circa actus humanos, circa quae versantur judicia Astrologiae, de nativitatibus, interrogationibus, & electionibus: & haec pars Astrologiae, ut quidam Doctores [i.e., Nicole Oresme, from whose work d'Ailly is borrowing here] dicunt, falsa est, superstitiosa, & inutilis, ac etiam impossibilis sciri."

28. Ibid., col. 602: ". . . plures hominum suas passiones corporales sequuntur; et ideo actus eorum disponuntur, ut in pluribus, secundum inclinationem coelestium corporum," quoting Thomas Aquinas, *Summa theologiae*, IIa IIae, q. 95, a. 5, ad 2.

29. *De falsis prophetis II*, 1: col. 525: "Prophetia est inspiratio vel revelatio divina . . . per lumen intellectuale . . . naturalis rationis lumen excedens."

30. Ibid., col. 533: ". . . Ptolomaeus in prima propositione Centilogii, dicit quod aliqui propter vim animae in eis dominantem, licet non habeant multam scientiam artis astrorum, habent ex meliori parte cognitionem futurorum, & sunt propinquiores veritati quam Astrologus. Super quod dicit ibi Haly in Commento: . . . [quod] . . . haec via, cum pura fuerit, a Philosophis vocatur divina." The *Centiloquium* to which d'Ailly refers was wrongly assumed to have been a work by Ptolemy; it may have originated in ninth- or tenth-century Cairo, the work of one Ahmed ibn Yusuf. Similarly, the commentary d'Ailly cites is also wrongly attributed to "Haly." It may be the work of the same Ahmed ibn Yusuf. Both were translated into Latin in the twelfth century and were standard sources of astrological lore. See Helen Lemay, "The Stars and Human Sexuality: Some Medieval Scientific Views," *Isis* 71 (1980): 127; and S. J. Tester, *A History of Western Astrology* (Woodbridge, Suffolk: Boydell Press, 1987), p. 154.

31. *De falsis prophetis II*, 1: col. 550: ". . . secundum Astronomos; si Mercurius Soli corporaliter sit conjunctus in aliqua domorum Saturni, et Luna fuerit in domo Mercurii eiusdem triplicitatis; tunc virtus coelestis multum fortiter movet ad occultas et profundas cognitiones rerum" ("According to astrologers, if Mercury and the Sun are corporeally conjoined in any *domus* of Saturn, and the Moon is in a *domus* of Mercury in the same triplicity, then the celestial virtue very strongly moves [one] toward occult and profound cognition of things").

32. Ibid., col. 546: "Ubi dicit Haly in Commento [on *Centiloquium*, v. 3] quod Stellae tales sunt animabus, qualia sunt elementa corporibus. . . . intelligit quod sicut corpus virtutes elementorum participat diversimode, secundum quod diversa elementa dominantur in ipso; ita anima diversas proprietates adquirit, secundum dominium stellarum dominantium in infusione ipsius."

33. Ibid., col. 547: ". . . motores coelestes ipsarum stellarum, seu orbium, principaliter impressionem faciunt in animabus. Sic scilicet, quod si Mercurius fortis fuerit hora prin-

cipii generationis alicujus, vel hora nativitatis . . .; motor orbis Mercurii faciet fortis memoriae animam talis nati, & ad hoc dirigit corpus ejus virtus ipsius orbis Mercurii."

34. The Condemnations of 1277, no. 76 (in Mandonnet's numbering) addressed this issue: "Quod intelligentia motrix coeli influit in animam rationalem, sicut corpus coeli influit in corpus humanum." Pierre Mandonnet, *Siger de Brabant et l'averroïsme latin au XIIIme siècle*, 2d ed., 2 vols. in 1 (Louvain: Institut Supérieur de Philosophie, 1908 and 1911), p. 181. This was repeated by the university in 1398. Denifle and Chatelain, *Chartularium Universitatis Parisiensis*, 4:35.

35. E.g., "Nullum autem corpus potest imprimere in rem incorpoream," Thomas Aquinas, *Summa theologiae*, IIa IIae, q. 95, a. 5, ad 4, quoted in *De falsis prophetis II*, 1: col. 591.

36. *De falsis prophetis II*, 1: col. 590.

37. Ibid., col. 591: "Unde quia probatum est quod ex inspectione syderum, non accipitur praecognitio huiusmodi futurorum, nisi sicut ex causis praecognoscuntur effectus; consequens est quod illa futura ex syderibus non possunt praecognosci, quae substrahuntur ipsorum causalitate. . . . duplices effectus substrahuntur causalitate coelestium corporum. Primo quidem, omnes effectus per actus [Thomas has *per accidens*] contingentes . . . Secundo . . . actus liberi arbitrii . . .," following Thomas Aquinas, *Summa theologiae*, IIa IIae, q. 95, a. 5.

38. *De falsis prophetis II*, 1: col. 591–92, repeating some material from the Boethius commentary.

39. Ibid., col. 593: "Si quis ex consideratione astrorum utatur, ad praecognoscendum futuros casuales vel fortuitos eventus; aut etiam ad cognoscendum per certitudinem futura opera hominum: hoc procedet ex falsa & vana opinione, cui saepe operatio daemonis se immiscet, & ideo erit divinatio superstitiosa & illicita. Si vero quis utatur consideratione astrorum, ad praecognoscendum futura quae ex coelestibus causantur corporibus, puta siccitates ac pluvias, & hujusmodi, non erit illicita divinatio nec superstitiosa," quoting Thomas Aquinas, *Summa theologiae*, IIa IIae, q. 95, a. 5.

40. *De falsis prophetis II*, 1: col. 593, following Nicole Oresme, *Tractatus contra iudiciarios astronomos*, chap. 5.

41. For a summary of the two men's arguments, see chapter 2.

42. *De falsis prophetis II*, 1: col. 603: "Plurimos eorum audivi, novi multos: sed neminem in hoc errore diu fuisse recolo, in quo manus Domini condignam non exercuerit ultionem," quoting John of Salisbury, *Policraticus*, book 2, chap. 26 (D'Ailly probably took the quotation from Oresme, *Tractatus*, chap. 2).

43. Gentile prophets of Christ are frequently discussed in medieval Christian writings. The Erythraean Sibyl was believed to have composed verses whose first letters spelled out (when translated from Greek to Latin), "Jesus Christus filius dei salvator." Hermes Trismegistus, a mythical figure believed to have been a rough contemporary of Moses, was thought to have taught that God is one and good. Important sources for such discussions are Lactantius, *Liber divinarum institutionum*, 1:6 (Hermes Trismegistus and the Sibyl) and Augustine, *De civitate dei*, 8:23–24 (Hermes Trismegistus) and 18:23 (where he quotes the Erythraean Sibyl), and Vincent of Beauvais, *Speculum historiale*, 2:100–102 (Sibyls) and 4:10 (Hermes Trismegistus). D'Ailly cites Augustine and Lactantius in his sermon.

Ovid joins the list of gentile prophets of Christ via a work attributed to him (but written in the thirteenth century) entitled *De vetula*, on which see n. 83–85. Pseudo-

Ovid's prophecies of Christianity appear in book 3. The "Chaldeans" are the Magi of Matthew 2 who come in search of the Christ child. See, e.g., the commentary on Gratian's *Decretum*, causa 26, q. 5, by Guido de Baysio (Archidiaconus): "Magi non malefici sed philosophi Caldaeorum qui philosophantur de singulis." Cited in M. A. Screech, "The Magi and the Star (Matthew, 2)," in Olivier Fatio and Pierre Fraenkel, eds., *Histoire de l'exégèse au XVIe siècle*, Etudes de philologie et d'histoire, 34 (Geneva: Droz, 1978), p. 389.

44. *Sermo (tertius) de quadruplici adventu domini et specialiter de adventu ad iudicium*, in Pierre d'Ailly, *Tractatus et sermones* (Strasbourg: [Printer of Jordanus de Quedlinburg], 1490; repr., Frankfurt am Main: Minerva, 1971), fol. t2r: "Albumazar etiam doctrinam veterum caldeorum secutus. viii. [vi] maioris introductorii differentia prima de quadam virgine loquitur, et de illo puero eius quem gentes plurime Ihesum vocant."

45. Ibid.: "Sed et ante huius pueri id est Christi adventum per sex annos et paucos dies et horas facta est coniunctio maxima saturni et iovis circa principium arietis, cui prefuit mercurius dominans virginis. Per que omnia significabantur perspicue puerum de virgine nasciturum, qui foret maximus prophetarum. . . ."

46. Ibid.: "Sed nec solum pia sanctorum patrum devotio hunc prestolabatur adventum, quinimio inveterata gentilitas que a patre luminum deviarat, in ipsis etiam tenebris caligabat inter dum et ipsum licet umbratice sompniabat." D'Ailly shows basically the same impetus as that of Saint Anselm in the *Proslogion*—to strengthen Christians in their faith by showing how some of its fundamentals might be demonstrated by reason.

47. *Sermo (tertius) de adventu*, fol. t4r: ". . . sicut deus in productione mundi supernaturaliter operatus est sic in consummatione huius seculi supernaturaliter operabitur, quare . . . consummatio seculi aut eius tempus naturaliter non potest cognosci, nec per philosophorum rationes aut astrologorum speculationes." This same language of *naturaliter* and *supernaturaliter* will be used by d'Ailly in defense of astrology in the *De legibus et sectis* (1410). Here, however, d'Ailly was following Arnold of Villanova, *De tempore adventus Antichristi*, both in denying the possibility of astrological foreknowledge of the apocalypse and in attacking the particular astrological theory presented here. Lengthy excerpts from this treatise appear in Heinrich Finke, *Aus den Tagen Bonifaz VIII: Funde und Forschungen*, Vorreformationsgeschichtliche Forschungen, 2 (Münster: Aschendorff, 1902), pp. CXXIX–CLIX. In this sermon, d'Ailly follows the section of Arnold's treatise corresponding to pp. CXXXIV–CXXXV of Finke's edition. For more on Arnold's treatise and d'Ailly's use of it, see chapter 5.

48. *Sermo (tertius) de adventu*, fol. t4r: ". . . astrologi probent quod motus retardationis octave spere nequit citius quam in xxxvi milibus annorum compleri, licet etiam huius totius retardationis revolutio necessaria foret ut ipsorum quidam asserunt ad universalem perfectionem mundi . . .," following Arnold of Villanova. This period of 36,000 years was also linked with the concept of the *magnus annus*, the period of time after which the stars would all return to their original positions and, in the most fatalistic versions of the doctrine, all of history would repeat itself. For an explanation of precession and the "great year" theory, see chapter 1.

49. Ibid.: ". . . non tamen ex hoc sequitur presens seculum per tantum spacium duraturum. Nam quia deus suam potentiam et sapientiam non alligavit naturalibus causis, potens est motum orbium quantum sibi placet velocitare et revolutionem huiusmodi brevissimo tempore complere, ita ut revolutionem centum annorum uno anno vel dimidio perficiat," following Arnold of Villanova.

50. Ibid.: ". . . quod utique futurum esse circa finem mundi scriptura testatur dicens. Advenit dies domini sicut fur quo celi magno impetu transient. ii. Petri ultimo. Et idem testatur Sibilla asserens quod tunc minuetur anni sicut menses et menses sicut dies," following Arnold of Villanova (". . . that indeed this shall happen near the end of the world Scripture testifies, saying: The day of the Lord comes like a thief, in which the heavens shall go past with great speed [my trans.]. 2 Peter, end [3:10]. And the Sibyl testifies to the same, asserting that then the years shall be diminished like months, and the months like days").

51. *Sermo primus de adventu*, in Pierre d'Ailly, *Tractatus et sermones*, fol. s3v: "Deus omnipotens voluntatem habens universaliter efficacem potuit sine virili auxilio formare puerum in utero virginali." The sermon most likely dates from the years 1372–95.

52. Ibid.: "Huic enim generationi natura celi testimonium perhibuit celestia quoque hanc dei gloriam enarrarunt, nam in celestibus facta est mirabilis quedam coniunctio modicum priusquam Christus desideratus cunctis gentibus adveniret, aperte significans Jhesum ex virgine nasciturum, de qua coniunctione Albumazar doctrinam veterum caldeorum secutus, scilicet maioris introductorii differentia prima, loquitur multum plane, cuius sententia ponens Ovidius .iii. de vetula qui de hac coniunctione metrice multa scribit."

53. Ibid., fol. s4r: "tunc mirabilis novitas in celo apparuit et coniunctio alias inaudita sicut superius tangebatur, propter quod etiam magi venerunt ad adorandum eum [Christum]." On the miraculous quality of the star that appeared to the Magi, see also the biblical commentary of Nicholas of Lyra (ca. 1270–1340): "patet quod illa stella non erat de stellis existentibus in orbe, nec de stellis cometis." In *Biblia sacra cum glossa interlineari, ordinaria, et Nicolai Lyrani Postilla eiusdemque Moralitatibus, Burgensis Additionibus, et Thoringi replicis* (Venice: [publisher unknown], 1588), fol. 12r.

54. For a more detailed analysis of d'Ailly's defense of astrology, see chapter 2.

55. *Apologetica defensio astronomice veritatis*; *Alia secunda apologetica defensio eiusdem*; *Apologia astrologiae defensiva*.

56. *Epistola ad Papam Joannem XXIII* (*Incipit*: "Dudum . . ."), in Jean Gerson, *Opera omnia*, ed. Louis Ellies Dupin, 5 vols. (Antwerp: Sumptibus societatis, 1706), 2: cols. 880–881; *Sermo in die omnium sanctorum*, in Paul Tschackert, *Peter von Ailli: Zur Geschichte des grossen abendländischen Schisma und der Reformconcilien von Pisa und Constanz* (Gotha: Friedrich Andreas Perthes, 1877; repr., Amsterdam: Rodopi, 1968), app., pp. [43]–[45], [47]–[48], and [49].

57. *Sermo in die omnium sanctorum*, p. [44].

58. For an explanation of this doctrine, see chapter 1. On d'Ailly's use of the great conjunctions in historical writing, see chapter 4.

59. *Vigintiloquium (Concordantia astronomie cum theologia)*; *Concordantia astronomie cum hystorica narratione*; *Elucidarium*; *Pro declaratione decem dictarum figurarum*; and *De figura inceptionis mundi et coniunctionibus mediis sequentibus*.

60. *Concordantia astronomie cum hystorica narratione*, in d'Ailly 1490, chap. 61, fol. [d8r]. See my comments in chapter 6.

61. *Elucidarium*, in d'Ailly 1490, chap. 34. See my chapter 6.

62. *De persecutionibus ecclesie*, Marseilles, Bibliothèque Municipale, MS 1156, fols. 26v–27r. See chapter 6.

63. *De legibus et sectis*, in Pierre d'Ailly, *Tractatus de imagine mundi et varia ejusdem auctoris et Joannis Gersonis opuscula* (Louvain: Johann de Westphalia, ca. 1483), chap. 4.

64. In *Concordantia astronomie cum theologia (= Vigintiloquium)*; *Concordantia astronomie cum hystorica narratione*, both printed in d'Ailly 1490. Leopold of Austria was the author of a popular astrological *summa*. D'Ailly mentions other sources, but we cannot be sure he had read them yet. We can know he had looked at Albumasar and is not merely following Bacon again, because he cites text not found in Bacon; likewise for Alchabitius and Leopold of Austria. He also cites by name Ptolemy, Haly, Messahalla, and Albertus Magnus.

65. In the *Elucidarium*. For more on d'Ailly's thought on the great conjunctions, see chapter 4.

66. E.g., Ptolemy (and the pseudo-Ptolemy of the *Centiloquium*), "Haly," Albumasar, al-Kindi, Messahalla, Alchabitius, Henry Bate of Malines, Albertus Magnus, Avenesra, Johannes de Muris, John of Ashenden, Leopold of Austria, the Alfonsine tables, and Roger Bacon.

67. *De figura inceptionis mundi et coniunctionibus mediis sequentibus*, Vienna, Österreichische Nationalbibliothek, MS 5266, fols. 46r–50v. E.g., fol. 45v: "According to the figure made above, in the fourteenth year, with 229 days, 8 hours, and 36 minutes, from the beginning of the world, Jupiter and Saturn were conjoined in the fifteenth degree and 29 minutes of Taurus, according to their mean motions, with the twenty-fourth degree of Libra in the ascendant" ("Secundum prefactam figuram celi anno 14 cum 229 diebus 8 horas 36 minutis fuerunt coniuncti Jupiter et Saturnus in Thauro 14 gradu 29 minuto in medios motus ascendente 24 gradu Libre"). The conjunction thus was to have occurred 13 years, 229 days, 8 hours, and 36 minutes after the moment of Creation, set in this treatise at 5,328 years and 243 days before Christ.

68. *De concordia discordantium astronomorum*, in Pierre d'Ailly, *Tractatus de imagine mundi et varia ejusdem auctoris et Joannis Gersonis opuscula* (Louvain: Johann de Westphalia, ca. 1483).

69. John D. North, "Astrology and the Fortunes of Churches," *Centaurus* 24 (1980): 184–85.

70. *De legibus et sectis*, chap. 7. See also my discussion in chapter 2.

71. *Vigintiloquium*, v. 4, fol. a3v.

72. *Apologetica defensio astronomice veritatis (I)*, in Pierre d'Ailly, *Tractatus de imagine mundi et varia ejusdem auctoris et Joannis Gersonis opuscula* (Louvain: Johann de Westphalia, ca. 1483), fols. [gg6v]; [gg8r]–[gg8v].

73. *Concordantia astronomie cum hystorica narratione*, chap. 60; *Elucidarium*, chap. 34; *De persecutionibus ecclesie*, passim.

74. E.g., d'Ailly followed Aquinas in the Boethius commentary as well as in a 1419 letter to Jean Gerson.

75. In *Apologetica defensio astronomice veritatis (I)*, fol. [gg6v], he says that *De legibus et sectis*, *De concordia theologie et astronomie* (i.e., *Vigintiloquium*), a sermon from his youth, and a treatise on false and true prophets (i.e., *De falsis prophetis II*) all "reproved not the truth of astronomy, but the vanity of certain astronomers" ("ego autem non astronomie veritatem sed quorumdam astronomorum vanitatem secum reprobo").

76. For the *De legibus et sectis*, d'Ailly followed Roger Bacon, *Opus maius*, pars 4. The relevant passages are found in Roger Bacon, *The "Opus maius" of Roger Bacon*, ed. John Henry Bridges, 2 vols. (Oxford: Clarendon Press, 1897), vol. 1, pars 4, pp. 253–57, 261–69.

77. This theory dealt with the trepidational movement of the eighth sphere; d'Ailly took it from Albumasar, *De magnis coniunctionibus*, tr. 2, diff. 8 (as in Albumasar, *De magnis coniunctionibus, annorum revolutionibus, ac eorum profectionibus, octo continens tractatus* [Venice: Melchior Sessa, 1515]). D'Ailly, *Concordantia astronomie cum hystorica narratione*, chap. 60, fol. [d8r]: "For with the aforesaid conjunction and revolutions of Saturn there concurs the revolution or reversal of the superior orb, that is, of the eighth sphere, by which, just as by these others is recognized a mutation of sects. . . . as Albumasar notes more fully in *De magnis coniunctionibus*, tr. 2, diff. last, near the end" ("Nam cum predicta coniunctione et illis revolutionibus Saturni ad hoc concurret revolutio seu reversio superioris orbis id est octave spere per quam et per alia premissa cognoscitur sectarum mutatio. . . . Et plenius notavit Albumasar libro de magnis coniunctionibus tractatu .ii. differentia ultima circa finem").

78. "Nec dicta audivi, nec scripta legi," *Epistola ad novos Hebraeos*, chap. 4, quoted in Salembier 1886, p. 305. Excerpts from this work have been printed in Tschackert, *Peter von Ailli*, pp. [7]–[12].

79. *Apologeticus Hieronymianae versionis*, printed in Louis Salembier, "Une page inédite de l'histoire de la Vulgate," *Revue des sciences ecclésiastiques* 61 (1890): 500–13, and 62 (1890): 97–110. After composing the *Epistola ad novos Hebraeos*, there had come into d'Ailly's hands "magnus liber unius doctoris anglicani." Salembier, "Une page inédite," *Revue des sciences ecclésiastiques* 61:508. For the date of this work, see Max Lieberman, "Pierre d'Ailly, Jean Gerson et le culte de saint Joseph (II)," *Cahiers de Joséphologie* 14 (1966): 306. D'Ailly was criticizing Bacon's *De linguarum studii necessitate*. Salembier 1886, p. 309.

80. Lieberman, "Pierre d'Ailly, Jean Gerson et le culte," p. 307.

81. *Sermo (tertius) de adventu*, fol. t2r: "Albumazar etiam doctrinam veterum caldeorum secutus .viii. [.vi.] maioris introductorii differentia prima de quadam virgine loquitur et de illo puero eius quem gentes plurime Ihesum vocant."

82. The division of the signs into three "faces" of 10 degrees each is found in ancient Greek astrology and stems from an Egyptian tradition of dividing the 360 degrees of the ecliptic into "decans," 36 groups of 10 degrees each. The decans were associated with genii or demons, each of whom had its own name and face. These genii or demons came to be associated with the paranatella (the stars rising to the north and south of the zodiacal signs), and lists of such stars contain Egyptian constellations, glyphs, and sigils. Albumasar's virgin and child is a constellation described within such a list. Jean Seznec, *The Survival of the Pagan Gods: The Mythological Tradition and Its Place in Renaissance Humanism and Art*, trans. Barbara F. Sessions (New York: Pantheon Books, 1953), pp. 38–39.

83. The relevant passages are:
Albumasar, *Introductorium maius*, trans. John of Seville, tr. 6, diff. 1: "Virgo est duum corporum suntque ei tres species. Et ascendit in prima facie illius puella quam vocamus Celchuis Dorosthal, et est virgo pulchra atque honesta et munda prolixi capilli et pulchra facie, habens in manu sua duas spicas. Et ipsa sedet supra sedem stratam et nutrit puerum dans ei ad comedendum ius in loco qui vocatur Abrie, et vocant ipsum puerum quaedam gentium Ihesum cuius interpretatio arabice est Eiceh. Et ascendit cum ea vir sedens super ipsam sedem." Quoted by the editors in Albertus Magnus, *Speculum astronomiae*, ed. Stefano Caroti, Michela Pereira, Stefano Zamponi, and Paola Zambelli (Pisa: Domus Galilaeana, 1977), p. 72 (from Florence, Bibliotheca Medicea Laurenziana, MS Pluteo

29, 12). Albumasar's *Introductorium maius* was translated twice in the twelfth century, in 1133 by John of Seville, and in 1140 by Hermann of Carinthia. John's translation was used far more frequently, but only Hermann's translation has been printed. Richard Lemay, "The True Place of Astrology in Medieval Science and Philosophy: Towards a Definition," in Patrick Curry, ed., *Astrology, Science and Society: Historical Essays* (Woodbridge, Suffolk: Boydell Press, 1987), pp. 65–66.

Albertus Magnus, *Speculum astronomiae* (ed. Caroti et al.), chap. 12, pp. 36–37: "In tractatu namque sexto [of Albumasar's *Introductorium maius*], differentia prima, in capitulo de ascensionibus imaginum quae ascendunt cum virgine, invenitur: 'Et ascendit in prima facie illius (scilicet Virginis) puella quam vocat Celchuis Darostal; et est virgo pulchra atque honesta et munda prolixi capilli, et pulchra facie, habens in manu sua duas spicas, et ipsa sedet super sedem stratam, et nutrit puerum, dans ei ad comedendum ius in loco qui vocatur Abrie. Et vocant ipsum puerum quaedam gens Iesum, cuius interpretatio est arabice Eice. Et ascendit cum ea vir sedens super ipsam sedem. Et ascendit cum ea stella virginis etc.'"

Roger Bacon, *Opus maius* (ed. Bridges), vol. 1, pars 4, p. 257: "Et dicunt [mathematici], quod haec lex est prophetae nascituri de virgine, secundum quod omnes antiqui Indi, Chaldaei, Babylonii, docuerunt quod in prima facie Virginis ascendit virgo mundissima nutritura puerum in terra Hebraeorum, cui nomen Jesus Christus, ut dicit Albumazar in majori introductorio astronomiae." Note the reference to the Chaldeans, present in d'Ailly (see n. 81) and Bacon, but not in Albertus Magnus or Albumasar.

Pseudo-Ovid, *The Pseudo-Ovidian De vetula: Text, Introduction, and Notes*, ed. Dorothy M. Robathan (Amsterdam: Adolf M. Hakkert, 1968), book 3, lines 623–33: "namque / Hiis in imaginibus, que describuntur ab Indis / Et Caldeorum sapientibus ac Babilonis, / Dicitur ex veterum scriptis ascendere prima / Virginis in facie, prolixi virgo capilli, / Munda quidem magnique animi magnique decoris, / Pluris honestatis, et in ipsius manibus sunt / Spice suspensis et vestimenta vetusta. / Sede strata puerumque nutrit, puero ius / Ad comedendum dans, puerumque Ih'm vocat ipsum / Gens quedam, sedet et vir ibi sedem super ipsam." This poem is now generally thought to have been the work of Richard of Fournival (d. 1260). Note that there is a mention of the Chaldeans in this passage, but no reference to Albumasar (since the poem's pretended author would have lived long before the ninth-century Arab astrologer). This work has also been published as *Pseudo-Ovidius De vetula: Untersuchungen und Text*, ed. Paul Klopsch, Mittellateinische Studien und Texte, 2 (Leiden: E. J. Brill, 1967).

D'Ailly's source in this 1385 sermon thus appears to have been Roger Bacon's *Opus maius*. In the other Advent sermon discussed here, d'Ailly also cited this passage of Albumasar's, adding a reference to the *De vetula*. Again, mention of both Albumasar and the Chaldeans points to the *Opus maius* as d'Ailly's source. He may have read the *De vetula* himself, but he also could have become familiar with its contents by a reading of the *Opus maius*.

84. *Sermo (tertius) de adventu*, fol. t2r: "Sed et ante huius pueri id est Christi adventum per sex annos et paucos dies et horas facta est coniunctio maxima saturni et iovis circa principium arietis cui prefuit mercurius dominans virginis. Per que omnia significabantur perspicue puerum de virgine nasciturum, qui foret maximus prophetarum. . . ." For explanations of *coniunctio maxima* and of d'Ailly's use of conjunction theory in general, see my chapters 1 and 4.

85. *De vetula* (ed. Robathan), book 3, lines 611–15: "Una quidem talis [coniunctio] felice tempore nuper / Caesaris Augusti fuit, anno bis duodeno / A regni novitate sui, que significavit / Post annum sextum nasci debere prophetam / Absque maris coitu de virgine. . . ."

86. *Opus maius* (ed. Bridges), vol. 1, pars 4, p. 264: "Si enim revolvamus motus Saturni et Jovis ad tempus illud, inveniemus eos fuisse conjunctos per medios cursos suos ante nativitatem Christi per sex annos, quinque dies, et tres horas; et erat medius cursus utriusque in Ariete decem gradus, lvi minuta, lii secunda." I take the phrase "per medios cursos suos" to refer to the planets' mean and not their true motions (on which see chapter 4).

87. *Opus maius* (ed. Bridges), vol. 1, pars 4, p. 261: "Et ideo nusquam dominatur Mercurius tantum, sicut in Virgine. Nec aliquis planeta habet tot in ea dominia, propter quod appropriatur Virgini Mercurius. Et ideo ex hac causa dicunt legem Mercurialem debere esse sectam prophetae nascituri de virgine."

88. See Philippe de Mézières, *Le songe du vieil pelerin*, ed. G. W. Coopland, 2 vols. (Cambridge: Cambridge University Press, 1969). See esp. chaps. 140–62, vol. 1, pp. 594–621, for the treatment of astrology.

89. Of course, d'Ailly could have encountered Oresme's works earlier as a student in the College of Navarre, where Oresme had been a master.

90. Bernard Guenée, *Entre l'Eglise et l'Etat: Quatre vies de prélats français à la fin du Moyen Age (XIIIe–XVe siècle)* (Paris: Gallimard, 1987), pp. 260–66. This quiet period may have been a deliberate retreat from public life. Following the Council of Pisa, Boniface Ferrer had composed a treatise in which he accused d'Ailly of using the circumstances of the Schism to advance his own career. D'Ailly was stung by his criticisms and composed a response in 1412. Boniface Ferrer, *Tractatus pro defensione Benedicti XIII*, in Edmund Martène and Ursin Durand, eds., *Thesaurus novus anecdotorum* (Paris: Delaulne, 1717; repr., New York: Burt Franklin, 1968), 2: cols. 1436–1529, esp. 1447–48, 1464–65, 1496.

91. On the *Imago mundi*, see *Imago mundi de Pierre d'Ailly. Texte latin et traduction française des quatre textes cosmographiques de d'Ailly et des notes marginales de Cristophe Colomb. Etude sur les sources de l'auteur par Edmond Buron*, 3 vols. (Paris: Maisonneuve Frères, 1930); and Pauline Moffit Watts, "Prophecy and Discovery: On the Spiritual Origins of Christopher Columbus's 'Enterprise of the Indies,'" *American Historical Review* 90 (1985): 73–102.

92. *Exhortatio super Kalendarii correctione* (1411); *Compendium cosmographie*; *De vero cyclo lunari*; *Cosmographie tractatus duo*. All of these works appear in Pierre d'Ailly, *Tractatus de imagine mundi et varia ejusdem auctoris et Joannis Gersonis opuscula* (Louvain: Johann de Westphalia, ca. 1483).

93. He composed astrological treatises in Basel and Cologne, based, one may surmise, upon readings he encountered in his travels.

94. For more on the *Elucidarium* and other works dealing with the great conjunctions, see chapter 4.

95. Including Leopold of Austria, Albumasar, Alchabitius, (pseudo)-Ptolemy, Avenesra, Henry Bate of Malines, Johannes de Muris, John of Ashenden, Guido Bonatti, and Messahalla.

96. D'Ailly's views on the Schism and its relation to the apocalypse are discussed more fully in chapters 5 and 6.

97. Richard Lemay, "The Teaching of Astronomy in Medieval Universities, Principally at Paris in the Fourteenth Century," *Manuscripta* 20 (1976): 200–202.

98. *Elucidarium*, chap. 40, fol. [g7v]: "Scimus autem quosdam nobis obiecisse quod professionem nostram et similiter etatem magis decuisset circa theologica quam circa illa mathematica studia occupari."

99. *Sermo (tertius) de adventu*, fol. [t6r]: "De hac crudeli seditione presentis scismatis omnino timendum est ne . . . ipsa sit illa horrenda divisio et persecutio scismatica post quam antichristi seva persecutio celeriter est futura."

100. *De falsis prophetis II*, 1: col. 517: ". . . huis temporibus, quae fini mundi propinqua esse videntur."

101. This summary is based on the excellent account put forth by Bernard Guenée, *Entre l'Eglise et l'Etat*, pp. 212–22. For d'Ailly's difficulties at Cambrai, see also Tschackert, *Peter von Ailli*, pp. 97–101; Salembier 1886, pp. 43–53; and Salembier, *Le cardinal Pierre d'Ailly, chancelier de l'Université de Paris, évêque du Puy et de Cambrai 1350–1420* (Tourcoing: Imprimerie Georges Frère, 1932), pp. 116–25 (hereafter cited as Salembier 1932).

102. André Combes, however, has attacked the notion that the year 1408 was one so harsh for d'Ailly that he required letters of consolation from his friends. He cites a 1408 letter from d'Ailly's friend Nicolas of Clamanges in which Nicolas reports his pleasure at hearing that things were going so well for d'Ailly. Combes, "Sur les 'lettres de consolation' de Nicolas de Clamanges à Pierre d'Ailly," *Archives d'histoire doctrinale et littéraire du moyen âge* 15–17 (1940–42): 359–89. (He cites Nicolas of Clamanges's letter at pp. 381–87.)

103. *De legibus et sectis*, chap. 7, fol. [f5r]: ". . . omnes leges vel secte quantum ad illa que in eis naturalia sunt vel naturaliter fiunt astronomice potestati seu constellationi vel dispositioni celesti aliqualiter subesse possunt"; and fol. [f5v]–[f6r]: ". . . leges vel secte divine id est non humanitus sed divinitus inspirate qualis est lex Christi et qualem fuisse credimus legem Moysi inquantum a divina voluntate libera supernaturaliter et miraculose procedunt nullatenus astrorum legibus seu eorum constellationibus subiecte sunt." Bacon had urged the use of astrology with other means to predict Antichrist's advent, as d'Ailly here acknowledges, but rejects, in chap. 4, fol. f2v: "Unde concludendo [i.e., Bacon] dicit. Scio quod si ecclesia . . . iuberet considerari vias astronomie inveniretur sufficiens suspicio vel magis certitudo de tempore antichristi. Sed his videtur obviare dictum Christi: Non est vestrum nosse tempora vel momenta. . . ."

104. *Apologia concilii pisani contra tractatum domini Bonifacii*, printed in Tschackert, *Peter von Ailli*, pp. [31]–[41]. On Boniface's attack and d'Ailly's reply, see Max Lieberman, "Chronologie Gersonienne. VIII. Gerson et d'Ailly (III)," *Romania* 81 (1960): 74–76; Guenée, *Entre l'Eglise et l'Etat*, p. 270; Salembier 1886, pp. 83–84; Salembier 1932, pp. 254–59; and Tschackert, *Peter von Ailli*, pp. 162–63.

105. Lieberman, "Chronologie Gersonienne," pp. 79–80. Gerson similarly turned to mysticism after 1400; Guenée, *Entre l'Eglise et l'Etat*, p. 231. Among the spiritual works d'Ailly composed in 1414 are: *Devota meditatio super psalmum: Judica me Deus* (June 10, 1414); *Oratio dominica anagogice exposita* (July 6, 1414); *Devota meditatio super Ave Maria* (July 12, 1414); and *Devota meditatio seu expositio super psalmum: In te Domine speravi* (July 14, 1414). Salembier 1886, p. xviii.

106. *Concordantia astronomie cum hystorica narratione*, chap. 60.

107. *De persecutionibus ecclesie*, fol. 6v: ". . . de tempore antichristi cuius adventus adhuc non apparet propinquus."

108 Ibid., fol. 27r–v: ". . . erit complementum decem revolutionum saturnalium anno Christi millesimo septingentesimo octuagesimo nono. . . . dico quod si mundus usque ad illa tempora duraverit . . . probabiliter concluditur quod forte circa illa tempora veniet Antichristus cum lege sua vel secta dampnabili."

109. *Apologia astrologiae defensiva*, in Gerson, *Oeuvres complètes*, 2:219: "Nam sicut illos dico superstitiosos astronomos qui contra theologicam veritatem astrologiam ultra id quod potest nimis extollunt, sic et illos superstitiosos theologos qui contra philosophicam rationem astronomiae potestatem nimis deprimunt vel penitus tollunt." D'Ailly had defended astrology in 1410 under the guise of attacking "superstitious astrologers." Now, in a neat and conscious symmetry, he guarded the science against "superstitious theologians."

· **Chapter Four**
Astrology and the Narration of History

1. An outline of conjunction theory appears in chapter 1. For the two authors' versions of the theory of the great conjunctions, see Roger Bacon, *The "Opus maius" of Roger Bacon*, ed. John Henry Bridges, 2 vols. (Oxford: Clarendon Press, 1897), vol. 1, pars 4, pp. 263–64; and Albumasar, *De magnis coniunctionibus, annorum revolutionibus, ac eorum profectionibus, octo continens tractatus* (Venice: Melchior Sessa, 1515), tr. 1, diff. 1. Albumasar's treatise was also published by Erhard Ratdolt (Augsburg, 1489).

2. Albumasar, *De magnis coniunctionibus*, tr. 1, diff. 4.

3. See Bacon, *Opus maius* (ed. Bridges), vol. 1, pars 4, pp. 255–57. D'Ailly gives this fully in the *De legibus et sectis*, in Pierre d'Ailly, *Tractatus de imagine mundi et varia ejusdem auctoris et Joannis Gersonis opuscula* (Louvain: Johann de Westphalia, ca. 1483), chaps. 2–3. He attacks parts of the theory there, but uses it with apparent approval in the *Concordantia astronomie cum hystorica narratione*, in d'Ailly 1490, chap. 62, to strengthen the notion that no sect will come after Muhammad save that of Antichrist.

Bacon may have taken his discussion of the astrological basis of religion from the work that circulated under Ovid's name entitled *De vetula*, which he cites. For an edition, see *The Pseudo-Ovidian De vetula. Text, Introduction, and Notes*, ed. Dorothy M. Robathan (Amsterdam: Adolf M. Hakkert, 1968). The poet (probably Richard de Fournival) discusses the six principal religions in book 3, lines 522–93; and the great conjunctions of Saturn and Jupiter in book 3, lines 594–654. The poem also appears in *Pseudo-Ovidius De vetula: Untersuchungen und Text*, ed. Paul Klopsch, Mittellateinische Studien und Texte, 2 (Leiden: E. J. Brill, 1967).

The passage in Albumasar upon which Bacon and the *De vetula* ultimately rely is *De magnis coniunctionibus*, tr. 1, diff. 4.

4. See William Adler, *Time Immemorial: Archaic History and Its Sources in Christian Chronography from Julius Africanus to George Syncellus*, Dumbarton Oaks Studies, 26 (Washington, D.C.: Dumbarton Oaks, 1989), esp. pp. 2, 18, 47–48; Bernard Guenée, *Histoire et culture historique dans l'occident médiéval* (Paris: Aubier Montaigne, ca. 1980), pp. 147–54; Roderich Schmidt, "Aetates mundi: Die Weltalter als Gliederungsprinzip der Geschichte," *Zeitschrift für Kirchengeschichte* 67 (1955): 288–317; Anna-Dorothee v. den

Brincken, "Beobachtungen zum Aufkommen der retrospektiven Inkarnationsära," *Archiv für Diplomatik Schriftgeschichte Siegel- und Wappenkunde* 25 (1979): 1–8; and Richard Landes, "Lest the Millennium Be Fulfilled: Apocalyptic Expectations and the Pattern of Western Chronography 100–800 CE," in Werner Verbeke, Daniel Verhelst, and Andries Welkenhuysen, eds., *The Use and Abuse of Eschatology in the Middle Ages* (Louvain: Leuven University Press, 1988), pp. 137–211.

5. Luke 3:23 yields the information that at the time of his baptism and the beginning of his ministry, Jesus entered the thirtieth year of his life, pointing to a Creation date for Eusebius of some 5,199 years before his birth ("Et ipse Iesus erat incipiens quasi annorum triginta"). Eusebius declined, however, to date events in terms of years from Creation, citing the problems involved in establishing an accurate chronology beginning from Adam. His chronological tables put forth dates in years from Abraham and from the reign of Ninus. Adler, *Time Immemorial*, pp. 69–70.

6. See Bede, *De temporum ratione*, chap. 66, in *Bedae venerabilis opera*, pars 6, 2, Corpus Christianorum, Series Latina, 123B (Turnhout: Brepols, 1977), p. 495.

7. Vincentius Bellovacensis (Vincent of Beauvais), *Speculum quadruplex, sive speculum maius: naturale, doctrinale, morale, historiale*, 4 vols. (Douai: Baltazar Bellerus, 1624; repr., Graz, Austria: Akademische Druck, 1965), vol. 4 (*Speculum historiale*), book 2, chap. 115, p. 84: "Ita sunt variae et discrepantes de temporibus sententiae, ut vix in hac re aliquid certi statui possit." See also book 6, chap. 88, pp. 203–4, at which Vincent places Christ's birth in the year 3953 but also cites the figure of 5199.

8. *Concordantia astronomie cum theologia* (=*Vigintiloquium*), in d'Ailly 1490, prologue, fol. a2v: "De concordia theologie et astronomie aliquid scribere ab amico in utraque perito transmisse mihi littere occasionaliter induxerunt. . . . [Se] asserebat admirari nos fideles ex hebreorum et septuaginta interpretum sentenciis apparenter dissonantibus tot in lege numerosas contrarietates indeterminatas reliquisse ubi inter alia occurrit materia quod ab adam usque ad christum annorum millenaria historiographi varie descripserunt. . . . Que diversitas utique in cronicis non contingisset si legum et cronicarum scriptores magnarum tempora coniunctionibus in significandis scriptis suis veridice supposuissent."

D'Ailly may also have become familiar with some of the issues involved in calculating the time of Creation through Roger Bacon. Bacon, *Opus maius* (ed. Bridges), vol. 1, pars 4, pp. 187–210; *Opus tertium*, chap. 44 (in Roger Bacon, *Opera quaedam hactenus inedita*, ed. J. S. Brewer, Rerum Britannicarum Medii Aevi Scriptores [London: Longman, 1859], 1: 204–15).

9. *Vigintiloquium*, v. 2. Details of d'Ailly's defense of astrology appear in chapter 2.

10. *Vigintiloquium*, v. 3.

11. Ibid., v. 3, fol. a3r: ". . . Augustinus .v. de civitate dei describens fatum secundum estimationem illorum qui vocant fatum vim positionis siderum et constellationum secundum quam omnia in his inferioribus necessario eveniunt: sic diffinit quod fatum nihil sit. . . . A quibus non discordat peritissimus astronomorum Ptolomeus ubi ait quod vir prudens dominabitur astris."

12. Ibid., v. 4, fol. a3v: ". . . inter varias utilitates una esset theologicis prophetis astronomica iudicia respectu quorumdam futurorum eventuum coaptare."

13. Ibid., v. 5, fol. a4r: "Et ideo licet Noe illud diluvium precognoverit per revelationem propheticam tamen probabile videtur quod aliqua constellatio astronomica illum

effectum presignaverit." The 1483 edition of the *Vigintiloquium* adds the phrase "et ipsius aliqualiter partialis causa fuerit"; see Pierre d'Ailly, *Tractatus de imagine mundi et varia ejusdem auctoris et Joannis Gersonis opuscula* (Louvain: Johann de Westphalia, ca. 1483), fol. aa3r.

14. *Vigintiloquium*, in d'Ailly 1490, v. 20.

15. Ibid., v. 5, fols. a4r–a4v: ". . . benedicta Christi incarnatio et nativitas licet in multis fuerit miraculosa et supernaturalis tamen . . . natura tamquam famula domino suo et creatori subserviens divine omnipotentie cooperari potuit et in his per celi et astrorum virtutem concurrere cum virtute naturali virginis matris eius."

16. Ibid., fol. a4v: ". . . dei sapientia humanitati Christi hypostatice unita bonitatem constellationis augere potuit in melius ipsiusque dispositionem in malum penitus suspendere."

17. Ibid., prologue, fol. a2v: "Preter Methodium quem de magnitudine numeri excusant aliqui .5228. posuisse videantur Eusebius Cesariensis .6000. Beda vero numero minori .5200. minus uno." The figure of 5,199 is not Bede's (he places Creation, as noted, at 3952 B.C.) but is compatible with Eusebius's 5,228 from Adam to the start of Christ's ministry. D'Ailly may have had in mind a verse cited in Vincent of Beauvais, *Speculum historiale*, book 6, chap. 88, p. 204, which attributes to Bede the figure of 5,200 years minus 1 from Adam to Christ: "Unum tolle datis ad milia quinque ducentis; / Nascenti Domino tot Beda dat a Protoplasto." In the *Vigintiloquium*, v. 10, d'Ailly notes that the figure of 6,000 years falsely has been attributed to Eusebius.

18. *Vigintiloquium*, in d'Ailly 1490, v. 6 and 7. D'Ailly notes that Roman years were originally of 300 days; this was corrected by Numa to a figure of 354, and by Julius Caesar to 365¼ days; astronomers consider the year to be a fraction of an hour shorter.

19. Ibid., v. 8.

20. Ibid., v. 10–11.

21. Ibid., v. 12, fol. br. The era of the Flood is the earliest of the various epochs put forth in the Alfonsine tables. Expressed sexagesimally, as are all dates in the Alfonsine tables, the era of the Flood there is said to be 5,14,42,39 days before the Incarnation [that is, $(5 \times 216,000) + (14 \times 3,600) + (42 \times 60) + 39$ days], equal to 3,101 years, 319 days, or 6½ days more than d'Ailly's interval. Emmanuel Poulle, ed. and trans., *Les tables alphonsines avec les canons de Jean de Saxe: Edition, traduction et commentaire* (Paris: Editions du C.N.R.S., 1984), p. 108.

22. Hindu astronomers maintained that there were periodic mean conjunctions of all of the planets at Aries 0° at various long intervals. In Albumasar's lost *Book of the Thousands* there is such a conjunction in 3102 B.C., the epoch of the Flood, halfway through a *Yuga* of 360,000 years. David Pingree, *The Thousands of Abu Ma'shar*, Studies of the Warburg Institute, 30 (London: Warburg Institute, 1968), pp. 27–35. See also John D. North, "Chronology and the Age of the World," in Wolfgang Yourgrau and Allen D. Beck, eds., *Cosmology, History, and Theology* (New York: Plenum Press, 1977), pp. 315–17; Edward S. Kennedy, "Ramifications of the World-Year Concept in Islamic Astrology," *Proceedings of the Tenth International Congress of the History of Science* (1962): 23–43; and Otto Neugebauer, "Hindu Astronomy at Newminster in 1428," reprinted in Neugebauer, *Astronomy and History: Selected Essays* (New York: Springer-Verlag, 1983), pp. 425–32.

23. *Vigintiloquium*, v. 15 and 16.

24. Ibid., v. 15, fol. b2r: ". . . si earum [maximarum coniunctionum] tempora nota-verimus et ad ea historias applicemus inveniemus profecto circa huiusmodi tempora mag-nas et mirandas mutationes in hoc seculo evenisse."

25. Ibid., v. 16, fol. b2v: ". . . probabiliter temptare magis quam certitudinaliter trac-tare proponimus."

26. On the *thema mundi* and in particular on its representation in art, see Kristin Lip-pincott, "Giovanni di Paolo's 'Creation of the World' and the Tradition of the 'Thema Mundi' in Late Medieval and Renaissance Art," *Burlington Magazine* 132 (1990): 460–68.

27. Iulius Firmicus Maternus, *Matheseos Libri VIII*, ed. W. Kroll and F. Skutsch (Leipzig: B. G. Teubner, 1897), book 3, chap. 1; Macrobius, *Commentary on the Dream of Scipio*, trans. William Harris Stahl (New York: Columbia University Press, 1952), book 1, chap. 21, pp. 179–80; also see John of Salisbury, *Policraticus*, book 2, chap. 19, in Joannes Saresberiensis, *Opera omnia*, ed. J. A. Giles (Oxford: J. H. Parker, 1848), 3:103.

28. *Vigintiloquium*, in d'Ailly 1490, v. 19. Although d'Ailly cites the *De magnis coniunc-tionibus*, tr. 4, Albumasar does not present a horoscope of Creation there, but notes that according to Hindu astronomy the world began on a Sunday at sunrise ("Et estimaverunt indi quod principium fuit die dominica sole ascendente"); see *De magnis coniunctionibus* (Augsburg: Erhard Ratdolt, 1489), tr. 4, diff. 12, fol. [F8v]. D'Ailly's chart has the sun in Aries and Aries in the midheaven, not the ascendant. Alchabitius's list of planetary exalta-tions appears in Alchabitius, *Libellus isagogicus abdilazi id est servi gloriosi dei qui dicitur Alchabitius ad magisterium iudiciorum astrorum interpretatus a Joanne hispalensi . . ., in Alchabi-tius cum commento. Noviter impresso* (Venice: Melchior Sessa, 1512), diff. 1, fol. 2v.

29. *Vigintiloquium*, in d'Ailly 1490, v. 18, following William of Auxerre's *Summa aurea*, book 2, tr. 7. Guillermus Altissiodorensis (William of Auxerre), *Summa aurea* (Paris: Philippe Pigouchet, 1500; repr., Frankfurt: Minerva, 1964), fol. 55r.

30. *Vigintiloquium*, in d'Ailly 1490, v. 20.

31. Given d'Ailly's figures for Saturn's (30 years) and Jupiter's (12 years) average paths around the zodiac, Saturn would have completed 10⅔ such circuits in 320 years and Jupiter 26⅔. D'Ailly gives Saturn's original position as Aquarius 21° and Jupiter's as Sagittarius 15°. That puts Saturn at Libra 21° and Jupiter at Leo 15° in the year 320.

32. *Vigintiloquium*, in d'Ailly 1490, v. 20, fol. [b5r]: "hic non procedemus secundum calculationem equatam et precisam sed veritati proximam quorundam astronomorum opinionem satis probabilem insequendo."

33. Ibid., v. 20. D'Ailly later rejected the occurrence of a conjunction two years before the Flood. It is not clear what his source was for this conjunction; it does not appear in the writings of the major Arab astrologers. According to Kennedy, Ibn Hibinta quotes Masha'allah (Messahalla) as placing a conjunction 20 years before the Flood in Scorpio 1°24'. Perhaps this was the conjunction d'Ailly had in mind, although it would have to come from a different source; Ibn Hibinta's work does not exist in Latin and was thus unknown to d'Ailly. Kennedy, "Ramifications of the World-Year Concept," p. 26. D'Ailly later suggested that his conjunction 2 years before the Flood might have been that of 22 years before the Flood in Aries 6°, via a misreading or a scribal error. *Elucidarium*, in d'Ailly 1490, chap. 27.

34. *Vigintiloquium*, in d'Ailly 1490, v. 20, fol. [b6r]: "Unde valde expediens videretur quod ecclesia auctoritate concilii generalis vel summi pontificis plures peritos theologos et astronomos ad hoc committeret ut ex auctoritativa ecclesie declaratione et determina-

tione in premissis decor et confirmatio fidei sequeretur et tolleretur occasio scandali eorum qui detrahunt catholice veritati et improperant discordiam que inter hebreos et 70 interpretes sanctosque catholicos doctores in annorum mundi computatione reperitur."

35. North, "Chronology," pp. 317–23.

36. *Concordantia astronomie cum hystorica narratione*, chaps. 47 and 48, citing Albumasar, *De magnis coniunctionibus*, tr. 2, diff. 8. D'Ailly also could have taken this information from Roger Bacon. Bacon, *Opus maius* (ed. Bridges), vol. 1, pars 4, p. 265.

37. Friedrich von Bezold says he displays evidence of "eines recht dürftigen historischen Wissens." F. von Bezold, "Astrologische Geschichtsconstruction im Mittelalter," *Deutsche Zeitschrift für Geschichtswissenschaft* 8 (1892): 59.

38. *Concordantia astronomie cum hystorica narratione*, chap. 3.

39. Ibid., chap. 5, fol. cr: ". . . ipse modicum ante secundam coniunctionem maximam natus fuit."

40. Ibid., chap. 6, fol. cr: ". . . circa hoc tempus facta est ante diluvium tercia coniunctio maxima quam mirabiles mundi alterationes secute sunt inter quas precipua fuit diluvii inundatio cuius dictam coniunctionem parcialem causam extitisse annotavimus."

41. Ibid., chap. 10.

42. Ibid., chap. 12.

43. Ibid.

44. Ibid., chaps. 14–17.

45. Ibid., chap. 17, fol. c3v: "Ad multa igitur de his mirabilibus mundi que hic incidentaliter tetigimus plurimaque alia de quibus sacra narrat historia que hic gratia brevitatis subticemus concurrere potuerunt alique coniunctiones maiores revolutionesque Saturni que a quarta usque ad quintam maximam coniunctionem intervenerunt. Quas etiam verisimile est eversionis Troiane partiales causas fuisse. . . ."

46. Ibid., chaps. 19–20.

47. Ibid., chap. 22.

48. Ibid., chap. 23.

49. Ibid., chaps. 27–28. See esp. chap. 28, fol. [c7r]: "Ante hanc autem et post proximas fuisse magnas et miras mutationes iam dicta et infra dicenda manifestant."

50. Ibid., chap. 47, fol. d4r: "Et similiter mutant hoc citius vel tardius secundum proprietates planetarum dominantium regnis diversis." And chap. 48, fol. d4r: "Et possunt dictarum revolutionum significationes citius aut tardius compleri vel debilius vel fortius secundum exegentiam huiusmodi coniunctionum ut ex dictis patet."

51. Ibid., chap. 47.

52. Ibid., chap. 49.

53. Ibid., chaps. 60–64. For a full discussion of this prediction, see my chapter 6.

54. For the date, see Salembier, p. xviii. Ashenden's work was printed as Johannes Eschuid, *Summa astrologiae (Summa astrologiae iudicialis de accidentibus mundi quae anglicana uulgo nuncapatur)* (Venice: Iohannes Lucilius Santritter, 1489). D'Ailly borrows heavily from the prologue, and tr. 1, diff. 1, chaps. 1–4, where Ashenden examines the question of the age of the world. Ashenden's determination of the world's age is discussed in North, "Chronology and the Age of the World," pp. 317–20. For information on his career, see Hilary M. Carey, *Courting Disaster: Astrology at the English Court and University in the Later Middle Ages* (New York: St. Martin's Press, 1992), pp. 66, 73–77.

55. There follow five chapters of technical astronomy, dealing with the length of the year, the moon's orbit, etc. (*Elucidarium*, chaps. 35–39).

56. *Elucidarium*, chap. 1. D'Ailly notes that the change of triplicity to the fiery (Aries) triplicity would have occurred some 120 years after this conjunction signifying the Flood, that is, some 159 years before the Flood, and not 2 years prior as he had suggested in the *Vigintiloquium*.

57. *Elucidarium*, chap. 26.

58. Ibid.

59. Ibid., fol. [f6v]: "Et forte hanc imprecisionem voluit deus ut secreta celi absconderet ab indignis ne eis ex nimia familiaritate vilescerent."

60. For an explanation of terms such as "epicycle" and "equant," see chapter 1. On *medius motus*, see John D. North, *Richard of Wallingford: An Edition of His Writings with Introductions, English Translation and Commentary by John D. North*, 3 vols. (Oxford: Clarendon Press, 1976), 3:173–75; and Geoffrey Chaucer, *The Equatorie of the Planetis, Edited from Peterhouse MS. 75.I*, ed. Derek J. Price (Cambridge: Cambridge University Press, 1955), pp. 97–102.

61. To follow the procedure for the Alfonsine tables, see Poulle, *Les tables alphonsines*, chaps. 17–21 of John of Saxony's explanatory canons.

62. *Elucidarium*, chap. 13. D'Ailly's major discussion of true versus mean motions occurs in *Elucidarium*, chaps. 12 and 13.

63. See Poulle, *Les tables alphonsines*, chap. 25 of John of Saxony's canons.

64. If d'Ailly could not (or would not) calculate true conjunctions, he apparently would be in good company. A marginal note on a manuscript (Oxford, Bodleian MS Digby 176) of conjunction prognostications written by John of Ashenden states that the astronomer William Rede had made the calculations (of true conjunctions) on which Ashenden's predictions rested. See Bernard R. Goldstein and David Pingree, "Levi ben Gerson's Prognostication for the Conjunction of 1345," *Transactions of the American Philosophical Society* 80, pt. 6 (1990): 8. Hilary Carey has argued that this information does not necessarily mean that Ashenden was incapable of such calculations, but rather suggests that Rede was an expert in the area of calculations and Ashenden in the field of judgments. Carey, *Courting Disaster*, p. 66.

65. E.g., *Elucidarium*, chap. 13. D'Ailly notes that the 1345 conjunction of Saturn, Jupiter, and Mars was a true conjunction. Thus one cannot use it as a basis for calculating backward or forward to other (mean) conjunctions, but rather must use a mean conjunction for that purpose. For details of this conjunction, see n. 103.

66. There is an intermingling of both in chap. 33, which is a table of mean conjunctions with some true ones thrown in.

67. E.g., Albumasar's incorrect prediction of the demise of Islam was based on mean, not true, motions. Ibid., chap. 24.

68. *Concordantia astronomie cum hystorica narratione*, chap. 48, fol. d4r: "Horum dubiorum solutio non est clara nec tamen hic sed alibi declaranda."

69. *Elucidarium*, chap. 18, fol. f3r.

70. Expounded in chaps. 19–23.

71. Otto Neugebauer has suggested that the 360-year period corresponds to the Hindu "year of the gods" of 360 years. Neugebauer, "Hindu Astronomy at Newminster," p. 432. This theory corresponds exactly with that of the "mighty fardarat" outlined in Albumasar's lost *Book of the Thousands*. See Pingree, *The Thousands of Abu Ma'shar*, pp. 60–70; and Kennedy, "Ramifications of the World-Year Concept," p. 29. Pingree gives a list of mighty fardarat from an Arabic astrological history based upon Abu Ma'shar and Masha'allah; it corresponds almost exactly to d'Ailly's list of *magni orbes* from chap. 23 of

the *Elucidarium*. D'Ailly most likely derived his list from information in John of Ashenden's *Summa*, tr. 1, diff. 1, chap. 3.

72. *Elucidarium*, chap. 19. The signs and planets alternate their reigns according to their customary order.

73. Ibid., fols. f3–f3v, following John of Ashenden, *Summa*, tr. 1, diff. 1, chap. 3. Although this conjunction was commonly dated to 279 years before the Flood, d'Ailly notes that Henry Bate of Malines had found a different date for this conjunction and beginning of the *orbis*, using astrological tables he (Bate) had corrected. Bate was a thirteenth-century astrologer and doctor of theology. He translated astrological works by the twelfth-century Jewish author Abraham ibn Ezra (Avenesra) and composed a number of original works on the stars. Thorndike, *HMES*, 2:926–29. D'Ailly rather abruptly dismisses Bate's (and Albumasar's) figure of 287 years, saying simply that the opinion of Aomar (279 years) is the more common. Aomar ('Umar ibn al-Farrukhan al-Tabari, fl. 800) was an astrologer and translator from Persian into Arabic. Kennedy, "Ramifications of the World-Year Concept," p. 26.

74. *Elucidarium*, chaps. 25–26.

75. Ibid., chaps. 29–33.

76. Ibid., chap. 33.

77. Ibid., chap. 23.

78. Ibid., chap. 34, fol. g3v: "Hic etiam notare volui quamdam prenosticationem ex imaginaria coniectura magis quam ex ratione certa ab aliquibus posita." D'Ailly's text follows almost verbatim that of an anonymous *De antichristo* from a prophetic anthology compiled in the thirteenth century by Pierre de Limoges. The text is printed and discussed in Nicole Bériou, "Pierre de Limoges et la fin des temps," *Mélanges de l'Ecole Française de Rome. Moyen Age—Temps modernes* 98 (1986): 65–107. For d'Ailly's use of this anonymous *De antichristo*, see my chapter 6.

79. *Elucidarium*, chaps. 30–31.

80. A fact that d'Ailly acknowledges; ibid., chap. 33, fol. g2v.

81. Ibid., chap. 33, fols. g3–g3v—e.g., the two conjunctions described by Johannes de Muris in his *Epistola ad Clementem 6m*. On Johannes de Muris, see Thorndike, *HMES*, 3:294–323; Guy Beaujouan, "Observations et calculs astronomiques de Jean de Murs (1321–1344)," *Proceedings of the Fourteenth International Congress of the History of Science* (1974), no. 2; and, for his role in the construction of the Alfonsine tables, Emmanuel Poulle, "The Alfonsine Tables and Alfonso X of Castille," *Journal for the History of Astronomy* 19 (1988): 97–113. The text of his prognostication for the conjunctions of 1345 is edited with commentary in Goldstein and Pingree, "Levi ben Gerson's Prognostication," pp. 35–39 and 52–55; excerpts from his letter to Clement appear in Pierre Maurice Marie Duhem, *Le système du monde: Histoire des doctrines cosmologiques de Platon à Copernic*, 10 vols. (Paris: Hermann: 1913–59), 4:35–37.

82. *Elucidarium*, chap. 33, fol. g2v.

83. There have been other minor corrections to the first two works: a reevaluation of the horoscope drawn up for Creation (chaps. 14–16, 28); a listing of other conjunctions before the Flood (chap. 27); and a whole host of caveats about prognosticating from conjunctions (chap. 24).

84. *Elucidarium*, prologue, fol. ev: ". . . hunc terciam tractatum . . . *illis coniungere* dignum duxi tamquam necessarium ad perfectum eorum complementum. Nec verebor in re tam ardua fateri me ibi aliqua minus perfecte minusque complete scripsisse *que hic opus est elucidare perfectius et completius explicare*." Emphasis added.

85. Ibid., chap. 29, fol. [f8v].

86. *Apologetica defensio astronomice veritatis (I),* in Pierre d'Ailly, *Tractatus de imagine mundi et varia ejusdem auctoris et Joannis Gersonis opuscula* (Louvain: Johann de Westphalia, ca. 1483), fol. [gg8v]: "rationes quantum ad hoc magis probare videntur sciendi difficultatem quam scientie impossibilitatem."

87. *Pro declaratione decem dictarum figurarum; De figura inceptionis mundi et coniunctionibus mediis sequentibus; De concordia discordantium astronomorum.* For the dates of these works, see appendix 2.

88. *Pro declaratione decem dictarum figurarum,* Cracow, Biblioteka Jagiellonska, MS 575, fol. 110v: "Nota quod quandocumque coniunctio Saturni et Jovis pervenit (?) ad primi signum coniunctionum eorum semper erunt permutaciones vel commotiones in populis. Exemplum de coniunctione significante secta machometi que fuit in triplicitate aquaea scilicet in signo Scorpione in qua post 60 annos quando rediit significabat obitum machometi." Cf. Albumasar, *De magni coniunctionibus* (Augsburg: Erhard Ratdolt, 1489), tr. 2, diff. 8, fol. [D6r]: "Et fit istud sicut permutationis coniunctionis significatio de libra ad scorpionem et ad triplicitatem eius super nativitatem prophete sive machometi et quia [quando] rediit coniunctio post 60 annos ad scorpionem significat obitum eius. . . .; deinde rediit coniunctio ad scorpionem post 120 annos et fuit in illo anno diruptio templi eius . . . Et quotiens rediit coniunctio ad sagittarium [scorpionem] significavit aliquid accidere."

89. *De figura inceptionis mundi et coniunctionibus mediis sequentibus,* Vienna, Österreichische Nationalbibliothek, MS 5266, fol. 46r.

90. *De figura inceptionis mundi,* fol. 47v.

91. This is noted in Kennerly Woody, "Dante and the Doctrine of the Great Conjunctions," *Dante Studies* 95 (1977): 129, n. 11.

92. *De figura inceptionis mundi,* fol. 48r: "Sciendum autem quod constellationes celestes secundum quod fiunt ex coniunctionibus tardioribus etiam principales suos effectus tardius producunt. Et primo quosdam preambullos quasi flores postremo alios completos quasi fructus."

93. Ibid: "Alia fuit 748 annis post Christum in principio Sagittarius. Et dicitur significare novum prophetam et novum regnum cuius effectus debet apparere anno Christi 1600." Another later manuscript of this treatise has the new prophet appearing "ante annum 1600." Vienna, Österreichische Nationalbibliothek, MS 5318, fol. 95v. According to Tuckerman's tables, there was in fact a conjunction of Saturn and Jupiter in A.D. 749; on December 6 of that year, Saturn was at Scorpio 28.68° and Jupiter at Scorpio 28.78° (less than 2 degrees from the beginning of Sagittarius). Bryant Tuckerman, *Planetary, Lunar, and Solar Positions A.D. 2 to A.D. 1649 at Five-day and Ten-day Intervals,* Memoirs of the American Philosophical Society, 59 (Philadelphia: American Philosophical Society, 1964), p. 392.

94. *De concordia discordantium astronomorum,* in Pierre d'Ailly, *Tractatus de imagine mundi et varia ejusdem auctoris et Joannis Gersonis opuscula* (Louvain: Johann de Westphalia, ca. 1483), fol. [hh8v].

95. In the treatise *De persecutionibus ecclesie,* Marseilles, Bibliothèque Municipale, MS 1156, fol. 27v. Excerpts from this treatise have been printed in Noël Valois, "Un ouvrage inédit de Pierre d'Ailly, le *De persecutionibus ecclesiae,*" *Bibliothèque de l'Ecole des Chartes* 65 (1904): 557–74.

96. *Sermo in die omnium sanctorum,* in Paul Tschackert, *Peter von Ailli: Zur Geschichte des*

grossen abendländischen Schisma und der Reformconcilien von Pisa und Constanz (Gotha: Friedrich Andreas Perthes, 1877; repr., Amsterdam: Rodopi, 1968), p. [44]. He calls God "supremus ille astronomus almusque siderum conditor."

97. *Elucidarium*, in Pierre d'Ailly, *Tractatus de imagine mundi et varia ejusdem auctoris et Joannis Gersonis opuscula* (Louvain: Johann de Westphalia, ca. 1483), chap. 40, fol. [gg6r]: "Hic autem operi finem imponentes deo gratias agimus quia .8. tractatus satis utiles ex sapientum dictis ipso donante collegimus." In Erhard Ratdolt's 1490 edition, mention is made only of the three treatises printed therein: *Elucidarium*, fol. [g7v].

98. These works appear together in the 1483 incunabulum and in some manuscripts. See Lynn Thorndike, "Four British Manuscripts of Scientific Works by Pierre d'Ailly," *Imago mundi* 16 (1962): 157–60.

99. *Vigintiloquium*, in d'Ailly 1490, v. 1, fol. a2v: "Quia secundum philosophum omne verum omni vero consonat necesse est veram astronomie scientiam sacre theologie concordare."

100. Ibid., v. 20, fol. [b5v]: "Et in hoc verificatur computatio secundum .70. interpretes quam sequitur ecclesia."

101. *Concordantia astronomie cum hystorica narratione*, prologue, fol. [b6r]: "Quia vero ibi generaliter diximus quod [si] maximarum coniunctionum tempora notaverimus ad ea hystorias applicantes inveniemus circa huiusmodi tempora magnas et mirandas alterationes et mutationes in hoc seculo evenisse ideo nunc hoc specialiter proponimus declarare."

102. *Elucidarium*, in d'Ailly 1490, chap. 1, fol. e2v, following Aomar, Albumasar, and Henry Bate of Malines.

103. Ibid., chap. 10, fols. [e7v]–[e8r]. According to d'Ailly, the conjunction was followed by an eclipse of the moon and a great conjunction of Saturn and Jupiter in the same month. He notes that "astrologers" held that the three together signified great effects. Since Aquarius was a "fixed" sign, the effects of the conjunction therein were stronger, more universal, and more durable than those of other conjunctions. D'Ailly probably took this information from John of Ashenden, *Summa*, tr. 1, diff. 1, chap. 4. See also Goldstein and Pingree, "Levi ben Gerson's Prognostication," for other astrologers' interpretations of this event. Goldstein and Pingree point out that according to the Alfonsine tables, Mars and Jupiter were in conjunction on March 1 at Aquarius 14°32′, Mars and Saturn were in conjunction on March 4 at Aquarius 17°0′, and Jupiter and Saturn were in conjunction at Aquarius 18°45′ on March 21 (p. 52). Furthermore, Levi ben Gerson had predicted a lunar eclipse for March 18 (p. 43). It is easy to see how all this information might become confused and become a triple conjunction followed by a lunar eclipse and a great conjunction, which d'Ailly describes.

104. *Elucidarium*, in d'Ailly 1490, chap. 10: "Et quia istud signum semper est inimicum religioni ideo dixerunt aliqui quod illa coniunctio valde significativa fuit presentis scismatis," fol. [e8r]. According to Tuckerman, on October 23, 1365, Saturn and Jupiter were both in the sixth degree of Scorpio, while the sun was at Scorpio 8°. Tuckerman, *Planetary, Lunar, and Solar Positions*, p. 700.

105. *Elucidarium*, in d'Ailly 1490, chap. 11, fol. [e8v]. In A.D. 571 in the fifth degree of Scorpio there was a great conjunction of Saturn and Jupiter, with Venus in the first degree of Scorpio, signifying the sect of Muhammad, which followed. See Albumasar, *De magnis coniunctionibus*, tr. 2, diff. 8; Edward S. Kennedy and David Pingree, *The Astrological History of Masha'allah*, Harvard Monographs in the History of Science (Cambridge,

Mass.: Harvard University Press, 1971), p. 98. The Franks' strength was signified in a conjunction of Saturn and Jupiter in the sixth degree of Sagittarius, 808 years, 193 days, and 23 hours after Christ. The mendicant orders are said to have begun under a conjunction of Saturn and Jupiter in the third degree of Aquarius, March (or April) 4, 1226; d'Ailly notes that at the same time the ruler of the Tartars gained in strength. According to Tuckerman, Saturn and Jupiter were both in Aquarius 3° on March 7, 1226. Tuckerman, *Planetary, Lunar, and Solar Positions*, p. 631.

106. On the contrary, he gives two examples of wrong predictions: Albumasar's incorrect guess about the duration of Islam (*Elucidarium*, in d'Ailly 1490, chap. 24), and a conjunction of 1405 that was said to signify the end of the Schism (chap. 10).

107. *Vigintiloquium*, in d'Ailly 1490, v. 12.

108. This theme is discussed at length in chapter 6.

109. For details about these various periodizations, their interconnectedness, their origins, and their development, see Schmidt, "Aetates mundi"; Elizabeth Sears, *The Ages of Man: Medieval Interpretations of the Life Cycle* (Princeton, N.J.: Princeton University Press, 1986), pp. 54–70, and Guenée, *Histoire et culture*, pp. 147–50.

110. Bezold, "Astrologische Geschichtsconstruction im Mittelalter," 29–72.

111. Bezold, "Astrologische Geschichtsconstruction," p. 30: "so lässt sich indem wir in den äusserlich noch der kirchlichen Weltanschauung unterworfenen Zeiten auf astrologische Erklärung des geschichtlichen Lebens stossen, eine gewisse Parallele mit den Bemühungen des modernen Positivismus um die Geschichte und ihre Gesetze kaum von der Hand weisen."

112. Ibid., pp. 65 and 72, citing Jean Bodin, who wanted to avoid church periodization and followed the "new" style of ancient history, but became involved with the theory of the great conjunctions.

113. A brief, but more balanced overview of astrological historical writing occurs in Martin Haeusler, *Das Ende der Geschichte in der mittelalterlichen Weltchronistik* (Cologne and Vienna: Böhlau Verlag, 1980), chap. 10, pp. 142–55.

114. Kennedy and Pingree, *The Astrological History of Masha'allah*, p. vii.

115. Bezold, "Astrologische Geschichtsconstruction," p. 48 (Nebuchadnezzar, Moses, Alexander the Great, Jesus, and Muhammad); Haeusler, *Das Ende der Geschichte*, p. 144; Thorndike, *HMES*, 2:896–98.

116. Bezold, "Astrologische Geschichtsconstruction," pp. 45–47; see also Haeusler, *Das Ende der Geschichte*, p. 142, for attribution of Villani as first to organize history in this way. Giovanni's brother Matteo later attacked Giovanni for his use of astrology (Bezold, p. 47). For a discussion of Villani's *Cronica*, see Louis Green, *Chronicle into History: An Essay on the Interpretation of History in Florentine Fourteenth-Century Chronicles* (Cambridge: Cambridge University Press, 1972), pp. 9–43. Villani's astrology is treated at pp. 29–35. Like d'Ailly, Villani insisted upon God's ability to act both naturally (through causes like the stars) and supernaturally, according to his pleasure; see Green, p. 34.

117. Green, *Chronicle*, p. 33; Bezold, "Astrologische Geschichtsconstruction," pp. 47, 49; Bezold argues that Cecco was burned not merely for being an astrologer but for these more outrageous predictions. Thorndike suggests, however, that Cecco was prosecuted by his numerous enemies and not so much for extreme astrological teachings. Thorndike, *HMES*, 2:966–68.

118. Bezold, "Astrologische Geschichtsconstruction," pp. 57–63; Haeusler, *Das Ende der Geschichte*, p. 145.

119. When he drew up a horoscope of Christ, d'Ailly specifically left it to the experienced astrologers to interpret; e.g., *Elucidarium*, in d'Ailly 1490, chap. 2, fol. e4r: "de huius vero figure significatione ad astronomorum iudicia remitto. . . ."

120. Bezold, "Astrologische Geschichtsconstruction," pp. 67–72; Haeusler, *Das Ende der Geschichte*, pp. 151–54.

121. See Richard Lemay, "The Teaching of Astronomy in Medieval Universities, Principally at Paris in the Fourteenth Century," *Manuscripta* 20 (1976): 198–209; Lemay argues that in the universities of the fourteenth and fifteenth centuries, astrology was taught mainly by the faculties of medicine. From a survey of existing manuscripts, Hilary Carey has concluded that in later medieval England, astrology was a subject of subsidiary interest in the universities (outside medicine), but attracted a specialist's attention from a number of individual scholars. Carey, *Courting Disaster*, pp. 39–49.

122. Bezold, "Astrologische Geschichtsconstruction," p. 45.

123. Ibid., p. 65.

124. Roger Bacon, *Opus maius* (ed. Bridges), vol. 1, pars 4, p. 389: "Et potest homo revolvere historiam ad tempora praeterita, et considerare effectus coelorum a principio mundi, ut sint diluvia, terrae motus, pestilentiae, fames, cometae, monstra, et alia infinita, quae contingerunt tam in rebus humanis quam in naturalibus. Quibus comparatis, revolvat tabulas et canones astronomiae, et inveniet constellationes proprias singulis effectibus respondere. Deinde consideret per tabulas consimiles constellationes in futuro tempore propinquo vel remoto sicut vult; et poterit tunc pronuntiare in effectibus, qui consimiles erunt sicut fuerunt in praeterito, quia posita causa ponitur effectus."

125. Jacques Le Goff, "Au moyen âge: Temps de l'Eglise et temps du marchand," *Annales. Economies, sociétés, civilisations* 15 (1960): 417–33.

126. Tullio Gregory, "Temps astrologique et temps chrétien," in Jean-Marie Leroux, ed., *Le temps chrétien de la fin de l'antiquité au Moyen Age: IIIe–XIIIe siècles*, Colloques internationaux du Centre National de la Recherche Scientifique, 604 (Paris: Editions du C.N.R.S., 1984), pp. 557–73.

127. Ibid., p. 560.

128. E.g., Gregory (ibid., pp. 564–65) cites Aquinas on the stars' power to incline men to action; only the truly wise men can resist (*Summa theologiae*, IIa IIae, q. 95, a. 5; and Ia, q. 115, a. 4, ad 2; and Ia IIae, q. 9, a. 5). According to Gregory, the view of history as a series of community events controlled by the stars was expressed by Aquinas, Bonaventure, and Roger Bacon, among others.

129. He cites d'Ailly as an example. Even though the astrological view brought Christian eschatology into question, d'Ailly is an example of the attempt to make apocalyptic doctrine absorb astrological teaching. Gregory, "Temps astrologique," pp. 566–67.

130. Ibid., pp. 567–68.

131. Krzystoff Pomian, "Astrology as a Naturalistic Theology of History," in Paola Zambelli, ed., *"Astrologi hallucinati": Stars and the End of the World in Luther's Time* (Berlin and New York: Walter de Gruyter, 1986), pp. 29–43.

132. "Every chronosophy is . . . dependent upon some procedure supposed to predict the future, sometimes even a very distant one, with a reasonable if not absolute certainty." Pomian, ibid., p. 29.

133. Ibid., p. 32.

134. Ibid., pp. 39–43.

135. Arnaldo Momigliano similarly has argued that scholars err when they make a

great distinction between Jewish and Greek views of time in antiquity. Momigliano, "Time in Ancient Historiography," in Momigliano, *Essays in Ancient and Modern Historiography* (Middletown, Conn.: Wesleyan University Press, 1977), pp. 179–204.

136. *Vigintiloquium*, in d'Ailly 1490, v. 7; *Elucidarium*, in d'Ailly 1490, chap. 4. E.g., the sun's *anni maximi* are 1,461; *anni maiores*, 120; *anni medii*, 39 1/2; and *anni minores*, 19. (*Elucidarium*, chap. 4.)

137. See esp. Kennedy, "Ramifications of the World-Year Concept," pp. 23, 37–38. The Persian system of planetary "thousands" presupposes that the world will endure only 12,000 years (or less—some versions have the beginning of planetary motion taking place only in the third or fourth millennium). See Kennedy and Pingree, *The Astrological History of Masha'allah*, pp. 69–75.

138. Gregory, "Temps astrologique," p. 561.

139. *Concordantia astronomie cum hystorica narratione*, chap. 61.

140. Ibid., chap. 28.

141. *Elucidarium*, in d'Ailly 1490, chap. 26. The conjunction nearest the Flood was in the 1,498th year of the world.

142. *Vigintiloquium*, in d'Ailly 1490, v. 9, citing Augustine, *De civitate dei*, book 22. See R. A. Markus, *Saeculum: History and Society in the Theology of Saint Augustine*, rev. ed. (Cambridge: Cambridge University Press, 1970).

143. This is treated as one aspect of the Christian view of time in Le Goff, "Temps de l'Eglise et temps du marchand," p. 423.

144. *Concordantia astronomie cum hystorica narratione*, chap. 46. This tradition found its source in the prophecies of Daniel 7.

145. Guenée, *Histoire et culture*, p. 163 and, on the medieval "mastery of time," in general, pp. 147–63.

146. E.g., Donald J. Wilcox, *The Measure of Times Past: Pre-Newtonian Chronologies and the Rhetoric of Relative Time* (Chicago: University of Chicago Press, 1987), esp. p. 188.

147. Anna-Dorothee v. den Brincken, "Beobachtungen zum Aufkommen der retrospektiven Inkarnationsära," pp. 1–20; see also Adalbert Klempt, *Die Säkularisierung der universalhistorischen Auffassung zum Wandel des Geschichtsdenkens im. 16. und 17. Jahrhundert*, Göttinger Bausteine zur Geschichtswissenschaft, 31 (Göttingen: Musterschmidt, 1960), pp. 85–87.

148. A. T. Grafton and N. M. Swerdlow, "Technical Chronology and Astrological History in Varro, Censorinus and Others," *Classical Quarterly* 35 (1985): 456–60.

149. Keith Thomas, *Religion and the Decline of Magic* (New York: Scribner's, 1971), pp. 335–38.

150. Alexander Murray, *Reason and Society in the Middle Ages* (Oxford: Clarendon Press, 1978), pp. 201–6.

151. *Concordantia astronomie cum hystorica narratione*, chap. 59, fol. [d7v]: "alioquin formidandum est ne istud sit illud magnum scisma quod esse debet preambulum adventus antichristi."

152. Ibid.: ". . . secundum aliquos astronomos prenosticatum est ex figura celi anni presentis quod retrogradatio iovis in principio anni in prima domo significat destructionem religionis et pacem in ecclesia adhuc non firmari." According to Tuckerman, Jupiter was retrograde from roughly the end of March through the beginning of August 1414. Tuckerman, *Planetary, Lunar, and Solar Positions*, p. 725. (The astrologers' year began with the vernal equinox.)

Chapter Five
The Great Schism and the Coming of the Apocalypse

1. Louis B. Pascoe, "Pierre d'Ailly: Histoire, Schisme et Antéchrist," in Jean Favier, ed., *Genèse et débuts du Grand Schisme d'occident* (Paris: Editions du C.N.R.S., 1980), pp. 615–22.

2. Sir Richard Southern has stressed the extent to which all of these components were used in the study of the future. See R. W. Southern, "Aspects of the European Tradition of Historical Writing: 3. History as Prophecy," *Royal Historical Association. Transactions*, 5th ser., 22 (1972): 166. On medieval apocalypticism in general, see Richard Kenneth Emmerson, *Antichrist in the Middle Ages: A Study of Medieval Apocalypticism, Art, and Literature* (Seattle: University of Washington Press, 1981); Bernard Guenée, *Histoire et culture historique dans l'occident médiéval* (Paris: Aubier Montaigne, ca. 1980); Robert E. Lerner, "Refreshment of the Saints: The Time after Antichrist as a Station for Earthly Progress in Medieval Thought," *Traditio* 32 (1976): 97–144; Lerner, *The Powers of Prophecy: The Cedar of Lebanon Vision from the Mongol Onslaught to the Dawn of the Enlightenment* (Berkeley: University of California Press, 1983); Bernard McGinn, *Visions of the End: Apocalyptic Traditions in the Middle Ages* (New York: Columbia University Press, 1979); Marjorie Reeves, *The Influence of Prophecy in the Later Middle Ages: A Study in Joachimism* (Oxford: Clarendon Press, 1969); Roberto Rusconi, *L'Attesa della fine: Crisi della società, profezie ed Apocalisse in Italia al tempo del grande scisma d'Occidente (1378–1417)*, Studi storici, 115–18 (Rome: Instituto Storico Italiano per il medio evo, 1979); Ernst Wadstein, *Die eschatologische Ideengruppe: Antichrist—Weltsabbat—Weltende und Weltgericht, in den Hauptmomenten ihrer christlich-mittelalterlichen Gesamtentwickelung* (Leipzig: O. R. Reisland, 1896); the papers in the volume of the *Mélanges de l'Ecole Française de Rome. Moyen Age* 102 (1990), entitled *Les textes prophétiques et la prophétie en Occident (XIIe–XVIe siècle)*; and the excellent bibliography given by R. E. Kaske in Kaske, Arthur Groos, and Michael W. Twomey *Medieval Christian Literary Imagery: A Guide to Interpretation* (Toronto: University of Toronto Press, 1988), pp. 151–64 ("Eschatology"). In these chapters I use the following definitions, following Richard Landes, "Lest the Millennium Be Fulfilled: Apocalyptic Expectations and the Pattern of Western Chronography 100–800 CE," in Werner Verbeke, Daniel Verhelst, and Andries Welkenhuysen, eds., *The Use and Abuse of Eschatology in the Middle Ages* (Louvain: Leuven University Press, 1988), pp. 205–6. "Eschatology" is defined as the belief that time will have an end and, by consequence, refers to any study of the end of time. "Apocalypticism," by contrast, refers to the belief that the end of the world is imminent. "Millenarianism" is the belief that the final events will usher in a reign of peace here on earth (the millennium).

3. Lerner, "Refreshment," p. 116.

4. Ibid., p. 101; Emmerson, *Antichrist*, p. 45.

5. Pierre d'Ailly in *Concordantia astronomie cum hystorica narratione*, in d'Ailly 1490, chap. 59, fol. [d7v]: "de quo scismate ipse apostolus Paulus ubi de adventu antichristi loquitur antea prophetasse videtur dicens illum non esse venturum nisi venerit dissentio primum etc., quod exponunt sapientes de dissentione id est scismatica divisione vel subtractione obedientie Romane ecclesie seu Romani imperii." Compare the *glossa ordinaria*: "*Quoniam nisi venerit.* Occulte loquitur de destructione imperii Romani, ne irritaret eos ad persecutionem ecclesiae. Vel hoc dicit de spirituali imperio Romanae ecclesiae, vel discessione a fide." In *Biblia sacra cum glossa interlineari, ordinaria, et Nicolai Lyrani Postilla*

eiusdemque Moralitatibus, Burgensis Additionibus, et Thoringi replicis (Venice: [publisher un-known], 1588). Note that d'Ailly reverses the order in the *glossa*, implying that a schism in the church is a more likely interpretation of Paul's text.

6. Lerner, "Refreshment," p. 103. Augustine insisted that this passage "relaxes the fingers of all who do sums on them about this matter." Augustine, [*De civitate dei*] *The City of God against the Pagans*, ed. and trans. William M. Green, Loeb Classical Library (Cambridge, Mass.: Harvard University Press, 1960–72), book 18, chap. 53: "Omnium vero de hac re calculantium digitos resolvit et quiescere iubet ille qui dicit: Non est vestrum scire tempora, quae Pater posuit in sua potestate."

7. Landes, "Lest the Millennium Be Fulfilled." The first chronology, dating Creation at 5500 B.C. was that of Hippolytus (p. 146). As we have noted in chapter 4, this was replaced in its latter years by Eusebius's calculation (in his *Chronicle*) of Creation at 5,228 years before Christ's ministry, rejuvenating the world by almost three centuries (ibid., p. 149). Bede, in the 5900s of that era, introduced a third dating of Creation at 3952 B.C., based on the Hebrew calculations, as well as the move to date events by A.D. rather than by years from Creation. Both of these were anti-apocalyptic moves. (Ibid., pp. 175–78.)

8. According to Landes, Charlemagne and his advisors must have been aware of the fact that his coronation in A.D. 800 coincided with the world's 6,000th year. Ibid., pp. 197–201.

9. *City of God*, book 22, chap. 30; Augustine was an opponent of the equation of these ages with set lengths of time and argued vigorously against would-be calculators of the End.

10. Felix Hemmerlin, *Registrum querele*, quoted in Wadstein, *Die eschatologische Ideen-gruppe*, p. 33.

11. Printed by Ernst Sackur, ed., *Sibyllinische Texte und Forschungen. Pseudomethodius, Adso und die Tiburtinische Sibylle* (Halle: Niemeyer, 1898); also see the discussion of this figure in Reeves, *Influence*, pp. 299ff.; and Emmerson, *Antichrist*, pp. 47–49. Excerpts in English translation appear in McGinn, *Visions*, pp. 70–76.

12. Pseudo-Methodius (ed. Sackur), chaps. 11 and 13, pp. 80, 89–90.

13. Reeves, *Influence*, p. 301.

14. A change that Southern attributes to the enormous influence of the writings of Joachim of Fiore. Southern, "History as Prophecy," pp. 166, 170, 173.

15. Ibid., p. 166. On the various relationships between astrology and prophecy in the later Middle Ages, see also Jean-Patrice Boudet, "Simon de Phares et les rapports entre astrologie et prophétie à la fin du Moyen Age," *Mélanges de l'Ecole Française de Rome. Moyen Age* 102 (1990): 617–48.

16. E.g., the so-called Toledo Letter, which first appeared in 1184; its different ver-sions predicted the End for 1186, 1229, 1345, 1395, and 1516. See Emmerson, *Antichrist*, p. 54; Lerner, *The Powers of Prophecy*, p. 20; and also M. Gaster, "The Letter of Toledo," *Folk-Lore* 13 (1902): 115–34. An English translation appears in McGinn, *Visions*, p. 152. From the same period dates the *Oraculum angelicum*, supposedly dictated by an angel to a Carmelite friar named Cyril. It has been edited by Franz Ehrle, in *Archiv für Litteratur und Kirchengeschichte des Mittelalters* 2 (1886): 327ff. One should note also the famous Cedar of Lebanon prophecy, studied by Robert Lerner in *The Powers of Prophecy*. This text circu-lated throughout Europe from the thirteenth through the seventeenth centuries.

17. The "*spiritualis intellectus*." Such a gift was cited to justify the nontraditional exege-sis of a number of thinkers of the High Middle Ages and later. See Robert E. Lerner,

"Ecstatic Dissent," *Speculum* 67 (1992): 33–57; and André Vauchez, "Les théologiens face aux prophéties à l'époque des papes d'Avignon et du Grand Schisme," *Mélanges de l'Ecole Française de Rome. Moyen Age* 102 (1990): 577–88. On Joachim in general, see esp. Reeves, *Influence,* and her *Joachim of Fiore and the Prophetic Future* (London and New York: S.P.C.K., 1976); Morton Bloomfield, "Joachim of Flora: A Critical Survey of His Canon, Teachings, Sources, Biography, and Influence," *Traditio* 13 (1957): 249–311; and Lerner, "Refreshment," which tempers Reeves's estimate of Joachim's influence. Selections from Joachim's works in English translation appear in McGinn, *Visions,* pp. 126–42.

18. This summary is based on Reeves, *Influence,* pp. 30–39.

19. Ibid., pp. 59–61.

20. Ibid., pp. 126, 135–46.

21. There exists also a number of alchemical works attributed to Arnold. On his life and career, see Michael R. McVaugh, ed., *Arnaldi de Villanova Opera Medica Omnia,* 2: *Aphorismi de gradibus* (Granada and Barcelona: Seminarium Historiae Medicae Granatensis, 1975), pp. 76–82; Thorndike, *HMES,* 2: chap. 68, pp. 841–61; and Lerner, "Ecstatic Dissent," pp. 42–46. Lengthy excerpts from Arnold's treatise are printed in Heinrich Finke, *Aus den Tagen Bonifaz VIII: Funde und Forschungen,* Vorreformationsgeschichtliche Forschungen, 2 (Münster: Aschendorff, 1902), pp. CXXIX–CLIX. See also McGinn, *Visions,* pp. 222–24.

22. Arnold of Villanova, *De tempore adventus Antichristi,* ed. Finke, p. CXLII: "Non ergo per illa verba respondit Christus ad interrogationem factam de tempore Antichristi, set de consummatione seculi, vel ultima conversione Judeorum ad Christum."

23. Ibid., p. CXXXII: "Quod nobis tamen hic sufficit, istud est, scilicet quod completis mille CC.XC. annis a tempore, quo populus Judaicus amisit possessionem terre illus [*sic*], stabit, ut ait dominus, abhominatio desolationis, scilicet Antichristus in loco sancto, quod erit circa septuagesimum octavum annum centenarii sequentis videlicat [*sic*] quarti decimi a salvatoris adventu."

24. Ibid., p. CXXXIV: ". . . sicut in productione mundi fuit supernaturaliter operatus, sic et in consummatione huius seculi supernaturaliter operabitur. Et si totius retardationis revolutio necessaria foret, ut asserunt [astrologi], ad universalem perfectionem, nichilominus Deus est potens motum orbium velocitare, quantum placuerit, et revolutionem complere brevissimo tempore. . . ." Pierre d'Ailly would use this passage in his sermon for Advent in 1385. It is not clear whence Arnold took this theory about the 36,000 years, although it appears to be related to the theory of the *magnus annus* condemned in Paris in 1277. (On precession and the *magnus annus,* see chapter 1.) Condemned in 1277 was the notion that after 36,000 years the stars would return to their original positions and the same series of effects would recur, not that the world would then end. (See chapter 2.) Perhaps this was the argument "*rationibus naturalibus*" used in an anonymous treatise of 1288, which gave the world many thousands of years yet to come. (Cited in Wadstein, *Die eschatologische Ideengruppe,* p. 32.)

25. "The *Tractatus de Antichristo* by John of Paris: A Critical Edition, Translation, and Commentary," ed. Sara Beth Peters Clark (Ph.D. diss., Cornell University, 1981), p. 9; Franz Pelster, "Die Quaestio Heinrichs von Harclay über die zweite Ankunft Christi und die Erwartung des baldigen Weltendes zu Anfang des XIV Jahrhunderts," *Archivio italiano per la Storia della Pietà* 1 (1951): 35.

26. Pelster, "Quaestio," p. 44; on his mystical experiences, see Lerner, "Ecstatic Dissent," pp. 44–46.

27. The whole affair is treated in Pelster, "Quaestio," pp. 32–46, who prints Henry of Harclay's response at pp. 53–82.

28. "Non negat Dominus precognitionem centenarii annorum vel huiusmodi secundum probabilitatem vel coniecturam, sed solum secundum certitudinis comprehensionem." *Tractatus de Antichristo* (ed. Clark), p. 66. The treatise is also edited in *Abbas Joachim Magnus Propheta . . .Expositio Magni Prophete Joachim: In Librum Beati Cyrilli . . . Item Tractatus de Antechristo Magistri Joannis Parisiensis Ordinis Predicatorum* (Venice: Laca de Soardis, 1516); Clark's is the preferable text.

29. John of Paris, *Tractatus de Antichristo* (ed. Clark), p. 44.

30. Ibid., p. 45.

31. Ibid., pp. 46–47. John assumes both that the sun was created in its exaltation of Aries 15° and that, at the time of Creation, its *aux* was in the same location as the sun. He takes Ptolemy's figures for precession (1 degree per 100 years) and for the position of the sun's *aux* in Ptolemy's time (Gemini 6°) to determine that 5,100 years must have passed between Creation and Ptolemy's writing in A.D. 130. The method is similar in spirit to that which d'Ailly will use in the *Vigintiloquium* (see chapter 4).

32. John of Paris, *Tractatus de Antichristo* (ed. Clark), p. 61. See chapter 1.

33. Ibid., pp. 53–54.

34. E.g., the author of the mid-fourteenth century French treatise *On the Natural Causes of Certain Future Events and Tribulations and Other Changes in the World*, who commented on current trials such as the plague and the disasters of Crécy and Calais. He felt that Antichrist's advent was imminent. (Cited by Lerner, *The Powers of Prophecy*, p. 104.) See also Lerner, "The Black Death and Western European Eschatological Mentalities," *American Historical Review* 86 (1981): 533–52.

35. Cited in Reeves, *Influence*, pp. 93–94; see also Lerner, *The Powers of Prophecy*, p. 93. His collection included the Sibyls, pseudo-Methodius, Merlin, Hildegard, and much, real and spurious, of Joachim's work. Many of Henry's texts stress that the apocalypse is near: one prophet says there will be only four more popes; another that Antichrist is already six years old in 1349.

36. In 1356, for example, the author of an anonymous English treatise *On the Last Age of the Church* predicted the appearance of Antichrist in 1400. (Wadstein, *Die eschatologische Ideengruppe*, pp. 93–94.) Similarly, an "unnamed soothsayer" known to Henry of Langenstein (d. 1397) used the prophecies of Daniel 12 to calculate that Christ would come to slay the son of perdition in 1400. Henry of Langenstein, *Liber adversus Thelesphori eremitae vaticinia de ultimis temporibus*, in Bernhard Pez, *Thesaurus anecdotorum novissimus* (Augsburg: Veith, 1721), vol. 1, pt. II, col. 541: "Ille innominatus . . . vaticinator, cujus nuper tractatulum perlegi . . . invenit suo modo calculando, quod utique, ut dicit, adventus Christi pro interficiendo Antichristum . . . erit anno Domini 1400."

37. On Roquetaillade, see Reeves, *Influence*, pp. 225–28, 321–23; Emmerson, *Antichrist*, p. 69; Lerner, "Refreshment," p. 132; and the articles appearing under the head "Figures de prophètes aux derniers siècles du Moyen Age: Autour de Jean de Roquetaillade," in *Mélanges de l'Ecole Française de Rome. Moyen Age* 102 (1990): 297–401. Excerpts of his works in translation appear in McGinn, *Visions*, pp. 231–33.

38. See, for example, Reeves's eloquent statement: "The Great Schism brought into the sharpest possible focus all the various elements of the prophetic tradition that we have been tracing: the forces of Antichrist creating schism and persecution in the church, the expectation of terrible tribulation and judgement, the prophetic summons of the Pope

back to Rome to fulfill the full destiny of the *renovatio ecclesie*. Above all, it was the fact of the Great Schism itself which set the seal of truth on the prophets from Joachim and St. Francis to Jean de Roquetaillade." Reeves, *Influence*, p. 422. See also Rusconi, *L'Attesa della fine*; Vauchez, "Les théologiens face aux prophéties"; and Hélène Millet, "Ecoute et usage des prophéties par les prélats pendant le Grand Schisme d'Occident," *Mélanges de l'Ecole Française de Rome. Moyen Age* 102 (1990): 425–55.

39. Lerner, "Refreshment," p. 139.

40. "Credo inconcussa certitudine, tempora magni iudicij in ianuis esse: atque adeo in ianuis, ut usque ad articulum diei illius nimium tremendi vix existimen triennii moram superesse, sed nec de biennio satis mihi constat." *De Antichristo et ortu eius, vita, moribus et operibus*, in Nicolaus de Clemingiis (Nicolas of Clamanges), *Opera omnia* (Lyons: Johannes Balduinus, 1613; repr., Farnborough, England: Gregg Press, 1967), p. 357.

41. Emmerson, *Antichrist*, p. 52; Millet, "Ecoute et usage," p. 426.

42. Millet describes one such anthology, probably a compilation by Simon de Bosc ("Ecoute et usage," pp. 428–38); see also Matthew Tobin, "Une collection de textes prophétiques du XVe siècle: Le manuscrit 520 de la Bibliothèque Municipale de Tours," *Mélanges de l'Ecole Française de Rome. Moyen Age* 102 (1990): 417–23. The prophetic texts in the collection Tobin describes demonstrate a preoccupation with the Great Schism and the Hundred Years' War.

43. See Millet, "Ecoute et usage," pp. 438–41. It is worth noting that Benedict XIII found her services so worthwhile that he vouchsafed her an annual income of sixty florins (ibid., p. 439).

44. Vauchez, "Les théologiens face aux prophéties," pp. 578–80, 586. For example, Vauchez speculates that Henry of Langenstein's condemnation of Telesphorus (on whom, see subsequent discussion) and the Joachimitic tradition undergirding his treatise was due in large part to Telesphorus's adamantly pro-French and pro-Avignon stance. Similarly, he notes that Gerson, who authored a treatise on the art of discerning true revelation (*De distinctione verarum revelationum a falsis*), enthusiastically maintained that Joan of Arc's was a divine mission.

45. Reeves, *Influence*, pp. 325–27; Rusconi, *L'Attesa della fine*, pp. 171–84; Lerner, "Refreshment," p. 132. Excerpts in English appear in McGinn, *Visions*, pp. 249–50.

46. The *Liber adversus Thelesphori eremitae vaticinia*. See also Thorndike, *HMES*, 3:506–7; Millet, "Ecoute et usage," pp. 441–48; and Vauchez, "Les théologiens face aux prophéties," pp. 580–86.

47. Henry of Langenstein, *Liber adversus Thelesphori eremitae vaticinia*, col. 517: "Unde his et aliis tandem confusus exivit monasterium praedictum, et habitu religionis ibidem relicto, in vili seculari tunica per silvas Monasterii adjacentes vagabatur." (The whole story appears in cols. 516–17.)

48. The most comprehensive study of the saint's career remains that of Henri Fages, *Histoire de Saint Vincent Ferrier*, 4 vols. in 2 (Louvain: Uystpruyst, and Paris: Picard, 1901–5). Fages also produced an edition of Vincent's works. See also P. Sigismund Brettle, *San Vincent Ferrer und sein literarischen Nachlass*, Vorreformationsgeschichtliche Forschungen, 10 (Münster: Aschendorff, 1924); and Etienne Delaruelle, "L'Antéchrist chez S. Vincent Ferrier, S. Bernardin de Sienne et autour de Jeanne d'Arc," in Delaruelle, *La piété populaire au Moyen Age* (Torino: Bottega d'Erasmo, 1975), pp. 329–54; and Rusconi, *L'Attesa della fine*, pp. 218–33. Excerpts from Vincent's writings in English translation appear in McGinn, *Visions*, pp. 256–58.

49. The complete text appears in Fages, *Ferrier*, 4:213–24. Brettle, *Ferrer*, prints extensive excerpts at pp. 167–72.

50. Fages, *Ferrier*, 4:213–14; Brettle, *Ferrer*, p. 170. "Prima conclusio est, quod tempus Antichristi, et finis mundi in eodem coincidunt temporaliter. Ratio est, propter brevitatem temporis durationis mundi post mortem Antichristi. Quoniam per Sacram Scripturam non invenitur tempus majoris durationis hujus mundi post mortem Antichristi, quam quadraginta quinque dierum . . ." (Fages, *Ferrier*, 4:213).

51. "Quarta conclusio est, quod tempus Antichristi, et finis mundi erunt cito, et bene cito, ac valde breviter." Fages, *Ferrier*, 4:220; Brettle, *Ferrer*, p. 171.

52. "Nam cum ego predicarem in partibus Lombardie prima vice (modo jam sunt novem anni completi), venit ad me de Tuscia ille vir, missus (ut dicebat) a quibusdam sanctissimis eremetis in partibus Tuscie. . . . annuncians quod eisdem viris expresse revelationes divinitus facte fuerant, quod Antichristus jam erat natus. . . ." Fages, *Ferrier*, 4:222; Brettle, *Ferrer*, p. 171.

53. "Quinto, patet eadem conclusio per quamdam aliam revelationem expressam, quam in Pedemontio audivi, relatu cujusdam mercatoris Venetiarum . . . dicentis quod cum ipse esset ultra mare in quodam Monasterio Fratrum Minorum, et audiret Vesperas . . . duo parvi Novitii ejusdem Monasterii . . . visibiliter rapti per magnum spatium temporis, tandem concorditer et terribiliter clamaverunt: 'Hodie hac hora natus est Antichristus mundi destructor.' . . . Ego autem exquirens, et interrogans de tempore hujus visionis, manifeste reperi, quod jam sunt novem anni completi. . . ." Fages, *Ferrier*, 4:222; Brettle, *Ferrer*, p. 171.

54. "Nam mihi per mundum predicando, discurrenti per diversas regiones . . . frequenter occurrerunt diverse persone devote et spirituales, narrantes et referentes certitudinaliter de tempore Antichristi et fine mundi diversimode et multifarie suas revelationes, juxta ea que dicta sunt unanimiter concordantes." Fages, *Ferrier*, 4:222; Brettle, *Ferrer*, p. 171.

55. ". . . dies iudicii in brevi veniet. . . . Et quod cito veniat, probat Apostolus Paulus dicens: 'Nisi venerit discessio primum.' Glossa: recessus universalis ab ecclesia et imperio, sicut modo est." *Sermo de Antichristo* (preached in Freiburg, March 10, 1404), in Brettle, *Ferrer*, p. 181. For the text from the *glossa ordinaria*, see n. 5.

56. Ibid., p. 178. (The reference to multiple wives was perhaps a slur on Islam.)

57. Ibid., p. 179: "primo omnia bona temporalia auferet a te. Item interficiet pueros et amicos in presentia parentum. Item de hora in horam, de die in diem faciet a te abscindi unum membrum post aliud, non simul et semel, sed per plura tempora continuando."

58. Ibid., p. 181.

59. See Palémon Glorieux, "Les années d'études de Pierre d'Ailly," *Recherches de théologie ancienne et médiévale* 44 (1977): 145–47; Francis Oakley, *The Political Thought of Pierre d'Ailly: The Voluntarist Tradition* (New Haven: Yale University Press, 1964).

60. Louis Salembier dates the sermon as pre-1379 (Salembier 1886, p. xiv). In his 1910 bibliography of d'Ailly's works he gives the date 1379: Louis Salembier, "A propos de Pierre d'Ailly, évêque de Cambrai. Biographie et bibliographie," *Mémoires de la Société d'Emulation de Cambrai* 64 (1910): 111. I follow Glorieux, who gives the sermon the date August 4, 1379 ("Les années d'études," p. 135). The sermon in printed in Pierre d'Ailly, *Tractatus et sermones* (Strasbourg: [Printer of Jordanus de Quedlinburg], 1490; repr., Frankfurt am Main: Minerva, 1971), fols. C3r–C5r.

61. *Sermo de beato Dominico confessore*, in *Tractatus et sermones*, fol. C3v: "Ecce iam insurgunt scismatici . . . Adversus quos noster Dominicus in celo pugnat."

62. Ibid.: "Sed esto victi sint aut ad concordia reducti scismatici . . . quod quanto erit nescio deus scit. Numquid ideo cessabunt persecutiones civitatis dei. . . . Nam ut in preallegato libro [*De civitate dei*] diffinit Augustinus, non est diffiniendus numerus persecutionum quibus exerceri oportet ecclesiam."

63. Ibid., fol. C4v: "Ex quibus verbis datur intelligi circa finem mundi maiores persecutiones debere fieri."

64. Ibid.

65. "Hoc anno (1380) Mathias Parisiensis natione Bohemus valde prolixum librum de Antichristo edidit; docuitque publice eum iam natum esse: ab eo omnes Universitates et eruditorum collegia seducta fuisse, ita ut iam nihil, quod sanum esset et vere catholicum docerent." Wadstein, *Die eschatologische Ideengruppe*, p. 87, citing Bullay, *Historia Universitatis Parisiensis*, IV.

66. Although Salembier dates the sermon to 1382, following the indication in *Tractatus et sermones*, it is preferable to use Glorieux's dating of October 4, 1380. Palémon Glorieux, "L'oeuvre littéraire de Pierre d'Ailly. Remarques et précisions," *Mélanges de science religieuse* 22 (1965): 70. Glorieux points out that d'Ailly states that the sermon was preached in Paris before the university; since d'Ailly was in Noyon in 1382, this would have been impossible. "Les années d'études," p. 135.

67. *Sermo de beato Francisco confessore*, in *Tractatus et sermones* (Strasbourg: [Printer of Jordanus de Quedlinburg], 1490; repr., Frankfurt am Main: Minerva, 1971), fol. [C6v]: (the church's vices are so bad) "ut vix synagoga iudeorum aut secta aliqua paganorum vel hereticorum ab evangelica Christi regula tam unquam in moribus oberrasse credatur quantum hodie malignantium ecclesia clericorum."

68. Ibid., fol. [C7r]: "Nam secundum prophetiam beate hyldegardis in fine illius muliebris temporis quod super quartum esse meminimus gravissimi scismatis et confusionis laqueus super omnem clerum et ordinem ecclesiasticum extendetur. . . . Ex quo satis intelligi datur quod illud quintum tempus cui primum ve correspondet vel iam inceptum est vel de propinquo inchoandum imminet."

69. Ibid., fol. [C7v]: "Ideo vestre caritati proposui ut . . . contra previsa persecutionum iacula cautiores reddamur."

70. He cites the pseudo-Joachim, *Super hieremiam*. Ibid., fol. [C7v].

71. The *Sermo de beato Bernardo* appears in Paris, Bibliothèque Nationale, MS Lat. 3122, fols. 92v–97v. According to Glorieux, this manuscript contains works from the years 1372–93 ("Les années d'études," p. 127). Since the sermon mentions the Schism, we can date it to the years 1378–93. Tschackert prints excerpts from the sermon; however, he omits the sections discussed here. Paul Tschackert, *Peter von Ailli: Zur Geschichte des grossen abendländischen Schisma und der Reformconcilien von Pisa und Constanz* (Gotha: Friedrich Andreas Perthes, 1877; repr., Amsterdam: Rodopi, 1968), pp. [21]-[23].

72. *Sermo de beato Bernardo*, fol. 97r.

73. Ibid., fol. 97r–v (emphasis added): "Et ideo de tempore determinato [?] nec possumus nec debemus certitudinaliter diffinire. Sed tamen ex scripturis quas legimus et experienciis quas videmus antichristi propinquitatem et finis mundi vicinitatem *probabilibus coniecturis* valemus."

74. Ibid., fol. 97v: "Primum est habundancia et inflammatio mondane iniquitatis; secundum est inopia et refrigeratio christiane caritatis; tertium est scismatica divisio catholice unitatis." Matthew 24:12 taught that in the final times "quoniam abundavit iniquitas, refrigescet charitas multorum" ("because iniquity shall abound, the love of many shall wax cold").

75. Ibid. (emphasis added): "ante diem domini asserit futuram esse quamdam dissessionem sive divisionem ecclesie dei *immediate* preambulum antichristi."

76. Ibid.: "Et hec divisio per presens scisma verisimiliter creditur esse facta. Qua propter timendum est iuxta prophetiam beati cirilli et sermoniam abbatis ioachim in commento eiusdem ne hec sit illa horrenda divisio et persecutio scismatica post quam antichristi persecutio celeriter est futura." He refers to the pseudo-Joachim commentary on the *Oraculum cyrili*. Again, d'Ailly reads the *glossa ordinaria* as pointing mainly to a schism in the church.

77. *Sermo (tertius) de quadruplici adventu domini et specialiter de adventu ad iudicium*, in *Tractatus et sermones* (Strasbourg: [Printer of Jordanus de Quedlinburg], 1490; repr., Frankfurt am Main: Minerva, 1971), fol. t3v–t4r.

78. Ibid., fol. [t5r].

79. For the view that d'Ailly believed the *potentia absoluta* would not be used in the present order, see William Courtenay, "Covenant and Causality in Pierre d'Ailly," *Speculum* 46 (1971): 94–119. For the view that d'Ailly believed God exercised his *potentia absoluta* in the case of miracles, see Francis Oakley, *Political Thought*, pp. 27–28; Oakley, "Christian Theology and the Newtonian Science: The Rise of the Concept of the Laws of Nature," *Church History* 30 (1961): 433–57; and Oakley, *Omnipotence, Covenant, and Order: An Excursion in the History of Ideas from Abelard to Leibniz* (Ithaca, N.Y.: Cornell University Press, 1984), p. 56. For a similar view, see Leonard A. Kennedy, *Peter of Ailly and the Harvest of Fourteenth-Century Philosophy*, Studies in the History of Philosophy, 2 (Lewiston, N.Y.: Edwin Mellen Press, 1986). Kennedy maintains that d'Ailly believed God would act *de potentia absoluta* in this world, but that miracles were included within the purview of *potentia ordinata*. I tend to side with Courtenay—that d'Ailly saw *potentia absoluta* as incorporating potential actions by God, not the power involved in suspensions of the ordinary course of events in miracles. For a fuller discussion of this distinction, see chapter 7.

80. *Sermo (tertius) de adventu*, fol. t4r. D'Ailly apparently has not yet read John of Paris, for it seems inconceivable that if he had, he would omit to mention the astronomical arguments against the 36,000-year theory. One would think that d'Ailly would have known about "Thabit"'s theory of trepidation and the other figures for the motion of the eighth sphere at this time. Perhaps he had not yet written his commentary on Sacrobosco's *Sphere* in 1385.

81. Ibid., fol. t4v: "cum iam de .xiiii. centenario a nativitate christi fluxerint plusquam anni lxxvii. cum nunc annum octogesimumquintum fere compleverimus. . . . consequens est secundum huius opinionis fictam divinationem quod iam esset antichristi persecutio consummata et seculi consummatio iam completa, que omnia falsa esse experientia manifestat."

82. For this passage and Roger Bacon as its source, see chapter 3.

83. Roger Bacon, *Opus maius* (ed. Bridges), vol. 1, pars 4, pp. 264 (conjunction before Christ); 261–62 (sect of the moon); 254 (foreknowledge of Antichrist's sect through the use of astrology).

84. *Sermo (tertius) de adventu*, fol. [t6r].

85. Ibid.: "Quid est enim erunt signa in sole et luna et stellis nisi quod ecclesie luminaria tenebrescent. Quod utique iam prochdolor factum esse conspicimus. . . . timeamus summi terroris futurum iudicium, quia prope est."

86. Ibid., fol. [t5r]: "Sic enim beatus cirillus, abbas ioachim et sancta hyldegardis de temporibus novissimis multa prophetasse creduntur."

87. Ibid: "Cum fuerint anni transacti mille trecenti / Et deni decies post partum virginis alme / Tunc antichristus regnabit demone plenus." Marjorie Reeves lists numerous versions of these verses, all giving different dates (Reeves, *Influence*, pp. 49–50). She does not list this prediction for 1400.

88. *Sermo (tertius) de adventu*, fol. [t5r]: ". . . a quibusdam traditur secundum revelationem factam eidem ioachim iam non restant plures quam anni quindecim usque ad regnum antichristi."

89. Ibid.: "Prima ergo consideratio est quod ibi christus loquitur per verbum de presenti, dicens non est vestrum, ubi ait glosa, non dicit non erit sed non est."

90. Bernard Guenée, *Entre l'Eglise et l'Etat: Quatre vies de prélats français à la fin du Moyen Age (XIIIe–XVe siècle)* (Paris: Gallimard, 1987), p. 227.

91. The questions of true versus false prophecy, the approach of Antichrist, and the current situation all merge, for example, in a passage from Gerson's *De distinctione verarum revelationum a falsis*: ". . . alia est quaestio, qualiter praesuppositis eis quae fidei sunt, cognoscere nos fideles poterimus, et secundum doctrinam Joannis probare spiritus si ex Deo sunt, ne fallamur. Et in hanc quaestionem sciens incidi propter illusiones plurimas quas nostro tempore cognovi contigisse; quas etiam in hoc senio saeculi, in hac hora novissima, in praecursore Antichristi, mundus tanquam senex delirus, phantasias plures et illusiones somniis similes pati habet . . . ." Jean Gerson, *Oeuvres complètes*, ed. Palémon Glorieux, 10 vols. (Paris: Desclée & Cie, 1960–73), 3:37–38. See Millet, "Ecoute et usage," p. 440; Vauchez, "Les théologiens face aux prophéties," pp. 580–88.

92. Both treatises appear in Paris, Bibliothèque Nationale, MS Lat. 3122, as well as in Jean Gerson, *Opera omnia*, ed. Louis Ellies Dupin, 5 vols. (Antwerp: Sumptibus societatis, 1706). According to Glorieux, the works in the manuscript date from 1372–93, although both the treatises *De falsis prophetis* mention the Schism and thus postdate 1378. The treatises appear in the manuscript (and in Dupin's edition, which follows it) in the traditional order. *De falsis prophetis I*, however, has a number of references back to *De falsis prophetis II*, indicating that the second treatise was, in fact, composed first. (That one of these references is in the future tense must be attributed to the correction of an astute scribe, for in the *De falsis prophetis II*, d'Ailly nowhere refers to *De falsis prophetis I*.) This discrepancy in numbering was brought to notice and fully discussed by Max Lieberman, "Chronologie Gersonienne. VIII. Gerson et D'Ailly (III)," *Romania* 81 (1960): 82–84. Lieberman suggests that the treatise numbered second is a product of the years before 1395, while the treatise numbered first may date from as late as 1410–13.

93. *De falsis prophetis II*, in Gerson, *Opera omnia* (ed. Dupin), 1: cols. 515–16: "sciendum est quod tres generales persecutiones passura est Ecclesia, ante specialem persecutionem antichristi: unam violentam, aliam fraudulentam, tertiam ex vi et dolo compositam. Prima facta est per tyrannos, tempore Apostolorum, et Martyrum. . . . Secunda sequuta est per apertos hereticos, . . . sequutura est tertia per falsos ypocritas, primo per fraudem ypocrisis fideles Christi seducentes, postea per vim. . . ."

94. Ibid., col. 516: "ipsi ypocritae sunt praecursores seu praeambuli illius antichristi famosi, qui dicitur *homo peccati, filius perditionis*. II. ad Thessalonicenses. II."

95. Ibid.: "Nam per tales sic multiplicatos, primo per fraudem ypocrisis, et simulationem sanctitatis, deinde per vim simulationis pietatis populum seducentes et Ecclesiam dividentes instar antichristi capitis sui, sicut praedictum est: tandem procuranda est divisio generalis, seu scissura aliarum Ecclesiarum ab Ecclesia Romana, . . . quae divisio erit via seu praeparatio antichristo venturo, de qua prophetavit Apostolus 2. Tessa. II."

96. Ibid., col. 517: "Et maxime apparet quod utilis sit tempore moderno, cum jam

appareat hujusmodi discessio in Ecclesia, id est divisio schismatica, de qua verisimiliter timendum est ne sit antichristi praeambulum."

97. This section is treated at length in chapter 3.

98. *De falsis prophetis I*, in Gerson, *Opera omnia* (ed. Dupin), 1: col. 497.

99. Ibid., col. 503: "Secundo, notandum quod miracula falsorum prophetarum, maxime his temporibus quae ad finem mundi appropinquare videntur magnopere debent esse suspecta, sicut declaratum fuit in prima parte primi articuli, de Arte cognoscendi falsos prophetas" (i.e., as was declared in *De falsis prophetis II*).

100. Ibid., col. 504.

101. Ibid., col. 508: "Secundo, notandum quod omnes illi qui praedicant vel publice docent, non a Deo missi sunt falsi prophetae, et pseudodoctores."

102. Pierre d'Ailly, *Oratio in Matthaeum*, quoted in Fages, *Ferrier*, 4:211–12: "Recessit lex a sacerdotibus, justicia a principibus, consilium a senioribus, fides a populo, a filiis reverentia, a subjectis charitas, a prelatis religio, a monachis devotio, a monialibus honestas, a juvenibus disciplina, a clericis doctrina, a magistris studium, a scholaribus timor, a servitoribus equitas, a judicibus integritas, a militibus fidelitas, concordia a civibus, communitas a rusticis, bonitas a artificibus, veritas a mercatoribus, largitas a divitibus, castitas a virginibus, virtus et meror a viduis, munditia ac fiducia a conjugatis.—Et nunc quid, fratres, nisi venire antechristum!" (This work is not listed in the bibliographies of Salembier or Glorieux; the date is unknown. It would seem to be in keeping with the spirit of the pre-1400 works discussed previously, however.)

Chapter Six
Astrology and the Postponement of the End

1. Pierre d'Ailly, *De reformatione ecclesiae* (= pt. III of his *Tractatus de materia concilii generalis*), ed. in Francis Oakley, *The Political Thought of Pierre d'Ailly: The Voluntarist Tradition* (New Haven: Yale University Press, 1964), p. 316. D'Ailly mentions that Joachim and Hildegard had foreseen the Schism, then adds, "Haec autem Deus misericordissimus, qui solus ex malis bona novit elicere, ideo permittere credendus est ut eorum occasione Ecclesia sua in melius reformetur. Quod nisi celeriter fiat, audeo dicere quod licet magna sint quae videmus, tamen in brevi incomparabiliter majora videbimus, et post ista tonitrua tam horrenda, alia horribiliora in proximo audiemus."

2. For further discussion of d'Ailly's defense of astrology and the doctrine of the great conjunctions, see chapters 2 and 4.

3. *De legibus et sectis contra superstitiosos astronomos*, in Pierre d'Ailly, *Tractatus de imagine mundi et varia ejusdem auctoris et Joannis Gersonis opuscula* (Louvain: Johann de Westphalia, ca. 1483), chap. 4, fol. f2v: "Scio quod si ecclesia vellet revolvere textum sacrum et prophetias sacras atque prophetias Sibille et Merlini aquile et Joachim et multorum aliorum, insuper historias et libros philosophorum atque iuberet considerari vias Astronomie inveniretur sufficiens suspicio vel magis certitudo de tempore antichristi," quoting Bacon's *Opus maius*, as in *The "Opus maius" of Roger Bacon*, ed. John Henry Bridges, 2 vols. (Oxford: Clarendon Press, 1897), vol. 1, pars 4, p. 269.

4. *De legibus et sectis*, chap. 7, fol. [f5v]: "Secunda conclusio est quod leges vel secte que vel humana adinventione vel diabolica suggestione introducte sunt sicut secta Machometi vel supradicte secte ydolatrie aut alie huiusmodi celesti dispositioni naturaliter subsunt."

5. Richard Landes, "Lest the Millennium Be Fulfilled: Apocalyptic Expectations and the Pattern of Western Chronography 100–800 CE," in Werner Verbeke, Daniel Verhelst, and Andries Welkenhuysen, eds., *The Use and Abuse of Eschatology in the Middle Ages* (Louvain: Leuven University Press, 1988), pp. 137–211. See my chapter 5.

6. *Sermo de beato Dominico confessore*, in Pierre d'Ailly, *Tractatus et sermones* (Strasbourg: [Printer of Jordanus de Quedlinburg], 1490; repr., Frankfurt am Main: Minerva, 1971) fol. c4v; *Sermo de beato Bernardo*, Paris, Bibliothèque Nationale, MS Lat. 3122, fol. 97r; *Sermo (tertius) de quadruplici adventu domini et specialiter de adventu ad iudicium*, in *Tractatus et sermones*, fols. t3v, t5r.

7. *Concordantia astronomie cum theologia* (= *Vigintiloquium*), in d'Ailly 1490, v. 11, fol. [a8v].

8. Ibid., v. 20, fol. [b5r]. For a fuller discussion of this treatise, see chapter 4.

9. See my chapters 1 and 4 for an explanation of the doctrine of the great conjunctions and the revolutions of Saturn and for the contents of this treatise.

10. *Concordantia astronomie cum hystorica narratione*, in d'Ailly 1490, chap. 57, fol. [d7r]: "Ipsa quoque multorum aliorum malorum causa fuit quorum hucusque remanent vestigia. . . . inter que mala non est silendum hoc horrendum scisma quod ab annis .36. usque hodie perseverat." For details of the actual astronomical situation in March 1345, see my chapter 4, n. 103.

11. *Concordantia*, chap. 58. Such a list ultimately derived from the author known as Telesphorus of Cosenza (on whom, see my chapter 5); the list of twenty-two schisms, however, circulated independently of Telesphorus's work, for example, in Paris, where d'Ailly could have encountered it. Hélène Millet, "Ecoute et usage des prophéties par les prélats pendant le Grand Schisme d'Occident," *Mélanges de l'Ecole Française de Rome. Moyen Age* 102 (1990): 443–45.

12. *Concordantia*, chap. 59, fol. [d7v]: "Et nichilominus secundum aliquos astronomos prenosticatum est ex figura celi anni presentis quod retrogradatio Jovis in principio anni in prima domo significat destructionem religionis et pacem in ecclesia adhuc non firmari."

13. Ibid.: "Sed deus est ille vere sapiens qui solus dominatur astris, cuius singulari auxilio huic malo conveniens poterit adhiberi remedium, alioquin formidandum est ne istud sit illud magnum scisma quod esse debet preambulum adventus antichristi."

14. Ibid., chap. 60, fols. [d7v]–[d8r]: "Inter dictam coniunctionem et illud complementum dictarum .10. revolutionum erit status octave spere circiter per annos .25. quod sic patet quia status octave spere erit anno .444. post situm augium que secundum tabulas astronomicas sunt adequate ad annum Christi 1320 perfectum. Et ideo anno Christi 1764 quibus annis si addas .25. fiunt anni .1789. quos prediximus. . . . Anno Christi 1765 diebus .136. completis erit statio motus accessus et recessus et habebimus pro motu eius .30.[90] gradus." According to the Alfonsine tables, the eighth sphere's trepidational movement of access and recess would attain precisely 90 degrees on May 16, 1766, at which point, presumably, *recessus* would begin. The *"status"* that d'Ailly describes would thus be analogous to the "stations" through which the planets passed as their motions shifted from direct to retrograde (or vice versa). The movement from 0 to 90 degrees took a period of time stretching approximately from the Incarnation through the mid-eighteenth century. No doubt d'Ailly delighted in placing Antichrist's advent at the end of an era beginning roughly around the time of Christ. See Emmanuel Poulle, ed., *Les tables alphonsines avec les canons de Jean de Saxe: Edition, traduction et commentaire* (Paris:

Editions du C.N.R.S., 1984), p. 199. John D. North cites the work of a sixteenth-century Frenchman, Pierre Turrel, who also calculated the end of the world using the trepidation of the eighth sphere. Turrel described four such "stations," at 1,750-year intervals, marked by the Flood, Exodus, the destruction of Jerusalem, and the end of the world. North, "Chronology and the Age of the World," in Wolfgang Yourgrau and Allen D. Beck, eds. *Cosmology, History, and Theology* (New York: Plenum Press, 1977), pp. 322–23. Turrel's work may have stemmed indirectly from d'Ailly's. North notes that Turrel's source was Jean de Bruges; Jean de Bruges, in turn, had borrowed from d'Ailly. Jean-Patrice Boudet, "Simon de Phares et les rapports entre astrologie et prophétie à la fin du Moyen Age," *Mélanges de l'Ecole Française de Rome. Moyen Age* 102 (1990): 639.

15. *Concordantia*, chap. 60, fol. [d8r]: "His itaque presuppositis dicimus quod si mundus usque ad illa tempora duraverit quod solus deus novit multe tunc et magne ac mirabiles alterationes mundi et mutationes future sunt et maxime circa leges et sectas."

16. Ibid., chap. 61, fol. [d8r]: "Unde ex his probabiliter concluditur quod forte circa illa tempora veniet antichristus cum lege sua vel secta damnabili."

17. Albumasar discusses all three predictors in the *De magnis coniunctionibus*. (On conjunctions and revolutions of Saturn, see my chapters 1 and 4.) In tr. 2, diff. 8, Albumasar discusses the motion of the eighth sphere. He points to a period of 640 years: eight degrees of access (or recess) at one degree per 80 years. He notes that the completion of such a period signifies great changes in the world, the "revolution of sects, and the permutation of kingdoms," particularly if it happens at the same time in which Saturn changes from one sign to another, which, presumably, was the case with Saturn's completion of its tenth revolution in 1789. ("Et illud est [*sic*] cum permutatur saturnus de signo ad signum significat esse accidentium in mundo et signa celestia et terrestria et revolutionem sectarum et permutationem regni de gente ad gentum. . . . Et iam fiet similiter mutatio communis maior cum compleverit motus orbis iste 9 [8] gradus accedendo vel recedendo et illud contingit in omnibus 664 [640] annis." *De magnis coniunctionibus, annorum revolutionibus, ac eorum profectionibus, octo continens tractatus* [Augsburg: Erhard Ratdolt, 1489], fol. [D6v].)

18. *Concordantia*, chap. 60, fol. [d8r]: "patet quod ab hoc anno .1414. usque ad statum octave spere erunt anni .253. [350] perfecte."

19. Ibid., chap. 63, fol. [d8v]: "primum est illa dissentio de qua in secunda ad thessalonicenses epistola dicit apostolus nisi venerit dissentio primum etc. Sed de hac iam aliquid breviter tetigimus."

20. Ibid.: "Secundum preambulum est quod . . . consurgent pro illis adversus Romanum imperium filii Israel [Ismael] filii Agar de quibus Daniel predixit. Et hoc erit in septime millenario annorum mundi. . . . Tercium preambulum est quod terram promissionis propter peccata inhabitantium in ea obtinebunt filii Ysmael id est sarraceni que deberet esse terra christianorum. Et adhuc applicat illud apostoli nisi venerit dissentio primum etc., dicens quod dissentio est disciplina vel correptio. . . ." D'Ailly offers an abridgment of pseudo-Methodius, *Revelationes*, chaps. 10–11 (pp. 78–85 in Ernst Sackur, ed., *Sibyllinische Texte und Forschungen: Pseudomethodius, Adso und die Tiburtinische Sibylle* [Halle: Niemeyer, 1898]).

21. *Concordantia*, chap. 63, following pseudo-Methodius, chap. 12–13 (pp. 86–93 in the Sackur ed.).

22. Heiko Oberman, thus, misunderstands d'Ailly's thought on the Schism when he characterizes him as showing increased anxiety about the apocalypse in the years *after* the

Council of Pisa and outright apocalypticism at the opening of the Council of Constance. Oberman, "The Shape of Late Medieval Thought: The Birthpangs of the Modern Era," in Charles Trinkaus and Heiko A. Oberman, eds., *The Pursuit of Holiness in Late Medieval and Renaissance Religion*, Studies in Medieval and Reformation Thought, 10 (Leiden: E. J. Brill, 1974), p. 11. He correctly notes that d'Ailly did not believe that the apocalypse was irreversibly impending (p. 16).

23. Pierre d'Ailly, *Epistola ad summum pontificem Joannem XXIII (Incipit:* "Dudum . . ."), dated June 18, 1414. The letter is printed in Jean Gerson, *Opera omnia*, ed. Louis Ellies Dupin, 5 vols. (Antwerp: Sumptibus societatis, 1706), 2: cols. 876–82; a copy in d'Ailly's own hand is reproduced in facsimile in *Le recueil épistolaire autographe de Pierre d'Ailly et les notes d'Italie de Jean de Montreuil*, ed. Gilbert Ouy, Umbrae Codicum Occidentalium, 9 (Amsterdam: North-Holland, 1966).

24. *Epistola ad Joannem XXIII (Incipit:* "Dudum . . ."), in Gerson, *Opera omnia* (ed. Dupin), 2: col. 876: "aliqua valde notanda, quae coram sanctae memoriae Domino Urbano Quinto, . . . praedicavit quidam solemnis sacrae Theologiae Doctor breviter recolligere, . . . et Vestrae Beatitudini praesentare decrevi."

25. Ibid., cols. 877–78.

26. Ibid., col. 879: "Quartus est error diffidentium, quibus non videtur in hoc posse dari remedium. Sed quod sicut aliqua Regna habuerent suam periodum, ita necesse est quod hujus Ecclesiae dominatio capiat finem. . . . Sed haec [scripture quoted earlier] non significant impossibilitatem, sed bene difficultatem; quia, Perversi difficulte corriguntur. Eccle. 1.15. Unde opinio de immutabilitate futurorum, non solum adversatur Theologiae, et Philosophiae; sed etiam Astronomiae."

27. Ibid., col. 879: "[paraphrasing the unnamed preacher:] licet, secundum praemissa, super Ecclesiam sit tribulatio praeparata; tamen si mores nostros in melius reformemus; . . . si corde et opere ad Deum revertamur; ipse modo inaestimibili nos juvabit, et a flagellatione cessabit. . . . [continuing in his own voice, now:] Haec, Beatissime Pater, ex dictis alienis colligere et allegare volui . . . quia cito post [this sermon], defuncto Domino Gregorio Papa XI. successore dicti Domini Urbani, secuta est tam horrenda tribulatio Ecclesiae. . . . [referring to his own writings on the Schism, including the sermon on St. Francis:] Unde quia non sunt secuti [i.e., the clergy] Regulam Evangelicam, non mansit pax Ecclesiastica super illos."

28. Ibid., col. 880: "licet nec Propheta, nec filius sum Prophetae; tamen sine temeraria assertione audeo dicere, quod nisi in Generali Concilio proxime celebrando contra haec damnabilia scandala praevisa fuerint remedia, hanc divisionem schismaticam penitus extirpando, et Ecclesiam tam multipliciter deformatam in melius reformando; verisimiliter credere debemus quod adhuc majora mala, et ampliora sint futura."

29. Ibid.: "In hiis vero, non in peccatori [as per MS Vatic. Regin. Lat. 689 A, f 352v in Ouy's facsimile; Dupin has *peccatorum*] fidem dari requiro; sed illis qui Spiritu sancto inspirati locuti sunt, inter quos non spernendi videntur ille venerabilis Abbas Joachim, de Calabria; et Sancta mulier Hyldegardis, de Alemania. . . ."

30. Ibid., cols. 880–81: "Sed etiam hiis concordant non solum testimonia Prophetica, verum et Astronomica judicia, quae non sunt a Catholicis penitus respuenda, sicut me notasse recolo in Tractatu quem nuper compilavi, *De concordia Astronomicae veritatis, et narrationis historicae*. . . ." He goes on to list the treatises "*De concordantia Theologiae et Astronomiae* [the *Vigintiloquium*], . . . *Super correctionem litteralem Kalendarii* [a work he composed urging calendar reform and anticipating the later Gregorian calendar]," and *De*

legibus et sectis, as well as making general reference to his various writings on the Schism. It was not unusual for a pope to be acquainted with astrological materials. The astrologer Johannes de Muris, for example, had given predictions based upon Saturn-Jupiter conjunctions in a letter to Pope Clement VI (1342–52). D'Ailly's defensive tone concerning astrology, here, reflects his more general belief that astrology and theology, though seemingly incompatible, could be harmonized.

31. *Elucidarium*, in d'Ailly 1490, chap. 6, fol. [e6v].

32. Ibid., chap. 6, fol. [e6v] (5,328 years to Christ); chap. 23, fol. f4v (5,344 years to Christ); chap. 28, fol. [f8r] (5,492 years and 223 days to Christ); and chap. 33, fol. g2v (5,343 years to Christ).

33. Ibid., chap. 26, fol. [f6v] (mean greatest conjunction in 1915); chap. 33, fol. g3v (mean greater conjunction in 3072); and chap. 12, fol. f1r (true greatest conjunction in 3469). On "true" vs. "mean" conjunctions, see my chapter 4.

34. *Elucidarium*, chap. 10, fol. [e8r]: "Fuit etiam alia [coniunctio] supradictam [coniunctionem] immediate subsequens in anno domini 1365 in .8. gradu Scorpionis. Et fuit media eorum scilicet Saturni et Jovis ac Solis coniunctio in Libra. Et ita debuisset mansisse in triplicitate aerea in qua sumus sed propter eorum directionem pervenit coniunctio eorum ad Scorpionem. Et quia istud signum semper est inimicum religioni ideo dixerunt aliqui quod illa coniunctio valde significativa fuit presentis scismatis." Both Jupiter and Saturn exhibited retrograde motion into Libra in the spring and summer of 1365 prior to their conjunction in October. See Bryant Tuckerman, *Planetary, Lunar, and Solar Positions A.D. 2 to A.D. 1649 at Five-day and Ten-day Intervals*, Memoirs of the American Philosophical Society, 59 (Philadelphia: American Philosophical Society, 1964), p. 700.

35. *Elucidarium*, chap. 10, fol. [e8r]: "Et post hanc immediate sequens fuit [coniunctio] anno 1405 .12. ianuarii in 27. Aquarii. Et hanc quidam reputant facere per terminationem scismatis et unione ecclesie propter coniunctionem duorum superiorum [i.e., Jupiter and Saturn] in Aquario in cuius triplicitate sumus quia fit per hanc coniunctionem reditus ad signum triplicitatis aerie licet immediate precedens fuerit in Scorpione que pretendebat scisma sed cum difficultate et tarditate propter signum fixum. . . ." D'Ailly notes that the effects of the 1365 conjunction were delayed somewhat because Scorpio was a "fixed sign." Since Aquarius was also considered one of the "fixed signs," one might anticipate the effects of a conjunction in that sign to be delayed as well.

36. Ibid., chap. 34.

37. Ibid., fol. g4v: "Ex premissis ergo concludunt isti quod sicut spacium celestis circuli quod est ab Ariete ad Libram est medietas illius circuli equalis altius medietati a Libra usque ad Arietem, sic quantum est spacium temporis lapsi ab Adam sive a principio mundi usque ad christum tantum naturaliter debet esse a christo usque ad finem mundi."

38. Paris, Bibliothèque Nationale, MS Lat. 15972, fols. 83r–84v. The work is edited and discussed in Nicole Bériou, "Pierre de Limoges et la fin des temps," in *Mélanges de l'Ecole Française de Rome. Moyen Age—Temps modernes* 98 (1986): 65–107 (the treatise appears on pp. 96–102). On the basis of codicological and paleographical evidence, Bériou identifies the anonymous *De antichristo* as a missing part of Pierre de Limoges's anthology of prophetic and Joachite texts (Paris, Bibliothèque Nationale, MS Lat. 16397). D'Ailly's use of this text has also been noted in Boudet, "Simon de Phares et les rapports entre astrologie et prophétie," pp. 638–39.

39. *De antichristo*, in Bériou, "Pierre de Limoges," pp. 100–01: "Sed anni ab inicio mundi usque ad tempus Domini sunt noti, igitur et quot sunt ab incarnacione Domini usque ad finem mundi."

40. Salembier 1886, p. 95. His retinue upon his entry into Constance included forty-four horsemen, befitting his special status as papal legate to Germany, site of the council. Bernard Guenée, *Entre l'Eglise et l'Etat: Quatre vies de prélats français à la fin du Moyen Age (XIIIe–XVe siècle)* (Paris: Gallimard, 1987), p. 276.

41. The sermon is printed in *Tractatus et sermones*, where it is given the erroneous date of 1417. That the sermon is rather from 1414 is shown by Salembier 1886, p. XVII; and Palémon Glorieux, "L'oeuvre littéraire de Pierre d'Ailly. Remarques et précisions," *Mélanges de science religieuse* 22 (1965): 70. An edition of the text is also found in Gerson, *Opera omnia* (ed. Dupin), 2: cols. 917–24.

42. Pierre d'Ailly, *Sermo secundus de adventu,* in *Tractatus et sermones* (Strasbourg: [Printer of Jordanus de Quedlinburg], 1490; repr., Frankfurt am Main: Minerva, 1971), fol. [s5v]: "Hec est ergo mirabilis mutatio dextere [dextera] excelsi quam in sole, luna et stellis nunc inchoari videmus. Dum hoc sacrum concilium ad hunc finem congregatum aspicimus ut sancta mater ecclesia iam dudum per scismaticam divisionem miserabiliter deformata, nunc per pacificam unionem feliciter reformetur."

43. Ibid.: "O beatos oculos qui hoc videre meruerint. O felicem locum hanc videlicet constantiensem civitatem." D'Ailly draws especial pleasure from the fact that Constance is within his jurisdiction as legate to Germany: "meque pariter cum ea iocunde exultationis participem, quam infra terminos commisse mihi legationis ad partes alamanie dominus ad hoc eligere dignatus est."

44. Ibid., fol. tr: "Si ergo contra hec mala hec sacra synodus non provident remedia congrua audacter affirmo quod post tam horrendas choruscationes quas videmus horribiliora sequentur tonitrua."

45. Guenée has stressed d'Ailly's reputation as a brilliant orator. He was, of course, extremely astute politically as well. Guenée, *Entre l'Eglise et l'Etat,* pp. 167–68.

46. *Epistola ad Joannem XXIII (Incipit: "Dudum . . ."),* col. 880: "Unde ex hiis omnibus praeambulis Antichristi, verisimiliter credere, et vehementer timere cogimur, quod nisi ad Deum, per veram poenitentiam convertamur, et ipse ad nos per piam misericordiam convertatur, fiet contra Praelatos et Clerum popularis seditio, et tribulatio talis et tanta, ut eam formidet animus cogitare, linguaque referre."

47. Marseilles, Bibliothèque Municipale, MS 1156, fols. 1–8, 11–30. Excerpts from this treatise have been printed in Noel Valois, "Un ouvrage inédit de Pierre d'Ailly, le *De persecutionibus ecclesiae,*" *Bibliothèque de l'Ecole des Chartes* 65 (1904): 557–74. I am grateful to the Bibliothèque Municipale in Marseilles for providing me with a copy of this important work.

48. Peter Aureoli, *Compendium Bibliorum.* Nicholas of Lyra expanded upon Auriol's interpretation in his *Postilla litteralis super totum Bibliam.* See R. E. Kaske, Arthur Groos, and Michael W. Twomey, *Medieval Christian Literary Imagery: A Guide to Interpretation* (Toronto: University of Toronto Press, 1988), pp. 19–23. D'Ailly may have been thinking of his own earlier interpretation of the persecutions of the church, as well. In March 1417 he recopied his sermon on Saint Francis (of 1380), in which he had discussed this theme. Glorieux, "L'oeuvre," p. 70.

49. Pierre d'Ailly, *De persecutionibus ecclesie,* fols. 4r–5r.

50. Ibid., fol. 5r: "Quintum tempus fuit a tempore heynrici predicti et durat secundum expositionem aureoli usque ad tempus antichristi et ultimi iudicii. In quo tempore ecclesia quandoque victoriam habuit *de babilone et muliere meretrice* id est de secta machometi sub godefrido et balduino regibus jherusalem." The reference is to the First Crusade, which succeeded in capturing Jerusalem in 1099; Godfrey de Bouillon (d. 1100) and his

brother Baldwin (d. 1118) were the first two rulers of the newly established kingdom of Jerusalem.

51. Ibid., fol. 6v (my emphasis): "Secunda ratio contra expositionem predictam est quia secundum eam omnia scripta in tribus capitulis, scilicet a xvii° capitulo usque ad xx^m, ab annis pluribus sunt completa. Littera vero sequens exponitur ab omnibus de tempore antichristi *cuius adventus adhuc non apparet propinquus*."

52. Ibid., fol. 6v–7r: "Et ideo quod de statu ecclesie quantum ad contingentia in tanto intermedio tempore nichil scripserit iohannes non videtur conveniens, maxime cum multa postea valde notabilia circa ecclesiam contingerint, et precipue illud magnum scisma quod iam quadraginta annis duravit [i.e., now in 1418] et amplius, de quo verisimile est iohannem in hoc libro prophetasse. . . . Cum insuper beatus cirillus, et venerabilis abbas ioachim, ac sancta hildegardis de hoc magno scismate multa predixisse legantur."

53. Ibid., fols. 7r–v: "Premissis igitur . . . presuppositis . . . et sine assertione temeraria dici potest quod in quinta visione predicta beatus Joannes describit persecucionem ecclesie sub magno scismate, ipsiusque scismatis decursum, et aliorum malorum inde sequencium usque ad tempus antichristi."

54. Ibid., fols. 8r (whore = schismatic church), 11v (Babylon = Rome and its obedience), and 12r (beast = temporal power).

55. Ibid., fol. 23v: "post illam victoriam regis grecorum sive romanorum de qua loquitur methodius, sequitur pax ecclesie et tranquilla concordia unde subdit *et vidi sedes* [Rev. 20:4]. Ubi notandum quod tunc sacerdocium et imperium adinvicem concordabunt et perfectam ecclesie reformationem ac per hoc veram eius pacem et unionem procurabunt."

56. Ibid., fol. 24r–v: "In prima huius capituli [Rev. 20] parte post ecclesie persecucionem scismaticam descripta est eius victoria et consolatio, postea vero circa medium capituli describitur eius ultima persecucio que fiet per antichristum." Louis Pascoe has noted the optimistic tone of d'Ailly's treatise and has related it to the influence of moderate Franciscan ideas about the apocalypse. Louis B. Pascoe, "Pierre d'Ailly: Histoire, Schisme et Antéchrist," in Jean Favier, ed., *Genèse et débuts du Grand Schisme d'Occident* (Paris: Editions du C.N.R.S., 1980), pp. 619–20.

57. *De persecutionibus ecclesie*, fols. 13v–16r.

58. Ibid., fol. 26r: "Sed ut dicit nicolaus de lira sumitur pro toto tempore christi quod currit usque ad antichristum et ponitur hic numeros determinatus pro indeterminato sicut sepe fit in divinis scripturis."

59. Ibid., fol. 26v: "Sciendum tamen est quod licet de adventus antichristi determinato tempore vel momento haberi non possit humanitus certitudo sicut alibi declaravi tamen indeterminate loquendo possint notari aliqua tempora circa que potest haberi probabilis coniectura de eius adventu et secta per astronomica iudicia."

60. Ibid., fol. 27v: "Unde ex hiis proba[bi]liter concluditur quod forte circa illa tempora veniet antichristus cum lege sua vel secta dampnabili." Following *Concordantia astronomie cum hystorica narratione*, chap. 61.

61. *Elucidarium*, chap. 12, fol. f1r (1761, although this is not a "true" *maxima* conjunction, being too far from the head of Aries—see my chapter 4); and chap. 26, fol. [f6v] (1915 for eighth *maxima* conjunction).

62. See Oakley, *Political Thought*, chap. 7 and app. 1.

63. The passage given by d'Ailly corresponds to p. 69, lines 1223–34, of the edition of John of Paris in "The *Tractatus de Antichristo* by John of Paris: A Critical Edition,

Translation, and Commentary," ed. Sara Beth Peters Clark (Ph.D. diss., Cornell University, 1981). See my chapter 5.

64. *De persecutionibus ecclesie*, fol. 28r: "Sed quia secundum [pseudo]Methodius ante adventum antichristi secta macometi totaliter destruetur ut super tactum est, ideo de huius secte duracione aliquid precognoscere videtur expedire."

65. Ibid., fol. 29v: "Et nichilominus constat quod incipiendo computacionem a nativitate macometi sive a morte eius sive a publicatione sue legis sive ab eiusdem legis reparacione emendacione et exposicione per suos successores et discipulis, iam multo plures anni transierunt quam sit maior numerus supradictus [693]." The various numbers appear at fols. 28r (666 of Revelation), 29v (693 and 584, according to Albumasar), 30v (1,141, according to Albumasar). The figures of 584 and 1,141 are obtained by reference to the "fridaric" years of the planet Venus, signifying the religion of Islam. Albumasar actually gave the number 1,151 for Venus's "maximum years," as d'Ailly reported in the *Elucidarium*, chap. 4 (probably following John of Ashenden and not Albumasar himself).

66. D'Ailly is following Albumasar's *De magnis coniunctionibus*, tr. 2, diff. 8. He does not comment on the fact that Christianity is here symbolized by the sun, whereas in most discussions of the great conjunctions Christianity is associated with Mercury. In the table of fridaric years d'Ailly reproduces in the *Elucidarium*, purportedly following Albumasar, he gives the figure 1,410.

67. *De persecutionibus ecclesie*, fol. 30v: "tunc lex christiana ab anno presenti qui est millesimus cccc[mus] decimus octavus non esset ultra annos quadraginta duos duratura."

68. Ibid.

69. Ibid.: "Ratio autem diversitatis in hoc forte potuit esse dispositio veneris seu habitudo ipsius ad gradus signi scorpionis tempore coniunctionis significantis dictam sectam et regnum arabum sicut notasse videtur albumazar loco praeallegato."

70. Ibid.: "Ex premissis igitur astronomicis coniecturis non videtur possibile aliquid certum concludere, sed tamen ex hiis et aliis supra scriptis probabilis haberi potest suspicio quod infra annum centesimum ab anno presenti magna fiet alteracio circa leges et sectas. Et specialiter circa legem et ecclesiam Christi. Et hoc aliqualiter precognoscere expedit ut ad tribulacionum pacienciam se constanter preparent christiani." Noël Valois, who edited this work in 1904, remarked that d'Ailly thus could be said correctly to have predicted both the French Revolution and the Reformation. Valois, "Un ouvrage inédit," p. 566.

71. Norman Cohn, *The Pursuit of the Millennium: Revolutionary Millenarians and Mystical Anarchists of the Middle Ages*, 3d rev. ed. (Oxford: Clarendon Press, 1970). For a critique of Cohn, see, e.g., Bernard McGinn, *Visions of the End: Apocalyptic Traditions in the Middle Ages* (New York: Columbia University Press, 1979), pp. 28–32. McGinn notes that many apocalyptic works were composed by well-placed clerics and tended to support the established institutions of medieval Christianity.

72. Robert E. Lerner, "Refreshment of the Saints: The Time after Antichrist as a Station for Earthly Progress in Medieval Thought," *Traditio* 32 (1976): 141.

73. *Epistola ad Joannem XXIII* (*Incipit*: "Dudum . . ."), col. 880.

74. *De falsis prophetis I*, in Gerson, *Opera omnia* (ed. Dupin), 1: col. 509: "Tales [i.e., illi qui sunt a Deo missi non immediate, sed per hominem] debint probare suam missionem, per canonica documenta: et in tali missione et eorum receptione, magna debet esse cautela et diligentia, propter eos qui importune se ingerunt, sine debita et Canonica electione."

75. *Errores sectae hominum intelligentiae et processus factus contra fratrem Wilhelmum de Hildernissem, ordinis beatae Mariae de monte Carmeli, per Petrum de Alliaco, episcopum Cameracensem, anno Christi MCCCCXI*, in Paul Fredericq, *Corpus documentorum inquisitionis haereticae pravitatis Neerlandicae*, 5 vols. (Ghent: J. Vuylsteke, 1889), 1:267–79. The work is also in Etienne Baluze, *Miscellanea novo ordine digesta, Tomus Secundus—Monumenta Sacra* (Lucca: Vincentius Junctinius, 1761), 2:288–92; and Charles du Plessis d'Argentré, *Collectio judiciorum de novis erroribus*, 3 vols. (Paris: Lambert Coffin, 1728–36), vol. 2, pt. 2:201–9. The name *homines intelligentiae* may have been meant to refer to Joachim's "gift of intelligence" for understanding Scripture.

76. *Errores sectae hominum intelligentiae*, in Fredericq, *Corpus documentorum*, pp. 271–72: "Et primo imitatores subscriptae sectae seu seductionis haereticae vocant se mutuo *homines intelligentiae* et habent duo capita. . . . Unum dictorum capitum vocatur frater Wilhelmus de Hildernissem carmelita. Aliud est quidam laicus illiteratus, forte sexagenarius, dictus Aegidius Cantoris. . . . Item dictus Aegidius habet modum specialem coeundi, non tamen contra naturam, quali dicit Adam in paradiso fuisse usum. . . . Item sibi invicem idioma fabricantes, actum carnalis copulae vocant delectationem paradisi vel alio nomine aclivitatem. . . . Item est quaedam maritata, quae non facit differentiam inter virum et virum, sed quemlibet indifferenter admittit suis tempore et loco. Et istud est quasi commune inter illas."

77. Ibid., p. 271: "Item dixit [Aegidius] quod diabolus finaliter salvabitur, . . . et quod finaliter omnes homines salvabuntur. . . . Item de statutis, praeceptis et ordinationibus ecclesiae non curant nec de orationibus, . . . Item nec curant de confessione. . . . Item nullum rigorem poenitentiae admittunt."

78. Ibid., pp. 276–77 (from the statement written by Wilhelmus in his own hand): "Item delatum est domino meo praedicto [i.e., d'Ailly] quod dicunt tempus veteris legis esse Patris et tempus novae legis esse Filii et pro nunc tempus esse Spiritus Sancti, quod dicunt esse tempus Heliae, quo removebuntur scripturae, ita ut quae prius tanquam vera habebantur, jam refutentur. . . . Hic dico quod talia nunquam audivi nec ab Aegidio nec ab aliis. Nec approbassem, si audivissem. . . ."

79. *Epistola ad Joannem XXIII (Incipit:* "Dudum . . ."), col. 880: "Nec desunt qui in odium ejus, errores contra Christi Fidem seminant, et haereses, et machinatur Sectarum novitates, sicut in hac mea Legatione, non sine cordis amaritudine, cognovi."

80. On Hus's trial and burning, see Gordon Leff, *Heresy in the Later Middle Ages*, 2 vols. (Manchester: Manchester University Press, 1967), 2:644–55. For the equation of pope with Antichrist, see Richard Kenneth Emmerson, *Antichrist in the Middle Ages: A Study of Medieval Apocalypticism, Art, and Literature* (Seattle: University of Washington Press, 1981), p. 71. Excerpts from Hus's writings dealing with Antichrist have been translated in McGinn, *Visions*, p. 263. For the ecclesiopolitical ramifications of Hus's ideas and of the council's actions, see Steven Ozment, *The Age of Reform 1250–1550: An Intellectual and Religious History of Late Medieval and Reformation Europe* (New Haven: Yale University Press, 1980), pp. 165–70.

81. Both Gerson's letter (*Epistola missa magistro Vincentio*) and d'Ailly's postscript appear in Jean Gerson, *Oeuvres complètes*, ed. Palémon Glorieux, 10 vols. (Paris: Desclée & Cie, 1960–73), 2:200–202.

82. Gerson, *Epistola missa magistro Vincentio*, p. 201: "Crede mihi, doctor emerite, multi multa loquuntur super praedicationibus tuis et maxime super illa secta se verberantium qualem constat praeteritis temporibus fuisse pluries et in locis variis reprobatam, quam nec approbas, ut testantur noti tui, sed nec efficaciter reprobas."

83. Jean Gerson, *Tractatus contra sectam flagellantium*, in Gerson, *Oeuvres complètes* (ed. Glorieux), 10:46–51.

84. Ibid., p. 48: "ordo hierarchicus confundatur . . si quilibet ad arbitrium suum posset instituere vel fovere novum ritum, sine duce, sine lege stabili, sine ordine, ubi sunt juvenes et virgines, senes cum junioribus, simul in unum dives et pauper."

85. Ibid., p. 51: "Proinde si praedicandum fuerit de finali judicio vel Antichristo, fiat hoc in generali, concludendo quod in morte quilibet habet suum judicium proximum et incertum." Recall that Saint Vincent Ferrer's sermons had specified that the Last Judgment was at hand and had described Antichrist's torments in vivid detail. Gerson went on to attack preachers' adducing of miracles to demonstrate the propinquity of Antichrist. This was perhaps also a dig at Vincent's supposed revelations of Antichrist's birth.

86. Ibid.: "Observandum praeterea summopere videtur, apud tales de societate quod nullo modo vivant otiose si laborare possint, ut sint in exemplar et animationem laborantium; ut praeterea nihil agant contentiose et velut ex vituperio et contemptu praelatorum majorum et minorum vel aliquorum generaliter clericorum; neque in praedicationibus faciendis neque in confessionibus audiendis."

87. Ibid.: "Tandem si forte senserit doctor insignis magister Vincentius, non posse convenienter super praemissis salubriter et efficaciter providere, videtur consultius ut ad tempus segregaret praesentiam suam, effugiens tantam societatem; quod fieri posset visitando sacrum Concilium vel altera occasione sibi sumpta."

88. Leff, *Heresy*, 2:491–93.

89. *Epistola ad Joannem XXIII (Incipit*: "Dudum . . ."), col. 880.

90. Based on unpublished research of Beatrice Hirsch-Reich, summarized in Marjorie Reeves, *The Influence of Prophecy in the Later Middle Ages: A Study in Joachimism* (Oxford: Clarendon Press, 1969), pp. 425–27. Michael H. Shank has suggested that Henry's disillusionment with prophecy may be parallel to his disappointment at his own failure to present an apology for Christianity sufficient to convert the Jews. Shank, *"Unless You Believe, You Shall Not Understand": Logic, University, and Society in Late Medieval Vienna* (Princeton, N.J.: Princeton University Press, 1988), p. 168. See also André Vauchez, "Les théologiens face aux prophéties à l'époque des papes d'Avignon et du Grand Schisme," *Mélanges de l'Ecole Française de Rome. Moyen Age* 102 (1990): 580–86. Vauchez proposes that disgust for the partisan prophecies of Telesphorus led Henry to reject the entire Joachimite tradition.

91. Henry of Langenstein, *Liber adversus Thelesphori eremitae vaticinia de ultimis temporibus*, in Bernhard Pez, *Thesaurus anecdotorum novissimus* (Augsburg: Veith, 1721), vol. 1, pt. ii, col. 518: "Quis sanus judicaret illum fuisse bonum Angelum, qui vanorum hominum vaticinia, vel eorum ingenio excogita, vel diabolico instinctu impressa, Spiritui Sancto attribui voluit?"

92. Ibid., cols. 521–22 (attack on pseudo-Joachim, *De semine scripturarum*, which attempted to forecast the time of the End by analogy with the Hebrew alphabet); 524: "Quod autem Joachim humana industria interpretatus sit verba illa ambigua libri adscripti Cyrillo, manifestum est."

93. Ibid., col. 544: "Itaque luce clarius est, quod illi calculatores ex astris vel alias undecunque, qui volunt vitas hominum quantificare, *et durationem mundi et legis Christi periodum mensurare*, sunt contrarii saluti humanae, et intentioni Divinae, et scripturis sacris, quae talia potius debere hominibus esse incerta patenter denuntiant, quam certa." Emphasis added.

94. Ibid., cols. 519–20: "Advertat ipse [Telesphorus] jam, quid profuerit fidelibus pro

pace Ecclesiae, sui somnii publicatio: consideret, quis inde motus sit, quis ab errore aversus, quis emendatus, aut quomodo res est inde propinquior facta ad pacem. . . . Sinamus ergo, et videamus, si ita erit sicut iste praedicit. . . . Vides ergo quomodo tales prophetiae, diabolica astutia interveniente, ad pacem non faciunt, sed eam retardantes, litem prolongant, et amplius discordiam in cordibus confirmant."

95. Ibid., col. 546: "Puto . . . quod omnia [preambles to Antichrist] praeterierint, non semel, sed forte pluries; ita quod nemo se jam ex hac parte assecurare poterit."

96. Reeves, *Influence*, pp. 425–26. Vauchez concurs with Hirsch-Reich's stress on the strong French bias in Telesphorus's work as a factor in Henry's reaction. Vauchez, "Les théologiens face aux prophéties," p. 586.

97. Henry of Langenstein, *Liber adversus Thelesphori eremitae vaticinia*, col. 532: ". . .verisimile videtur, quod sub illis [i.e., any line of popes instituted subsequent to the Great Schism] Ecclesia Christi continuo sit descensura in deterius, quemadmodum jam diu est. . . . Non videtur ergo, quod futura sit aliqua notabilis Ecclesiae reformatio usque post interfectionem manifestam Antichristi, per Jesum Christum Dominum nostrum."

98. *Henricus de hassia, mag. Paris. et Universitatis Vindobonens. professor, Petro de Alliaco, cancellario Paris.* . . . (Anno 1394, Vindobonae), in Heinrich Denifle and Emile Chatelain, eds., *Chartularium Universitatis Parisiensis*, 4 vols. (Paris: Delalain, 1894), 3:639 (no. 1695): "Sed quid, si scismate ablato infelici non redeant priorum aurea secula, nec forsitan ecclesia paci et unioni restituta erit in premissis emendacior, aut forte in virtutibus minor, quam ante scisma fuerat: nunquid si hoc undecunque verisimilius appareat, quempiam retrahere debet ut in negocio reunionis et pacis minus ferveat aut tardiori sollicitudine satagat? . . . Iterum pro verisimiliori habendum est, quod Deus, qui mala non permittit nisi gracia meliorum, ecclesiam suam de malo scismatice divisionis, qua modo concutitur, cum decreverit liberandam, utique in melius reformabit."

99. Henry of Langenstein, *Liber adversus Thelesphori eremitae vaticinia*, col. 513: "Inceperunt iam et alii sub hoc schismate loqui quasi Prophetae vel vates: sed verius divinatores . . . horum vaticiniis, verius vaniloquiis, omnes aures dederunt."

100. Pierre d'Ailly, *Concordantia astronomie cum hystorica narratione*, chap. 61, fol. [d8r]: "Nam licet de adventus sui determinato tempore vel momento haberi non possit humanitus certitudo sicut alibi declaravimus tamen indeterminate loquendo quod circa illa tempora venturus sit potest haberi probabilis coniectura et verisimilis suspicio per astronomica iudicia."

101. Ozment, *The Age of Reform*, p. 159.

102. Pierre d'Ailly, *Elucidarium*, fol. [g7v]: "Scimus autem quosdam nobis obiecisse quod professionem nostram et similiter etatem magis decuisset circa theologica quam circa illa mathematica studia occupari."

103. Ibid.: ". . . sciantur [i.e., these mathematical studies] ad fidei et theologie decorem ac utilitatem ecclesie pertinere."

Chapter Seven
The Concordance of Astrology and Theology

1. Bernard Guenée, *Entre l'Eglise et l'Etat: Quatre vies de prélats français à la fin du Moyen Age (XIIIe–XVe siècle)* (Paris: Gallimard, 1987), pp. 137 (powerful friends); 226 (contemporary view that d'Ailly chose sides to promote his career); and 281 (pleading ill health when his absence was politically expedient). D'Ailly's considerable political skills were

already obvious in his early years, in his use of the Blanchard affair as a springboard to prominence in the university. Alan E. Bernstein, *Pierre d'Ailly and the Blanchard Affair: University and Chancellor of Paris at the Beginning of the Great Schism* (Leiden: E. J. Brill, 1978), esp. pp. 67–69, 152.

2. See, e.g., d'Ailly's remark quoted in chapter 6 that people had objected to a person of his age and profession dabbling in astrological studies: *Elucidarium*, in d'Ailly 1490, fol. [g7v]: "Scimus autem quosdam nobis obiecisse quod professionem nostram et similiter etatem magis decuisset circa theologica quam circa illa mathematica studia occupari."

3. E.g., *Elucidarium*, chap. 2, fol. e4r: "De huius vero figure significatione ad astronomorum iudicia remitto."

4. E.g., in the *Elucidarium*, chap. 33, where d'Ailly quoted astrological predictions from the letter to Pope Clement VI by Johannes de Muris.

5. The prediction for 1789 occurs in the *Concordantia astronomie cum hystorica narratione*, chap. 60–61, and the *De persecutionibus ecclesie*. The forecast of a new religion to appear around 1600 is in d'Ailly's *De figura inceptionis mundi et coniunctionibus mediis sequentibus*. The warning of changes to come in the next one hundred years is also from the *De persecutionibus ecclesie*.

6. For d'Ailly's attempts to harmonize Christianity and astrology, see chapter 2.

7. Pierre d'Ailly, *Concordantia astronomie cum hystorica narratione*, in d'Ailly 1490, chap. 1, fol. [b8r]: "In his tamen lectorem cupimus esse premonitum quod nihil fatali necessitate ascribimus sed naturali causalitati et inclinationi cui liberum arbitrium in his que sue subiciantur facultati potest dei concurrente auxilio resistere ipsa quoque divina omnipotentia solo voluntatis imperio efficaciter obviare."

8. Pierre d'Ailly, *De persecutionibus ecclesie*, Marseilles, Bibliothèque Municipale, MS 1156, fol. 27r–v: "Hiis itaque presuppositis dico quod si mundus usque ad illa tempora duraverit quod solus deus novit, . . . ex hiis probabiliter concluditur quod forte circa illa tempora veniet antichristus cum lege sua vel secta dampnabili."

9. On the history of this distinction, see William J. Courtenay, *Capacity and Volition: A History of the Distinction of Absolute and Ordained Power* (Bergamo, Italy: Pierluigi Lubrina, 1990); and Francis Oakley, *Omnipotence, Covenant, and Order: An Excursion in the History of Ideas from Abelard to Leibniz* (Ithaca, N.Y.: Cornell University Press, 1984). See also Courtenay's review of Francis Oakley, *Omnipotence, Covenant, and Order*, in *Speculum* 60 (1985): 1008. In addition to these works, see esp. Courtenay, "Covenant and Causality in Pierre d'Ailly," *Speculum* 46 (1971): 94–119; and his "Nominalism and Late Medieval Religion," in Charles Trinkaus and Heiko A. Oberman, eds., *The Pursuit of Holiness in Late Medieval and Renaissance Religion*, Studies in Medieval and Reformation Thought, 10 (Leiden: E. J. Brill, 1974), pp. 37–39.

10. Pierre d'Ailly, *Quaestiones super libros sententiarum cum quibusdam in fine adjunctis* (Strasbourg: n.p., 1490; repr., Frankfurt: Minerva, 1968), I, q. 13, a. 1, D, fol. [t5r]: "Deum posse aliquid facere solet dupliciter intelligi. Uno modo secundum potentiam absolutam. Alio modo secundum potentiam ordinatam non quod in deo sint due potentie . . . sed deus dicitur illud posse de potentia absoluta quod simpliciter et absolute potest. . . . Sed deum aliquid posse de potentia ordinata potest dupliciter intelligi. Uno modo stricte quod potest stante sua ordinatione que eternaliter voluit se sic vel sic esse facturum et sic solum potest illa que ipse ordinavit se facturum. Alio modo potest intelligi magis large quod potest stante veritate legis seu scripture divine . . . aliquid est impossibile de potentia ordinata primo modo et non secundo modo."

11. Oakley, "Christian Theology and the Newtonian Science: The Rise of the Concept of the Laws of Nature," *Church History* 30 (1961): 455, n. 74, based upon *Quaestiones super libros sententiarum*, IV, q. 1, a. 2, N, fol. z4r (where d'Ailly addresses the question of whether a creature can create): "Sciendum est quod sicut dicitur quod deus aliquid potest de potentia absoluta quod non potest de potentia ordinata, ita dico de creatura, ideo concedo probabiliter quod licet creatura de potentia naturali seu naturaliter ordinata non possit creare vel annihilare ut dictum est, tamen ista potest de potentia simpliciter absoluta scilicet supernaturaliter seu miraculose." This passage does not seem to me to be an example of a miracle likely to happen in this world, but rather a hypothetical situation contrary to the laws of Aristotelian physics (i.e., the usual order). This is the sort of discussion, Edward Grant has argued, in which scholars used the language of *potentia absoluta* to justify speculation about events that were impossible in the Aristotelian cosmos. See Edward Grant, "The Condemnation of 1277, God's Absolute Power, and Physical Thought in the Late Middle Ages," *Viator* 10 (1979): 211–44; reprinted in Edward Grant, *Studies in Medieval Science and Natural Philosophy* (London: Variorum Reprints, 1981).

William Courtenay has stressed that in the most common use of the terms *potentia absoluta* and *potentia ordinata*, *potentia absoluta* represents the potential choices of action originally open to God, while *potentia ordinata* represents the order which God has actually chosen. *Potentia absoluta*, thus, does not represent a modality of divine action, such as is seen in the case of miracles. Confusion arose (and arises today) over whether God acts *de potentia absoluta* in such instances in part because of the imprecision of the term *potentia* and in part because canon lawyers in the thirteenth century applied these terms to describe different types of papal action. (Courtenay, *Capacity and Volition*, pp. 92–95.) Courtenay has argued, and I agree, that Oakley misinterprets Pierre d'Ailly in seeing d'Ailly as generally holding that miracles were an expression of the absolute power (review of Oakley's *Omnipotence, Covenant, and Order*, p. 1009). See Oakley, *Omnipotence, Covenant, and Order*, pp. 56–57. In his more recent reappraisal of the subject, Courtenay admits, however, that on occasion d'Ailly seems to adopt the canon lawyers' interpretation and to speak of *potentia absoluta* as a mode of divine action outside the (ordained) law (*Capacity and Volition*, p. 179). One must make careful note, however, of d'Ailly's own insistence "non quod in deo sint due potentie" ("that there not be two powers in God") (quoted in n. 10).

12. Courtenay describes such a conception as the "classic" formulation of the *absoluta/ordinata* distinction (*Capacity and Volition*, p. 74). My interpretation of d'Ailly's opinions, based upon his defense of astrology, owes much and is, I believe, in keeping with this view of d'Ailly, as set forth in Courtenay, "Covenant and Causality in Pierre d'Ailly," where Courtenay examines d'Ailly's interpretation of how the Sacraments work. Even if Oakley's interpretation is correct, and d'Ailly in the *Sentences* commentary did think miracles involved the absolute power, in the years between 1376 and the defense of astrology in 1410 he must have revised his view to place supernatural actions within God's ordained power. Leonard Kennedy, in his recent study of d'Ailly, is of the opinion that d'Ailly believed in the possibility of God exercising his absolute power to override the present world order. Even still he holds that d'Ailly placed miracles, and a supernatural order, within God's ordained power. Leonard A. Kennedy, *Peter of Ailly and the Harvest of Fourteenth-Century Philosophy* (Lewiston, N.Y.: Edwin Mellen Press, 1986), pp. 23, 27.

13. *Quaestiones super libros sententiarum*, I, q. 12, a. 3, HH, fol. t3r ("Deus potest de potentia absoluta rationali creature falsum dicere et eam decipere etiam per seipsum immediate et directe"); IV, q. 3, a. 3, G, fol. B1v ("Deus absolute posset manente obligatione ad penam eternam infundere gratiam et demittere culpam sine iniusticia sicut posset aliquem eternaliter punire sine demerito").

14. E.g., in the biographies of Salembier 1886, pp. 156–68, 195–256; and Paul Tschackert, *Peter von Ailli: Zur Geschichte des grossen abendländischen Schisma und der Reformconcilien von Pisa und Constanz* (Gotha: Friedrich Andreas Perthes, 1877; repr., Amsterdam: Rodopi, 1968), pp. 303–28. This is the philosophy formerly described under the disparaging term "nominalism," which is misleading in suggesting that the emphasis is on epistemology and not God's omnipotence, and that all philosophy between William of Ockham and Luther was basically the same. Courtenay gives an excellent overview of the developments in scholarship up to 1974 in "Nominalism and Late Medieval Religion."

15. See esp. Heiko A. Oberman, "The Shape of Late Medieval Thought: The Birthpangs of the Modern Era," in Charles Trinkaus and Heiko A. Oberman, eds., *The Pursuit of Holiness in Late Medieval and Renaissance Religion*, Studies in Medieval and Reformation Thought, 10 (Leiden: E. J. Brill, 1974), pp. 3–25; and Courtenay, "Nominalism and Late Medieval Religion," "Covenant and Causality," and *Capacity and Volition*.

16. Leonard Kennedy, *Peter of Ailly*, e.g., pp. 27, 181, 187, 193, 213 (God can use his absolute power in this world); and 1, 39–41 (d'Ailly's skepticism).

17. *De legibus et sectis contra superstitiosos astronomos*, in Pierre d'Ailly, *Tractatus de imagine mundi et varia ejusdem auctoris et Joannis Gersonis opuscula* (Louvain: Johann de Westphalia, ca. 1483), chap. 7, fol. [f6r], where d'Ailly gives examples "que facultatem nature et naturalem excellunt intellectum, sicut sancte et individue trinitatis eterne generationis ypostatice unionis, *conceptus virginalis*, et aliorum huiusmodi articulorum legis christiane" (my emphasis).

18. *Concordantia astronomie cum hystorica narratione*, chap. 59, fol. [d7v]: "Sed deus est ille vere sapiens qui solus dominatur astris, cuius singulari auxilio huic malo conveniens poterit adhiberi remedium, alioquin formidandum est ne istud sit illud magnum scisma quod esse debet preambulum adventus antichristi." D'Ailly is twisting the familiar maxim "Sapiens dominabitur astris," attributed to Ptolemy in the *Centiloquium*.

19. *Quaestiones super libros sententiarum*, I, q. 11, a. 2, N, fol. q6v, following Damascene, Augustine, Boethius, and others: "Unde sicut ea que nostris oculis sunt presentia intuitive videmus et per huius intuitionem certe iudicamus ubi et qualiter res visa existit . . . sic intellectualis et eternus oculus dei est quedam intuitio que immediate super quamlibet rem actualiter vel potentialiter existentem simul fertur, ideo tam circa futura quam circa presentia vel preterita omnium contingentium veritatum certum et distinctum ac verissimum habet iudicium." I am grateful to Professor Hester Gelber of Stanford University for an enlightening discussion of future contingents in the fourteenth century.

20. Ibid., q. 12, a. 3, HH, fol. t3r: "non est contra legem ordinatam imo possibile est *de potentia ordinata* quod deus decipiat aliquem tanquam causa mediata sine mediante creatura." Emphasis added.

21. Ibid., LL, fol. t4r: "Et in potestate antichristi futuri erit facere quod christus in aliquibus que predixit de ipso non fuerit propheta."

22. Calvin Normore, "Future Contingents," chap. 18 in Norman Kretzmann et al., eds., *The Cambridge History of Later Medieval Philosophy: From the Rediscovery of Aristotle to*

the Disintegration of Scholasticism, 1100–1600 (Cambridge: Cambridge University Press, 1982), p. 378.

23. In the *De reformatione ecclesie*.

24. On Ockham, see Normore, "Future Contingents," pp. 372–73. D'Ailly's views are in *Quaestiones super libros sententiarum*, I, q. 12, a. 3 (also see Normore, pp. 377–78).

25. *Quaestiones super libros sententiarum*, I, q. 12, a. 3, HH, fol. t3r: "Deus per seipsum frequenter fallit et decipit non directe sed indirecte et per accidens dicendo aliquid vel faciendo quod aliqui rationem non intelligentes crediderunt falsum, sicut patet in evangelio de illis quibus dixit Christus: Solvite templum hoc et in triduo reedificabitur illud; qui per hoc crediderunt falsum, et fuerunt decepti intelligentes de templo lapideo quod dixerat de templo corporis sui."

26. Jean-Patrice Boudet has argued that most astrologers of the later Middle Ages clearly distinguished their science from prophecy (defined as direct divine revelation), yet he notes also that the two "genres" mingled, particularly in popular works that did not come from the highest intellectual plane. He also notes that both types of prognostication were brought to bear upon the Great Schism, with astrology endowing prophecy with a sort of "scientific caution," while the general interest in eschatology gave certain astrologers a wider audience. Jean-Patrice Boudet, "Simon de Phares et les rapports entre astrologie et prophétie à la fin du Moyen Age," *Mélanges de l'Ecole Française de Rome. Moyen Age* 102 (1990): 617–48, esp. 626, 638, 642.

Appendix 1
A Note on the Availability of d'Ailly's Writings on Astrology

1. See, e.g., the descriptions offered by Lynn Thorndike in "Four British Manuscripts of Scientific Works by Pierre d'Ailly," *Imago mundi* 16 (1962): 157–60.

2. Ernst Zinner, *Verzeichnis der astronomischen Handschriften des deutschen Kulturgebietes* (Munich: C. H. Beck, 1925), pp. 341–422; Salembier 1886; Salembier, *Le cardinal Pierre d'Ailly, chancelier de l'Université de Paris, évêque du Puy et de Cambrai, 1350–1420* (Tourcoing: Imprimerie Georges Frère, 1932); Palémon Glorieux, "L'oeuvre littéraire de Pierre d'Ailly. Remarques et précisions," *Mélanges de science religieuse* 22 (1965): 61–78; Lynn Thorndike and Pearl Kibre, eds., *A Catalogue of Incipits of Medieval Scientific Writings in Latin*, rev. ed. (Cambridge, Mass.: Medieval Academy of America, 1963).

Appendix 2
A Chronology of d'Ailly's Works Dealing with Astrology

1. Palémon Glorieux, "Les années d'études de Pierre d'Ailly," *Recherches de théologie ancienne et médiévale* 44 (1977): 133. Glorieux earlier had specified a date of October 1375. Glorieux, "L'oeuvre littéraire de Pierre d'Ailly. Remarques et précisions," *Mélanges de science religieuse* 22 (1965): 62. The *Principium* preceded the lectures on the Bible. Tschackert (*Peter von Ailli*, p. 348) and Salembier (Salembier 1886, p. xiii) place the work in 1374, but Glorieux is to be followed instead because he situates it in its proper place in d'Ailly's education.

2. The work antedates September 26, 1396, date of transcription of Paris, Bibliothèque Nationale, MS Lat. 2831, which contains it. Glorieux, "L'oeuvre," p. 66. It would seem likely that it dates from d'Ailly's years of lecturing in the arts at Paris, for the

Meteorology of Aristotle was a required text, according to a university statute of 1366. (Lynn Thorndike, *University Records and Life in the Middle Ages* [New York: Columbia University Press, 1944], p. 247.) The years 1368–74 marked d'Ailly's years of "audition" in theology. Not yet permitted to lecture in theology, he probably incepted in arts while he was not attending lectures in theology. In a bibliography from 1910, Salembier gives the date of this work as September 26, 1396, apparently confusing the date of composition of the work with the date of the redaction of the manuscript containing it. Salembier, "A propos de Pierre d'Ailly, évêque de Cambrai. Biographie et bibliographie," *Mémoires de la Société d'Emulation de Cambrai* 64 (1910): 117.

3. Probable date given by Salembier 1886, p. xv. Lynn Thorndike suggests that the work may be from late in d'Ailly's career because of the level of expertise in astronomy. Thorndike, *The Sphere of Sacrobosco and Its Commentators* (Chicago: University of Chicago Press, 1949), pp. 38–39. I prefer to see the work as an early one, dating perhaps from the period in which d'Ailly lectured in the faculty of arts in Paris, and therefore I follow Salembier's dating. D'Ailly's commentary gives several mnemonic devices for remembering the signs of the zodiac, the triplicities, etc., suggesting that it was intended for students. Furthermore, d'Ailly treats several topics to which he returns in later works, and the *Questiones* show less expertise than in the later treatises. For example, on the problem of the size of the earth, he cites authors beyond Sacrobosco in his *Epilogus mappe mundi* of 1410, whereas the question on the appropriate section of Sacrobosco's *Sphera* cites none but the text itself.

4. Salembier 1886, p. xvi. This is the *Sermo primus de adventu domini* in Pierre d'Ailly, *Tractatus et sermones* (Strasbourg: [printer of Jordanus de Quedlinburg], 1490; repr., Frankfurt am Main: Minerva, 1971).

5. The dating of the *De anima* and of other philosophical works has been the subject of careful scrutiny in Marguerite Chappuis, Ludger Kaczmarek, and Olaf Pluta, "Die philosophischen Schriften des Peter von Ailly. Authentizität und Chronologie," *Freiburger Zeitschrift für Philosophie und Theologie* 33 (1986): 593–615. On the basis of internal evidence, they situate the work in the years 1377–81, between the time d'Ailly completed his *Sentences* commentary and the date of his receipt of the doctorate (p. 600). This dating must supersede the old date of 1372 assigned by Salembier (Salembier 1886, p. xiii) and Tschackert (*Peter von Ailli*, p. 348).

6. Chappuis, Kaczmarek, and Pluta assign this date due to internal evidence by the same reasoning as above ("Die philosophischen Schriften," pp. 602–4); the work was written while d'Ailly was a member of the faculty of arts and mentions the *De anima*. Again the old date of 1372 must be set aside. (This is the date given by Salembier 1886, p. xiii; and Tschackert, *Peter von Ailli*, p. 348.)

7. (D'Ailly composed two treatises bearing this title.) The work must be after 1378, because it mentions the Schism. The terminus ante quem comes from Glorieux: the treatise appears in Paris, Bibliothèque Nationale, MS Lat. 3122, which contains works from the years 1372–88 (or maybe 1394). Glorieux, "L'oeuvre," p. 67. Salembier places it between 1372 and 1395 (Salembier 1886, p. xv). Coopland's assertion that the work is post 1409 is simply wrong (G. W. Coopland, *Nicole Oresme and the Astrologers: A Study of His Livre de Divinacions* [Cambridge, Mass.: Harvard University Press, 1952], p. 11).

8. (= *Sermo [tertius] de adventu domini* in *Tractatus et sermones*.) The sermon bears the internal date of 1385; Glorieux supplies the specific date. "L'oeuvre," p. 70.

9. Salembier 1886, p. xvii.

10. This is the date given in the 1483 edition of the work and by Salembier 1886, p. xvii. Dupin's edition gives the erroneous date of December 24, 1416. Glorieux notes the discrepancy and follows Dupin, with the explanation that the work must be anterior to the *Epistola ad Joannem XXIII* (*Incipit*: "Dudum . . ."), which cites it. ("L'oeuvre," p. 67). Since this letter dates from 1414, as Glorieux himself states, and since other astrological works from the year 1414 also cite the *De legibus et sectis*, we must consider Glorieux's opinion about its date to be an oversight.

11. = *Concordantia astronomie cum theologia.*

12. = *Tractatus de concordia astronomice veritatis et narrationis hystorice.*

13. Salembier 1886, p. xviii.

14. The work discusses conjunctions mentioned in the *Elucidarium*. Since d'Ailly here follows the dating scheme used in that treatise, the *Pro declaratione* likely antedates the *De figura inceptionis mundi*, in which he offered revised times for these conjunctions.

15. It mentions the *Vigintiloquium*, the *Concordantia astronomie cum hystorica narratione*, and the *Elucidarium* and therefore must be considered to be later than September 1414. D'Ailly also promises that a treatise to follow will give more detail about the relation of the triplicities to the regions of the earth. This information comes in the *De concordia discordantium astronomorum* of January 1415. Glorieux suggests that the *De figura inceptionis mundi* may be the same as a treatise *Anni a principio mundi et gesta magis notanda cum figuris revolutionum et aliis*, Cambrai MS 955, fols. 124, 127–28 ("L'oeuvre," p. 66).

16. (= *Sermo secundus de adventu* in *Tractatus et sermones*.) The date (1417) given in *Tractatus et sermones* is incorrect. See Max Lieberman, "Chronologie Gersonienne. VIII. Gerson et d'Ailly (III)," *Romania* 81 (1960): 81; and Glorieux, "L'oeuvre," p. 70.

17. Salembier 1886, p. xviii.

18. Glorieux, "L'oeuvre," p. 71.

19. Date given in text.

20. (Written to Jean Gerson.) Date assigned in Glorieux, "L'oeuvre," p. 67. His 1960 edition of the text gives date "vers novembre 1419." Jean Gerson, *Oeuvres complètes*, ed. Palémon Glorieux, 10 vols. (Paris: Desclée & Cie, 1960–73), 2:218.

21. Gerson, *Oeuvres complètes* (ed. Glorieux), 2:222.

Select Bibliography

Works by Pierre d'Ailly (Petrus de Alliaco)

Apologeticus Hieronymianae versionis. Edited by Louis Salembier. In "Une page inédite de l'histoire de la Vulgate," *Revue des sciences ecclésiastiques* 61 (1890): 500–513; 62 (1890): 97–110.

Conceptus et Insolubiles: An Annotated Translation. Translated and edited by Paul Vincent Spade. Dordrecht and Boston: D. Reidel, 1980.

Concordantia astronomie cum theologia. Concordantia astronomie cum hystorica narratione. Et elucidarium duorum precedentium. Augsburg: Erhard Ratdolt, 1490.

De figura inceptionis mundi et coniunctionibus mediis sequentibus. Vienna. Österreichische Nationalbibliothek. MS 5266, fols. 46r–50v; MS 5318, fols. 92r–98r.

De persecutionibus ecclesie. Marseilles. Bibliothèque Municipale. MS 1156, fols. 1–8, 11–30.

Errores sectae hominum intelligentiae et processus factus contra fratrem Wilhelmum de Hildernissem, ordinis beatae Mariae de monte Carmeli, per Petrum de Alliaco, episcopum Cameracensem, anno Christi MCCCCXI. In Paul Fredericq, *Corpus documentorum inquisitionis haereticae pravitatis Neerlandicae,* 1:267–79. 5 vols. Ghent: J. Vuylsteke, 1889. Also in Etienne Baluze, *Miscellanea novo ordine digesta. Tomus Secundus—Monumenta Sacra,* 2:288–92. Lucca: Vincentius Junctinius, 1761. And Charles du Plessis d'Argentré, *Collectio judiciorum de novis erroribus,* vol. 2, pt. 2:201–9. 3 vols. Paris: Lambert Coffin, 1728–36.

Imago mundi. Translated by E. F. Keeber. Wilmington, N.C.: Linprint Co., 1948.

Imago mundi *by Petrus de Aliaco (Pierre d'Ailly) with Annotations by Christopher Columbus.* Facsimile edition. Boston: Massachusetts Historical Society, 1927.

Imago mundi *de Pierre d'Ailly: Texte latin et traduction française des quatre textes cosmographiques de d'Ailly et des notes marginales de Cristophe Colomb. Etude sur les sources de l'auteur par Edmond Buron.* 3 vols. Paris: Maisonneuve Frères, 1930.

Les oeuvres françaises du Cardinal Pierre d'Ailly, évêque de Cambrai, 1350–1420. Edited by Louis Salembier. Arras: Sueur-Charruey, 1907.

Pro declaratione decem dictarum figurarum. Cracow. Biblioteka Jagiellonska. MS 575, fols. 108r–111r; MS 584, fols. 56r–59r; and MS 586, fols. 66v–70r.

Quaestiones super libros Sententiarum cum quibusdam in fine adjunctis. Strasbourg, n.p.: 1490. Reprint. Frankfurt: Minerva, 1968.

Questiones Magistri Petri de Aylliaco cardinalis cameracensis super primum, tertium et quartum libros sententiarum. Paris: Jean Petit, 1505(?).

[*Questiones* on Sacrobosco's *Sphera*]. In Joannes de Sacrobosco, *Uberrimum sphere mundi commentum intersertis etiam questionibus domini Petri de Aliaco.* Paris: Jean Petit, 1498.

Le recueil épistolaire autographe de Pierre d'Ailly et les notes d'Italie de Jean de Montreuil. Edited by Gilbert Ouy. Umbrae Codicum Occidentalium, 9. Amsterdam: North-Holland, 1966.

Sermo de beato Bernardo. Paris. Bibliothèque Nationale. MS Lat. 3122. Fols. 92v–97v.

Tractatus de anima. Paris: Jean Petit, 1505.

Tractatus de imagine mundi et varia ejusdem auctoris et Joannis Gersonis opuscula. Louvain: Johann de Westphalia, ca. 1483.

Tractatus et sermones. Strasbourg: [Printer of Jordanus de Quedlinburg], 1490. Reprint. Frankfurt am Main: Minerva, 1971.

Tractatus Petri de Eliaco episcopi cameracensis super libros metheororum de impressionibus aeris. Ac de hijs quae in prima, secunda, atque tertia regionibus fiunt. Strasbourg: Johann Prüs, 1504.

Tractatus utilis super Boecii de consolatione philosophie. Paris, Bibliothèque Nationale. MS Lat. 3122. Fols. 110r–169r.

Le Traité de Pierre d'Ailly sur la Consolation de Boèce. Edited by Marguerite Chappuis. Bochumer Studien zur Philosophie. Amsterdam: B. R. Grüner, (forthcoming).

Other Primary Sources

Albertus Magnus. *Speculum astronomiae.* Edited by Stefano Caroti, Michela Pereira, Stefano Zamponi under the direction of Paola Zambelli. Pisa: Domus Galilaeana, 1977.

Albumasar (Abu Ma'shar). *De magnis coniunctionibus, annorum revolutionibus, ac eorum profectionibus, octo continens tractatus.* Augsburg: Erhard Ratdolt, 1489. Also Venice: Melchior Sessa, 1515.

————. *Flores albumasaris.* Augsburg: Erhard Ratdolt, 1488.

————. *Introductorium in astronomiam Albumasaris abalachi octo continens libros partiales.* Venice: Melchior Sessa, 1506.

Alchabitius (al-Kabisi, al-Qabisi). *Alchabitius cum commento. Noviter impresso.* Venice: Melchior Sessa, 1512.

Alkindus (al-Kindi). *De radiis.* Edited by M.-T. d'Alverny and F. Hudry. *Archives d'histoire doctrinale et littéraire du moyen âge* 41 (1974): 139–260.

Aquinas, Thomas. *Liber de veritate Catholicae fidei contra errores infidelium; qui dicitur Summa contra gentiles.* Edited by Petrus Marc, Ceslas Pera, and Petrus Caramello. 3 vols. Taurini: Marietti, 1961–67.

————. *Summa theologiae.* Latin text and English translation. Cambridge: Blackfriars, 1964.

Arnold of Villanova. *Tractatus de tempore adventus Antichristi.* Excerpts printed in Heinrich Finke. *Aus den Tagen Bonifaz VIII,* pp. CXXIX–CLIX. Vorreformationsgeschichtliche Forschungen, 2. Münster: Aschendorff, 1902.

Augustine. [*De civitate dei*] *The City of God against the Pagans.* With English translation by William M. Green. 7 vols. Loeb Classical Library. Cambridge, Mass.: Harvard University Press, 1960–72.

————. *Confessions.* With an English translation by William Watts. 2 vols. Loeb Classical Library. Cambridge, Mass.: Harvard University Press, 1912. Reprint. 1977.

Avenesra (Abraham ibn Ezra). *Abrahe Avenaris Judei astrologi peritissimi in re iudiciali opera: ab excellentissimo philosopho Petro de Abano post accuratam castigationem in latinum traducta.* Venice: Petrus Liechtenstein, 1507.

————. *Le Commence de la sapience des signes; Le livre des fondements astrologiques.* Edited and translated by Jacques Halbronn. Paris: Retz, 1977.

Bacon, Roger. *Opera quaedem hactenus inedita.* Vol. 1: *Opus tertium; Opus minus; Compendium philosophiae.* Edited by J. S. Brewer. Rerum Britannicarum Medii Aevi Scriptores. London: Longman, 1859.

————. *The "Opus maius" of Roger Bacon.* Edited, with introduction and analytical tables, by John Henry Bridges. 2 vols. Oxford: Clarendon Press, 1897.

Bede. *Bedae venerabilis opera.* Corpus Christianorum, Series Latina. Vol. 123B. Turnhout: Brepols, 1977.

Bernard Silvestris. *Cosmographia.* Edited by Peter Dronke. Leiden: E. J. Brill, 1978.

Biblia sacra cum glossa interlineari, ordinaria, et Nicolai Lyrani Postilla eiusdemque Moralitatibus, Burgensis Additionibus, et Thoringi replicis. Venice: [publisher unknown], 1588.

Bourgeois of Paris. *A Parisian Journal, 1405–1449. Translated from the Anonymous* Journal d'un Bourgeois de Paris. Translated by Janet Shirley. Oxford: Clarendon Press, 1968.

Brayer, Edith. "Notice d'un manuscrit 574 de la Bibliothèque municipale de Cambrai, suivie d'une édition des sermons français de Pierre d'Ailly." *Extraits des notices et extraits des manuscrits de la Bibliothèque Nationale et d'autres bibliothèques* 43 (1965): 145–342.

Chaucer, Geoffrey. *The Complete Works of Geoffrey Chaucer.* Edited by F. N. Robinson. Boston: Houghton Mifflin, 1933.

——. *The Equatorie of the Planetis, Edited from Peterhouse MS. 75.I.* Edited by Derek J. Price. Cambridge: Cambridge University Press, 1955.

Clemingiis, Nicolaus de (Nicolas of Clamanges). *Opera omnia.* Lyons: Johannes Balduinus, 1613. Reprint. Farnborough, England: Gregg Press, 1967.

Denifle, Heinrich, and Emile Chatelain, eds. *Chartularium Universitatis Parisiensis.* 4 vols. Paris: Delalain, 1889–97.

Du Plessis d'Argentré, Charles. *Collectio judiciorum de novis erroribus.* 3 vols. Paris: Lambert Coffin, 1728–36.

Eschuid, Johannes (John of Ashenden). *Summa astrologiae iudicialis de accidentibus mundi quae anglicana uulgo nuncapatur.* Venice: Iohannes Lucilius Santritter, 1489.

Ferrer, Boniface. *Tractatus pro defensione Benedicti XIII.* In Edmund Martène and Ursin Durand, eds., *Thesaurus novus anecdotorum,* 2: cols. 1436–1529. Paris: Delaulne, 1717. Reprint. New York: Burt Franklin, 1968.

Firmicus Maternus, Iulius. *Matheseos Libri VIII.* Edited by W. Kroll and F. Skutsch. Leipzig: B. G. Teubner, 1897.

Gerson, Jean. *Oeuvres complètes.* Edited by Palémon Glorieux. 10 vols. Paris: Desclée & Cie, 1960–73.

——. *Opera omnia.* Edited by Louis Ellies Dupin. 5 vols. Antwerp: Sumptibus societatis, 1706.

Gratian. *Decretum.* In J. P. Migne, ed., *Patrologiae cursus completus.* Series latina. Vol. 187. Paris: Garnier Fratres, 1844–64.

Guilelmus Arvernus (William of Auvergne). *De fide et legibus.* Augsburg: Gunther Zarnier, [1475–76].

Guillermus Altissiodorensis (William of Auxerre). *Summa aurea.* Paris: Philippe Pigouchet, 1500. Reprint. Frankfurt: Minerva, 1964.

Henry of Harclay. *Utrum astrologi vel quicumque calculatores possint probare secundum adventum Christi.* In Franz Pelster, "Die quaestio Heinrichs von Harclay über die zweite Ankunft Christi und die Erwartung des baldigen Weltendes zu Anfang des XIV Jahrhunderts," *Archivio italiano per la storia della pietà* 1 (1951): 25–82.

Henry of Langenstein (Henry of Hesse). *Liber adversus Thelesphori eremitae vaticinia de ultimis temporibus.* In Bernhard Pez, *Thesaurus anecdotorum novissimus,* vol. 1, pars ii, cols. 507–64. Augsburg: Veith, 1721.

Isidorus Hispalensis Episcopus (Isidore of Seville). *Etymologiarum sive originum libri XX.* Edited by W. M. Lindsay. 2 vols. Oxford: Clarendon Press, 1910.

Joannes Quidort Parisiensis (John of Paris). "The *Tractatus de Antichristo* of John of Paris: A Critical Edition, Translation, and Commentary." Edited and translated by Sara Beth Peters Clark. Ph.D. diss., Cornell University, 1981.

Joannes Saresberiensis (John of Salisbury). *Policraticus*. In Joannes Saresberiensis, *Opera Omnia*, edited by J. A. Giles, vols. 3–4. Oxford: J. H. Parker, 1848.

Lactantius. *Liber divinarum institutionum*. In J. P. Migne, ed., *Patrologiae cursus completus. Series Latina*. Vol. 6. Paris: Garnier Fratres, 1844–64.

Leopold of Austria. *Compilatio Leupoldi ducatus Austrie filii de astrorum scientia, decem continens tractatus*. Augsburg: Erhard Ratdolt, 1489.

Macrobius. *Commentary on the Dream of Scipio*. Translated by William Harris Stahl. New York: Columbia University Press, 1952.

Messahalla (Masha'allah). *Messahalae, antiquissimi ac laudatissimi inter arabes astrologi, Libri tres*. Nuremberg: Ioannes Montanus, 1549.

Mézières, Philippe de. *Le songe du vieil pelerin*. Edited by G. W. Coopland. 2 vols. Cambridge: Cambridge University Press, 1969.

Petrus Comestor (Peter the Eater). *Historia scholastica*. In J. P. Migne, ed., *Patrologiae cursus completus. Series Latina*. Vol. 198. Paris: Garnier Fratres, 1844–64.

Plotinus. *The Enneads*. Translated by Stephen McKenna. 2d ed. London: Faber and Faber, 1956.

Poulle, Emmanuel, ed. and trans. *Les tables alphonsines avec les canons de Jean de Saxe. Edition, traduction et commentaire*. Paris: Editions du C.N.R.S., 1984.

pseudo-Ovid. *The Pseudo-Ovidian De vetula: Text, Introduction, and Notes*. Edited by Dorothy M. Robathan. Amsterdam: Adolf M. Hakkert, 1968.

————. *Pseudo-Ovidius De vetula: Untersuchungen und Text*. Edited by Paul Klopsch. Mittellateinische Studien und Texte, 2. Leiden: E. J. Brill, 1967.

Ptolemaeus, Claudius (Ptolemy). *De praedictionibus astronomicis, cui titulum fecerunt Quadripartitum, Grece et Latine, Libri IIII. Philippo Melanthone interprete. Eiusdem Fructus librorum suorum, sive Centum dicta, ex conversione Ioviani Pontani*. Basil: Joannes Oporinus, 1553.

Sackur, Ernst, ed. *Sibyllinische Texte und Forschungen: Pseudomethodius, Adso und die Tiburtinische Sibylle*. Halle: Niemeyer, 1898.

Symon de Phares. *Recueil des plus célèbres astrologues et quelques hommes docts, faict par Symon de Phares du temps de Charles VIIIe*. Edited by Ernest Wickersheimer. Paris: Champion, 1929.

Thabit ibn Qurra. *The Astronomical Works of Thabit B. Qurra*. Edited by Francis J. Carmody. Berkeley: University of California Press, 1960.

————. "Thâbit ben Qurra 'On the Solar Year' and 'On the Motion of the Eighth Sphere.'" Edited by O. Neugebauer. *Proceedings of the American Philosophical Society* 106 (1962): 264–99.

Vincentius Bellovacensis (Vincent of Beauvais). *Speculum quadruplex, sive speculum maius: naturale, doctrinale, morale, historiale*. 4 vols. Douai: Baltazar Bellerus, 1624. Reprint. Graz, Austria: Akademische Druck, 1965.

Secondary Works

Adler, William. *Time Immemorial: Archaic History and Its Sources in Christian Chronography from Julius Africanus to George Syncellus*. Dumbarton Oaks Studies, 26. Washington, D.C.: Dumbarton Oaks, 1989.

Alverny, Marie-Thérèse d'. "Astrologues et théologiens au XIIᶜ siècle." In *Mélanges offerts à M.-D. Chenu*, pp. 31–50. Bibliothèque thomiste, 37. Paris: J. Vrin, 1967.

Amand, David. *Fatalisme et liberté dans l'antiquité grecque: Recherches sur la survivance de l'argumentation morale antifataliste de Carnéade chez les philosophes grecs et les théologiens chrétiens des quatre premiers siècles.* Louvain: Bibliothèque de l'Université, 1945.

Bauer, Georg-Karl. *Sternkunde und Sterndeutung der Deutschen in 9.-14. Jahrhundert.* Berlin: E. Ebering, 1937. Reprint. Nendeln: Kraus, 1967.

Beaujouan, Guy. "Observations et calculs astronomiques de Jean de Murs (1321–1344)." *Proceedings of the Fourteenth International Congress of the History of Science* (1974). No. 2.

Beaune, Colette. "La notion d'Europe dans les livres d'astrologie du XVᵉ siècle." In *La conscience Européene au XVᵉ et au XVIᵉ siècle*, pp. 1–7. Collection de l'Ecole Normale Supérieure de Jeunes Filles, 22. Paris: Ecole Normale Supérieure de Jeunes Filles, 1982.

Bègne, J.-Ph. *Exégèse et astrologie. A propos d'un ouvrage inédit de Pierre d'Ailly.* Extrait de la *Revue des sciences ecclésiastiques*. Lille: H. Morel, 1906.

Benjamin, Francis S., and G. J. Toomer. *Campanus of Novara and Medieval Planetary Theory.* Theorica planetarum, *edited with an introduction, English translation, and commentary.* Madison: University of Wisconsin Press, 1971.

Bériou, Nicole. "Pierre de Limoges et la fin des temps." *Mélanges de l'Ecole Française de Rome. Moyen Age—Temps modernes* 98 (1986): 65–107.

Bernstein, Alan E. *Pierre d'Ailly and the Blanchard Affair: University and Chancellor of Paris at the Beginning of the Great Schism.* Leiden: E. J. Brill, 1978.

Bezold, Friedrich von. "Astrologische Geschichtsconstruction im Mittelalter." *Deutsche Zeitschrift für Geschichtswissenschaft* 8 (1892): 29–72.

Bloomfield, Morton. "Joachim of Flora: A Critical Survey of His Canon, Teachings, Sources, Biography, and Influence." *Traditio* 13 (1957): 249–311.

Bober, Harry. "The Zodiacal Miniatures of the Très Riches Heures of the Duke of Berry—Its Sources and Meaning." *Journal of the Warburg and Courtauld Institutes* 11 (1948): 1–34.

Bois, Guy. *The Crisis of Feudalism: Economy and Society in Eastern Normandy, c. 1300–1550.* Cambridge: Cambridge University Press, 1984.

Boll, Franz Johannes, Carl Bezold, and Wilhelm Gundel. *Sternglaube und Sterndeutung: Die Geschichte und das Wesen der Astrologie.* Darmstadt: Wissenschaftliche Buchgesellschaft, 1974.

Bouché-Leclercq, Auguste. *L'astrologie grecque.* Brussels: Leroux, 1899. Reprint. Aalen: Scientia Verlag, 1979.

Boudet, Jean-Patrice. "Simon de Phares et les rapports entre astrologie et prophétie à la fin du Moyen Age." *Mélanges de l'Ecole Française de Rome. Moyen Age* 102 (1990): 617–48.

Brasewell, L. "Popular Lunar Astrology in the Late Middle Ages." *University of Ottawa Quarterly* 48 (1978): 187–94.

Brayer, Edith. "Un recueil de sermons, d'enseignments et d'exemples en ancien français, contenant des sermons de Pierre d'Ailly." *Academie des Inscriptions et Belles Lettres, Comptes rendus* (1959): 126–28.

Brettle, P. Sigismund. *San Vincent Ferrer und sein literarischer Nachlass.* Vorreformationsgeschichtliche Forschungen, 10. Münster: Aschendorff, 1924.

Brévart, Francis B. "The German *Volkskalender* of the Fifteenth Century." *Speculum* 63 (1988): 312–42.

Brincken, Anna-Dorothee v. den. "Beobachtungen zum Aufkommen der retrospektiven Inkarnationsära." *Archiv für Diplomatik Schriftgeschichte Siegel- und Wappenkunde* 25 (1979): 1–20.

Burnett, Charles, ed. *Adelard of Bath: An English Scientist and Arabist of the Early Twelfth Century.* London: Warburg Institute, 1987.

Campbell, Anna. *The Black Death and Men of Learning.* New York: Columbia University Press, 1931.

Capp, Bernard. *English Almanacs, 1500–1800: Astrology and the Popular Press.* Ithaca, N.Y.: Cornell University Press, 1979.

Carey, Hilary M. "Astrology at the English Court in the Later Middle Ages." In Patrick Curry, ed., *Astrology, Science and Society: Historical Essays,* pp. 41–56. Woodbridge, Suffolk: Boydell Press, 1987.

————. *Courting Disaster: Astrology at the English Court and University in the Later Middle Ages.* New York: St. Martin's Press, 1992.

Carmody, Francis J. *Arabic Astronomical and Astrological Sciences in Latin Translation: A Critical Bibliography.* Berkeley: University of California Press, 1956.

Caroti, Stefano. *La critica contro l'astrologia di Nicole Oresme e la sua influenza nel medioevo e nel Rinascimento,* pp. 545–684. Atti dell'Accademia Nazionale dei Lincei. Memorie. Classe di Scienze morali, storiche e filologiche. Ser. 8, vol. 23, Fasc. 6. 1979.

————. "Nicole Oresme: *Quaestio contra divinatores horoscopios.*" *Archives d'histoire doctrinale et littéraire du moyen âge* 43 (1976): 201–310.

————. "Nicole Oresme's Polemic against Astrology in His 'Quodlibeta.'" In Patrick Curry, ed., *Astrology, Science and Society: Historical Essays,* pp. 75–93. Woodbridge, Suffolk: Boydell Press, 1987.

Chappuis, Marguerite, Ludger Kaczmarek, and Olaf Pluta. "Die philosophischen Schriften des Peter von Ailly: Authentizität und Chronologie." *Freiburger Zeitschrift für Philosophie und Theologie* 33 (1986): 593–615.

Chappuis-Baeriswyl, Marguerite. "Notice sur le Traité de Pierre d'Ailly sur la Consolation de Boèce." *Freiburger Zeitschrift für Philosophie und Theologie* 31 (1984): 89–107.

Chenu, Marie-Dominique. "Astrologia praedicabilis." *Archives d'histoire doctrinale et littéraire du moyen âge* 31 (1964): 61–65.

————. "Nature and Man—The Renaissance of the Twelfth Century." In Marie-Dominique Chenu, *Nature, Man, and Society in the Twelfth Century: Essays on New Theological Perspectives in the Latin West.* Edited and translated by J. Taylor and Lester Little. Chicago: University of Chicago Press, 1968.

Clark, Charles. "The Zodiac Man in Medieval Medical Astrology." *Journal of the Rocky Mountain Medieval and Renaissance Association* 3 (1982): 13–38.

Cohn, Norman. *The Pursuit of the Millennium: Revolutionary Millenarians and Mystical Anarchists of the Middle Ages.* 3d rev. ed. Oxford: Clarendon Press, 1970.

Combes, André. "Sur les 'lettres de consolation' de Nicolas de Clamanges à Pierre d'Ailly." *Archives d'histoire doctrinale et littéraire du moyen âge* 15–17 (1940–42): 359–89.

Contamine, Philippe. "Les prédictions annuelles astrologiques à la fin du Moyen Age: Genre littéraire et témoin de leur temps." In *Histoire sociale, sensibilités collectives et mentalités: Mélanges Robert Mandrou,* pp. 191–204. Paris: Presses Universitaires de France, 1985.

Coopland, G. W. *Nicole Oresme and the Astrologers: A Study of His Livre de Divinacions.* Cambridge, Mass.: Harvard University Press, 1952.

Courtenay, William J. *Capacity and Volition: A History of the Distinction of Absolute and Ordained Power.* Bergamo, Italy: Pierluigi Lubrina, 1990.

―――. "Covenant and Causality in Pierre d'Ailly." *Speculum* 46 (1971): 94–119.

―――. "The Dialectic of Omnipotence in the High and Late Middle Ages." In Tamar Rudavsky, ed., *Divine Omniscience and Omnipotence in Medieval Philosophy,* pp. 243–69. Dordrecht: D. Reidel, 1985.

―――. "Nominalism and Late Medieval Religion." In Charles Trinkaus and Heiko A. Oberman, eds., *The Pursuit of Holiness in Late Medieval and Renaissance Religion,* pp. 26–59. Studies in Medieval and Reformation Thought, 10. Leiden: E. J. Brill, 1974.

―――. Review of Francis Oakley, *Omnipotence, Covenant, and Order. Speculum* 60 (1985): 1006–9.

Coville, A. "Ailly, Pierre de." In *Dictionnaire de biographie française.* Paris: Letouzey et Ané, 1933–.

Curry, Patrick. *Prophecy and Power: Astrology in Early Modern England.* Cambridge: Polity Press, 1989.

―――, ed. *Astrology, Science and Society: Historical Essays.* Woodbridge, Suffolk: Boydell Press, 1987.

Delaruelle, Etienne. "L'Antéchrist chez S. Vincent Ferrier, S. Bernardin de Sienne et autour de Jeanne d'Arc." In Etienne Delaruelle, *La piété populaire au Moyen Age,* pp. 329–54. Torino: Bottega d'Erasmo, 1975.

Desharnais, Richard P. "Reassessing Nominalism: A Note on the Epistemology and Metaphysics of Pierre d'Ailly." *Franciscan Studies* 34 (1974): 296–305.

Dijksterhuis, E. J. *The Mechanization of the World Picture.* Translated by C. Dikshoorn. Oxford: Oxford University Press, 1961.

Dinaux, Arthur. *Notice historique et littéraire sur le cardinal Pierre d'Ailly, évêque de Cambrai au XVe siècle.* Cambrai: S. Berthoud, 1824.

Duhem, Pierre Maurice Marie. *Medieval Cosmology: Theories of Infinity, Place, Time, Void, and the Plurality of Worlds.* Edited and translated by Roger Ariew. Chicago: University of Chicago Press, 1985.

―――. *Le système du monde: Histoire des doctrines cosmologiques de Platon à Copernic.* 10 vols. Paris: Hermann: 1913–59.

Eade, J. C. *The Forgotten Sky: A Guide to Astrology in English Literature.* Oxford: Clarendon Press, 1984.

Emmerson, Richard Kenneth. *Antichrist in the Middle Ages: A Study of Medieval Apocalypticism, Art, and Literature.* Seattle: University of Washington Press, 1981.

Fages, Henri. *Histoire de Saint Vincent Ferrier.* 4 vols. in 2. Louvain: Uystpruyst, and Paris: Picard, 1901–5.

Festugière, Le R. P. *La révélation d'Hermès Trismégiste.* Vol. 1: *L'astrologie et les sciences occultes.* Paris: Lecoffre, 1950.

Finke, Heinrich. *Aus den Tagen Bonifaz VIII: Funde und Forschungen.* Vorreformationsgeschichtliche Forschungen, 2. Münster: Aschendorff, 1902.

Flint, Valerie I. J. *The Rise of Magic in Early Medieval Europe.* Princeton, N.J.: Princeton University Press, 1991.

―――. "The Transmission of Astrology in the Early Middle Ages." *Viator* 21 (1990): 1–27.

Fontaine, Jacques. "Isidore de Séville et l'astrologie." *Revue des études latines* 31 (1953): 271–300.

Funkenstein, Amos. *Theology and the Scientific Imagination from the Middle Ages to the Seventeenth Century.* Princeton, N.J.: Princeton University Press, 1986.

Gandillac, Maurice P. de. "De l'usage et de la valeur des arguments probables dans les questions du Cardinal Pierre d'Ailly sur le 'Livre des Sentences.'" *Archives d'histoire doctrinale et littéraire du moyen âge* 8 (1933): 43–91.

Garin, Eugenio. *Astrology in the Renaissance: The Zodiac of Life.* Translated by Carolyn Jackson and June Allen. Translation revised by the author in conjunction with Clare Robertson. London and Boston: Routledge and Kegan Paul, 1983.

Gaster, M. "The Letter of Toledo." *Folk-Lore* 13 (1902): 115–34.

Gettings, Fred. *The Secret Zodiac: The Hidden Art in Mediaeval Astrology.* London and New York: Routledge and Kegan Paul, 1987.

Glorieux, Palémon. "Les années d'études de Pierre d'Ailly." *Recherches de théologie ancienne et médiévale* 44 (1977): 127–49.

———. "Deux élogues de la Sainte écriture par Pierre d'Ailly." *Mélanges de science religieuse* 29 (1972): 113–29.

———. "L'oeuvre littéraire de Pierre d'Ailly. Remarques et précisions." *Mélanges de science religieuse* 22 (1965): 61–78.

———. "Pierre d'Ailly et Saint Thomas." In *Littérature et religion: Mélanges offerts à Joseph Coppin,* pp. 45–54. Lille: Facultés Catholiques, 1966.

Goldstein, Bernard R., and David Pingree. "Levi ben Gerson's Prognostication for the Conjunction of 1345." *Transactions of the American Philosophical Society* 80, pt. 6 (1990).

Grafton, A. T., and N. M. Swerdlow. "Technical Chronology and Astrological History in Varro, Censorinus and Others." *Classical Quarterly* 35 (1985): 454–65.

Grant, Edward. "The Condemnation of 1277, God's Absolute Power, and Physical Thought in the Late Middle Ages." *Viator* 10 (1979): 211–44. Reprinted in Edward Grant, *Studies in Medieval Science and Natural Philosophy.* London: Variorum Reprints, 1981.

———. "Medieval and Renaissance Scholastic Conceptions of the Influence of the Celestial Region on the Terrestrial." *Journal of Medieval and Renaissance Studies* 17 (1987): 1–23.

———. "Science and Theology in the Middle Ages." In David C. Lindberg and Ronald L. Numbers, eds., *God and Nature: Historical Essays on the Encounter between Christianity and Science.* Berkeley: University of California Press, 1986.

———, ed. *Nicole Oresme and the Kinematics of Circular Motion: Tractatus de commensurabilitate vel incommensurabilitate motuum celi. Edited with an Introduction, English Translation, and Commentary.* Madison: University of Wisconsin Press, 1971.

Green, Louis. *Chronicle into History: An Essay on the Interpretation of History in Florentine Fourteenth-Century Chronicles.* Cambridge: Cambridge University Press, 1972.

Gregory, Tullio. "La nouvelle idée de nature et de savoir scientifique au XIIe siècle." In John E. Murdoch and Edith Sylla, eds., *The Cultural Context of Medieval Learning,* pp. 193–218. Boston Studies in the Philosophy of Science, 26. Dordrecht, Holland, and Boston: D. Reidel, 1975.

———. "Temps astrologique et temps chrétien." In Jean-Marie Leroux, ed., *Le temps chrétien de la fin de l'antiquité au Moyen Age: IIIe–XIIIe siècles,* pp. 557–79. Colloques internationaux du Centre National de la Recherche Scientifique, 604. Paris: Editions du C.N.R.S., 1984.

Guenée, Bernard. *Entre l'Eglise et l'Etat: Quatre vies de prélats français à la fin du Moyen Age (XIIIe–XVe siècle)*. Paris: Gallimard, 1987. English translation: *Between Church and State: The Lives of Four French Prelates in the Late Middle Ages*. Translated by Arthur Goldhammer. Chicago: University of Chicago Press, 1991.

―――. *Histoire et culture historique dans l'occident médiéval*. Paris: Aubier Montaigne, ca. 1980.

Guignebert, Charles. *De imagine mundi ceterisque Petri de Alliaco geographicis opusculis*. Paris: E. Leroux, 1902.

Gundel, Wilhelm. *Dekane und Dekansternbilder: Ein Beitrag zur Geschichte der Sternbilder der Kulturvölker*. Glückstadt und Hamburg: J. J. Augustin, 1936.

―――. *Sternglaube, Sternreligion und Sternorakel*. Leipzig: Verlag Quelle & Meyer, 1933.

Haeusler, Martin. *Das Ende der Geschichte in der mittelalterlichen Weltchronistik*. Cologne and Vienna: Böhlau Verlag, 1980.

Halbronn, Jacques. "L'itinéraire astrologique de trois italiens du XIIIe siècle: Pietro d'Abano, Guido Bonatti, Thomas d'Aquin." In Christian Wenin, ed., *L'homme et son univers au Moyen Age*, 2:688–74. Philosophes médiévaux, 26–27. Louvain: Institut Supérieur de Philosophie, 1986.

Hansen, Bert. *Nicole Oresme and the Marvels of Nature: A Study of His De causis mirabilium with Critical Edition, Translation, and Commentary*. Studies and Texts, 68. Toronto: Pontifical Institute of Mediaeval Studies, 1985.

―――. "The Complementarity of Science and Magic before the Scientific Revolution." *American Scientist* 74 (1986): 128–36.

Hartmann, Eduard. *Pierre d'Aillys Lehre von der sinnlichen Erkenntnis*. Fulda: Fuldaer Actiendruckerei, 1903.

Hartner, Willy. "The Mercury Horoscope of Marcantonio Michiel of Venice: A Study in the History of Renaissance Astrology and Astronomy." In Willy Hartner, *Oriens-Occidens: Ausgewählte Schriften zur Wissenschafts- und Kulturgeschichte. Festschrift zum 60. Geburtstag*, pp. [84]–[139]. Hildesheim: Georg Olms, 1968. Reprinted from A. Beer, ed. *Vistas in Astronomy*, 1:440–95. London and New York: Pergamon Press, 1955.

Heist, William W. *The Fifteen Signs before Doomsday*. East Lansing: Michigan State College Press, 1952.

Hissette, Roland. *Enquête sur les 219 articles condamnés à Paris le 7 mars 1277*. Philosophes médiévaux, 22. Louvain: Publications Universitaires, 1977.

Hübner, Wolfgang. *Zodiacus Christianus: Jüdisch-christliche Adaptationen des Tierkreises von der Antike bis zur Gegenwart*. Beiträge zur klassischen Philologie, 144. Königstein: Anton Hain, 1983.

Hüe, Denis. "Lire dans le ciel: Les *Pronostications*." In *Le soleil, la lune et les étoiles au Moyen Age*, pp. 159–75. Sénéfiance, 13. Aix-en-Provence: Publications du C.U.E.R.M.A., 1983.

Huizinga, Johan. *The Waning of the Middle Ages: A Study in the Forms of Life, Thought and Art in France and the Netherlands in the Dawn of the Renaissance*. Translated by F. Hopman. Garden City, N.Y.: Doubleday, 1954.

Jenks, Stuart. "Astrometeorology in the Middle Ages." *Isis* 74 (1983): 185–210.

Kaluza, Zénon. "Le Traité de Pierre d'Ailly sur l'Oraison dominicale." *Freiburger Zeitschrift für Philosophie und Theologie* 32 (1985): 273–93.

Kaske, R. E., Arthur Groos, and Michael W. Twomey. *Medieval Christian Literary Imagery: A Guide to Interpretation*. Toronto: University of Toronto Press, 1988.

Kennedy, Edward S. "Ramifications of the World-Year Concept in Islamic Astrology." *Proceedings of the Tenth International Congress of the History of Science* (1962): 23–43.

Kennedy, Edward S., and David Pingree. *The Astrological History of Masha'allah.* Harvard Monographs in the History of Science. Cambridge, Mass.: Harvard University Press, 1971.

Kennedy, Leonard A. *Peter of Ailly and the Harvest of Fourteenth-Century Philosophy.* Studies in the History of Philosophy, 2. Lewiston, N.Y.: Edwin Mellen Press, 1986.

Kibre, Pearl. "The Intellectual Interests Reflected in Libraries of the Fourteenth and Fifteenth Centuries." *Journal of the History of Ideas* 7 (1946): 257–97.

―――. *Studies in Medieval Science: Alchemy, Astrology, Mathematics, and Medicine.* London: Hambledon, 1984.

Kieckhefer, Richard. *Magic in the Middle Ages.* Cambridge Medieval Textbooks. Cambridge: Cambridge University Press, 1990.

Kinsman, Robert S., ed. *The Darker Vision of the Renaissance.* Berkeley: University of California Press, 1974.

Klempt, Adalbert. *Die Säkularisierung der universalhistorischen Auffassung zum Wandel des Geschichtsdenkens im. 16. und 17. Jahrhundert.* Göttinger Bausteine zur Geschichtswissenschaft, 31. Göttingen: Musterschmidt, 1960.

Klibansky, Raymond, Erwin Panofsky, and Fritz Saxl. *Saturn and Melancholy: Studies in the History of Natural Philosophy, Religion and Art.* London: Nelson, 1964. Reprint. Nendeln: Kraus, 1979.

Kren, Claudia. "Astronomical Teaching at the Late Medieval University of Vienna." *History of Universities* 3 (1983): 15–30.

―――. *Medieval Science and Technology: A Selected, Annotated Bibliography.* New York and London: Garland, 1985.

Kuhn, Thomas. *The Copernican Revolution: Planetary Astronomy in the Development of Western Thought.* Cambridge, Mass.: Harvard University Press, 1957.

Kunitzsch, Paul. *Mittelalterliche astronomisch-astrologische Glossare mit arabischen Fachausdrücken.* Bayerische Akademie der Wissenschaften. Philosophische-Historische Klasse. Setzungsberichte-Jahrgang 1977, Heft 5.

Kurze, Dietrich. *Joannis Lichtenberger: Eine Studie zur Geschichte der Prophetie und Astrologie.* Historische Studien, 379. Lübeck: Matthiesen, 1960.

Laistner, M.L.W. "The Western Church and Astrology during the Early Middle Ages." *Harvard Theological Review* 34 (1941): 251–75. Reprinted in M.L.W. Laistner, *The Intellectual Heritage of the Early Middle Ages: Selected Essays.* Edited by Chester G. Starr, pp. 57–82. Ithaca, N.Y.: Cornell University Press, 1957.

Landes, Richard. "Lest the Millennium Be Fulfilled: Apocalyptic Expectations and the Pattern of Western Chronography 100–800 CE." In Werner Verbeke, Daniel Verhelst, and Andries Welkenhuysen eds., *The Use and Abuse of Eschatology in the Middle Ages,* pp. 137–211. Louvain: Leuven University Press, 1988.

Leff, Gordon. *Heresy in the Later Middle Ages.* 2 vols. Manchester: Manchester University Press, 1967.

Le Goff, Jacques. "Au moyen âge: Temps de l'Eglise et temps du marchand." *Annales. Economies, sociétés, civilisations* 15 (1960): 417–33.

Lejbowicz, Max. "Chronologie des écrits anti-astrologiques de Nicole Oresme. Etude sur un cas de scepticisme dans la deuxième moitié du XIVe siècle." In Jeannine Quillet, ed., *Autour de Nicole Oresme*, pp. 119–76. Actes du Colloque Oresme organisé à l'Université de Paris, 12. Paris: J. Vrin, 1990.

Lemay, Helen. "The Stars and Human Sexuality: Some Medieval Scientific Views." *Isis* 71 (1980): 127–37.

Lemay, Richard. *Abu Ma'shar and Latin Aristotelianism in the Twelfth Century: The Recovery of Aristotle's Natural Philosophy through Arabic Astrology*. Beirut: American University, 1962.

——. "The Teaching of Astronomy in Medieval Universities, Principally at Paris in the Fourteenth Century." In *Science, Medicine, and the University: 1200–1500. Essays in Honor of Pearl Kibre, II*, edited by N. G. Siraisi and Luke Demaitre. *Manuscripta* 20 (1976): 197–217.

——. "The True Place of Astrology in Medieval Science and Philosophy: Towards a Definition." In Patrick Curry, ed. *Astrology, Science and Society: Historical Essays*, pp. 57–73. Woodbridge, Suffolk: Boydell Press, 1987.

Lerner, Robert E. "The Black Death and Western European Eschatological Mentalities." *American Historical Review* 86 (1981): 533–52.

——. "Ecstatic Dissent." *Speculum* 67 (1992): 33–57.

——. *The Powers of Prophecy: The Cedar of Lebanon Vision from the Mongol Onslaught to the Dawn of the Enlightenment*. Berkeley: University of California Press, 1983.

——. "Refreshment of the Saints: The Time after Antichrist as a Station for Earthly Progress in Medieval Thought." *Traditio* 32 (1976): 97–144.

Lieberman, Max. "Chronologie Gersonienne. VIII. Gerson et d'Ailly (III)." *Romania* 81 (1960): 44–98.

——. "Pierre d'Ailly, Jean Gerson et le culte de saint Joseph (I–III)." *Cahiers de Joséphologie* 13 (1965): 227–72; 14 (1966): 271–314; and 15 (1967): 5–113.

Lindbeck, George. "Nominalism and the Problem of Meaning as Illustrated by Pierre d'Ailly on Predestination and Justification." *Harvard Theological Review* 52 (1959): 43–60.

Lindberg, David C. *The Beginnings of Western Science: The European Scientific Tradition in Philosophical, Religious, and Institutional Context, 600 B.C. to A.D. 1450*. Chicago: University of Chicago Press, 1992.

——, ed. *Science in the Middle Ages*. Chicago: University of Chicago Press, 1978.

Lippincott, Kristin. "Giovanni di Paolo's 'Creation of the World' and the Tradition of the 'Thema Mundi' in Late Medieval and Renaissance Art." *Burlington Magazine* 132 (1990): 460–68.

Litt, Thomas. *Les corps célestes dans l'univers de Saint Thomas d'Aquin*. Louvain: B. Nauwelaerts, and Paris: Publications Universitaires, 1963.

Lloyd, G.E.R. *Magic, Reason, and Experience: Studies in the Origin and Development of Greek Science*. Cambridge: Cambridge University Press, 1979.

Luhrmann, T. M. *Persuasions of the Witch's Craft: Ritual Magic in Contemporary England*. Cambridge, Mass.: Harvard University Press, 1989.

Maierù, Alfonso. "Logique et théologie trinitaire: Pierre d'Ailly." In Zénon Kaluza and Paul Vignaux, eds. *Preuve et raisons à l'Université de Paris*, pp. 253–68. Paris: J. Vrin, 1984.

Mandonnet, Pierre. *Siger de Brabant et l'averroïsme latin au XIIIme siècle*. 2d ed. 2 vols. in 1. Louvain: Institut Supérieur de Philosophie, 1908 and 1911.

Markowski, Mieczyslaw. "Der Standpunkt der Gelehrten des späten Mittelalters und der Renaissance dem astrologischen Determinismus gegenüber." *Studia Mediewistycne* 23 (1984): 11–44.

Markus, R. A. *Saeculum: History and Society in the Theology of Saint Augustine.* Rev. ed. Cambridge: Cambridge University Press, 1970.

McGinn, Bernard. "Portraying Antichrist in the Middle Ages." In Werner Verbeke, Daniel Verhelst, and Andries Welkenhuysen, eds., *The Use and Abuse of Eschatology in the Middle Ages*, pp. 1–48. Louvain: Leuven University Press, 1988.

――――. *Visions of the End: Apocalyptic Traditions in the Middle Ages.* New York: Columbia University Press, 1979.

McGowan, John Patrick. *Pierre d'Ailly and the Council of Constance.* Washington, D.C.: Catholic University of America, 1936.

McVaugh, Michael R., ed. *Arnaldi de Villanova Opera Medica Omnia, 2: Aphorismi de gradibus.* Granada and Barcelona: Seminarium Historiae Medicae Granatensis, 1975.

Meller, Bernhard. *Studien zur Erkenntnislehre des Peter von Ailly.* Freiburg: Herder, 1954.

Millet, Hélène. "Ecoute et usage des prophéties par les prélats pendant le Grand Schisme d'Occident." *Mélanges de l'Ecole Française de Rome. Moyen Age* 102 (1990): 425–55.

Moeller, Bernd. "Piety in Germany around 1500." In Steven Ozment, ed., *The Reformation in Medieval Perspective*, pp. 50–75. Chicago: Quadrangle Books, 1971.

Momigliano, Arnaldo. "Time in Ancient Historiography." In Arnaldo Momigliano, *Essays in Ancient and Modern Historiography*, pp. 179–204. Middletown, Conn.: Wesleyan University Press, 1977.

Morison, Samuel Eliot. *Admiral of the Ocean Sea: A Life of Christopher Columbus.* Boston: Little, Brown, 1942.

Murdoch, John E., and Edith Sylla, eds. *The Cultural Context of Medieval Learning.* Boston Studies in the Philosophy of Science, 26. Dordrecht, Holland, and Boston: D. Reidel, 1975.

Murray, Alexander. *Reason and Society in the Middle Ages.* Oxford: Clarendon Press, 1978.

Neugebauer, Otto. "Astronomical and Calendrical Data in the Très Riches Heures." In Millard Meiss, *French Painting in the Time of Jean de Berry, The Limbourgs and Their Contemporaries*, text volume, pp. 421–32, 480–81. 2 vols. New York: G. Braziller, 1974. Reprinted in Otto Neugebauer, *Astronomy and History: Selected Essays*, pp. 507–20. New York and Berlin: Springer-Verlag, 1983.

――――. *The Exact Sciences in Antiquity.* 2d ed. New York: Dover, 1969.

――――. "The Study of Wretched Subjects." *Isis* 42 (1951): 111. Reprinted in Otto Neugebauer, *Astronomy and History: Selected Essays*, p. 3. New York and Berlin: Springer-Verlag, 1983.

Neugebauer, Otto, and Olaf Schmidt. "Hindu Astronomy at Newminster in 1428." *Annals of Science* 8 (1952): 221–28. Reprinted in Otto Neugebauer, *Astronomy and History: Selected Essays*, pp. 425–32. New York: Springer-Verlag, 1983.

Neugebauer, Otto, and H. B. Van Hoesen. *Greek Horoscopes.* Memoirs of the American Philosophical Society, 48. Philadelphia: American Philosophical Society, 1959.

Niccoli, Ottavia. *Prophecy and People in Renaissance Italy.* Translated by Lydia G. Cochrane. Princeton, N.J.: Princeton University Press, 1990.

Normore, Calvin. "Future Contingents." In Norman Kretzmann et al., eds., *The Cambridge History of Later Medieval Philosophy: From the Rediscovery of Aristotle to the Disinte-*

gration of Scholasticism, 1100–1600, chap. 18. Cambridge: Cambridge University Press, 1982.

North, John D. "Astrology and the Fortunes of Churches." *Centaurus* 24 (1980): 181–211.

————. "Celestial Influence—The Major Premiss of Astrology." In Paola Zambelli, ed., *"Astrologi hallucinati": Stars and the End of the World in Luther's Time*, pp. 45–100. Berlin and New York: Walter de Gruyter, 1986.

————. *Chaucer's Universe*. Oxford: Clarendon Press, 1988.

————. "Chronology and the Age of the World." In Wolfgang Yourgrau and Allen D. Beck, eds., *Cosmology, History, and Theology*, pp. 307–33. New York: Plenum Press, 1977.

————. *Horoscopes and History*. London: Warburg Institute, 1986.

————. "Medieval Concepts of Celestial Influence: A Survey." In Patrick Curry, ed., *Astrology, Science and Society: Historical Essays*, pp. 5–18. Woodbridge, Suffolk: Boydell Press, 1987.

————. *Richard of Wallingford: An Edition of His Writings with Introductions, English Translation and Commentary*. 3 vols. Oxford: Clarendon Press, 1976.

————. "Some Norman Horoscopes." In Charles Burnett, ed., *Adelard of Bath: An English Scientist and Arabist of the Early Twelfth Century*, pp. 147–61. London: Warburg Institute, 1987.

Oakley, Francis. "Christian Theology and the Newtonian Science: The Rise of the Concept of the Laws of Nature." *Church History* 30 (1961): 433–57.

————. "Gerson and d'Ailly: An Admonition." *Speculum* 40 (1965): 74–83.

————. *Omnipotence, Covenant, and Order: An Excursion in the History of Ideas from Abelard to Leibniz*. Ithaca, N.Y.: Cornell University Press, 1984.

————. "Pierre d'Ailly." In B. A. Gerrish, ed., *Reformers in Profile*, pp. 40–57. Philadelphia: Fortress Press, 1967.

————. "Pierre d'Ailly and the Absolute Power of God." *Harvard Theological Review* 56 (1963): 59–73.

————. *The Political Thought of Pierre d'Ailly: The Voluntarist Tradition*. New Haven: Yale University Press, 1964.

————. *The Western Church in the Later Middle Ages*. Ithaca, N.Y.: Cornell University Press, 1979.

Oberman, Heiko A. "Reformation and Revolution: Copernicus' Discovery in an Era of Change." In Heiko A. Oberman, *The Dawn of the Reformation: Essays in Late Medieval and Early Reformation Thought*. Edinburgh: T. & T. Clark, 1986. Reprinted from Owen Gingerich, ed., *The Nature of Scientific Discovery*, pp. 134–69. Washington, D.C.: Smithsonian Institution Press, 1975. And the expanded version in J. E. Murdoch and E. D. Sylla, eds., *The Cultural Context of Medieval Learning*, pp. 397–435. Boston Studies in the Philosophy of Science, 26. Dordrecht, Holland, and Boston: D. Reidel, 1975.

————. "The Shape of Late Medieval Thought: The Birthpangs of the Modern Era." In Charles Trinkaus and Heiko A. Oberman, eds., *The Pursuit of Holiness in Late Medieval and Renaissance Religion*, pp. 3–25. Studies in Medieval and Reformation Thought, 10. Leiden: E. J. Brill, 1974.

————. "Some Notes on the Theology of Nominalism." *Harvard Theological Review* 53 (1960): 47–76.

Ouy, Gilbert. "Ailly, Pierre d'." In *Lexikon des Mittelalters*. Munich: Artemis, 1980–.

Ozment, Steven. *The Age of Reform 1250–1550. An Intellectual and Religious History of Late Medieval and Reformation Europe.* New Haven: Yale University Press, 1980.

Pascoe, Louis B. "Pierre d'Ailly: Histoire, Schisme et Antéchrist." In Jean Favier, ed., *Genèse et débuts du Grand Schisme d'Occident*, pp. 615–22. Paris: Editions du C.N.R.S., 1980.

———. "Theological Dimensions of Pierre d'Ailly's Teaching on the Papal Plenitude of Power." *Annuarium historiae conciliorum* 11 (1979): 357–66.

Peters, Edward. *The Magician, the Witch and the Law.* Philadelphia: University of Pennsylvania Press, 1978.

Pingree, David. "Astrology." In *Dictionary of the History of Ideas.* New York: Charles Scribner's Sons, 1968, 1973.

———. *The Thousands of Abu Ma'shar.* Studies of the Warburg Institute, 30. London: Warburg Institute, 1968.

Pluta, Olaf. "Albert von Köln und Peter von Ailly." *Freiburger Zeitschrift für Philosophie und Theologie* 32 (1985): 261–71.

———. *Die philosophische Psychologie des Peter von Ailly· Ein Beitrag zur Geschichte der Philosophie des späten Mittelalters.* Bochumer Studien zur Philosophie, 6. Amsterdam: B. R. Grüner, 1986.

Pomian, Krzystoff. "Astrology as a Naturalistic Theology of History." In Paola Zambelli, ed., *"Astrologi hallucinati": Stars and the End of the World in Luther's Time*, pp. 29–43. Berlin and New York: Walter de Gruyter, 1986.

Pontes, José Maria da Cruz. "Astrologie et apologétique au Moyen Age." In Christian Wenin, ed., *L'homme et son univers au Moyen Age*, 2:631–37. Philosophes médiévaux, 26–27. Louvain: Institut Supérieur de Philosophie, 1986

Poulle, Emmanuel. "The Alfonsine Tables and Alfonso X of Castille." *Journal for the History of Astronomy* 19 (1988): 97–113.

———. *Les instruments de la théorie des planètes selon Ptolémée: Equatoires et horlogerie planétaire du XIIIᵉ au XVIᵉ siècle.* 2 vols. Geneva: Droz, and Paris: Champion, 1980.

Préaud, Maxime. *Les astrologues à la fin du Moyen Age.* Collection Lattès/Histoire. Groupes et sociétés Paris: J C. Lattès, 1984.

Pruckner, Herbert. *Studien zu den astrologischen Schriften des Heinrich von Langenstein.* Studien der Bibliothek Warburg, 14. Leipzig: Teubner, 1933.

Quillet, Jeannine. *La Philosophie politique du Songe du Vergier (1378).* Paris: J. Vrin, 1977.

———, ed. *Autour de Nicole Oresme.* Actes du Colloque Oresme organisé à l'Université de Paris, 12. Paris: J. Vrin, 1990.

Reeves, Marjorie. "History and Prophecy in Medieval Thought." *Mediaevalia et Humanistica*, n.s., 5 (1974): 51–75.

———. *The Influence of Prophecy in the Later Middle Ages: A Study in Joachimism.* Oxford: Clarendon Press, 1969.

———. *Joachim of Fiore and the Prophetic Future.* London: S.P.C.K., 1976.

Rusconi, Ruberto. *L'Attesa della fine: Crisi della società, profezie ed Apocalisse in Italia al tempo del grande scisma d'Occidente (1378–1417).* Studi storici, 115–18. Rome: Instituto Storico Italiano per il medio evo, 1979.

Sale, Kirkpatrick. *The Conquest of Paradise: Christopher Columbus and the Columbian Legacy.* New York: Knopf, 1990.

Salembıer, Louis. "A propos de Pıerre d'Ailly, évêque de Cambrai. Biographie et biblio-graphie." *Mémoires de la Société d'Emulation de Cambrai* 64 (1910): 101–26.

———. *Le cardinal Pierre d'Ailly, chancelier de l'Université de Paris, évêque du Puy et de Cambrai 1350–1420.* Tourcoing: Imprimerie Georges Frère, 1932.

———. *Petrus de Alliaco.* Lille: J. Lefort, 1886.

———. *Pierre d'Ailly et la découverte de l'Amérique.* Paris: Letouzey et Ané, 1912.

Saxl, Fritz, ed. *Verzeichnis der astrologischen und mythologischen illustrierten Handschriften des lateinıschen Mittelalters.* 4 vols. Heidelberg: C. Wınter, 1915–66.

Saxl, Fritz, and Hans Meier. *Catalog of Astrological and Mythological Illuminated Manuscripts of the Latin Middle Ages.* London, 1953.

Schmıdt, Roderich. "Aetates mundi: Die Weltalter als Gliederungsprinzip der Ge-schıchte." *Zeitschrift für Kirchengeschichte* 67 (1955): 288–317.

Schove, D. Justın. *Chronology of Eclipses and Comets, AD 1–1000.* Woodbridge, Suffolk: Boydell Press, 1984.

Schumaker, Wayne. *The Occult Sciences in the Renaissance: A Study in Intellectual Patterns.* Berkeley: University of California Press, 1972.

Screech, M. A. "The Magi and the Star (Matthew, 2)." In Olivier Fatio and Pierre Fraenkel, eds., *Histoire de l'exégèse au XVIe siècle,* pp. 385–409. Etudes de philologie et d'histoire, 34. Geneva: Droz, 1978.

Sears, Elizabeth. *The Ages of Man: Medieval Interpretations of the Life Cycle.* Princeton, N.J.: Princeton University Press, 1986.

Seznec, Jean. *The Survival of the Pagan Gods: The Mythological Tradition and Its Place in Renaissance Humanism and Art.* Translated by Barbara F. Sessions. New York: Pantheon Books, 1953.

Shank, Mıchael H. *"Unless You Believe, You Shall Not Understand": Logic, University, and Society in Late Medieval Vienna.* Princeton, N.J.: Princeton University Press, 1988.

Shore, Lys Ann. "A Case Study in Medieval Nonlıterary Translation: Scientific Texts from Latin to French." In Jeanette Beer, ed., *Medieval Translators and Their Craft,* pp. 297–327. Kalamazoo: Western Michigan University Press, 1989.

Siraisi, Nancy G. *Medieval and Early Renaissance Medicine: An Introduction to Knowledge and Practice.* Chicago: University of Chicago Press, 1990.

Le Soleil, la lune, et les étoiles au Moyen Age. Sénéfiance, 13. Aix-en-Provence: C.U.E.R.M.A., Université de Provence, 1983.

Southern, R. W. "Aspects of the European Tradition of Historical Wrıtıng: 3. History as Prophecy." *Royal Historical Association. Transactions.* 5th ser., 22 (1972): 159–80.

Steinmetz, Max. "Johann Virdung von Hassfurt, sein Leben und seine astrologischen Flugschrıften." In Hans Joachım Köhler, ed., *Flugschriften als Massenmedium der Refor-mationszeit,* pp. 353–72. Spätmittelalter und Frühe Neuzeit, 13. Stuttgart: Klett-Cotta, 1981.

Steneck, Nicholas. *Science and Creation in the Middle Ages: Henry of Langenstein (d. 1397) on Genesis.* Notre Dame, Ind.: University of Notre Dame Press, 1976.

Stillwell, Margaret B. *The Awakening Interest in Science during the First Century of Printing, 1450–1550.* New York: Bibliographical Socıety of America, 1970.

Stock, Brian. *Myth and Science in the Twelfth Century: A Study of Bernard Silvestris.* Prince-ton, N.J.: Princeton University Press, 1972.

Swerdlow, Noel M. "The Derivation and First Draft of Copernicus's Planetary Theory:

A Translation of the Commentariolus with Commentary." *Proceedings of the American Philosophical Society* 117 (1973): 423–512.

Swiezamski, Stefan. "La pensée philosophique du moyen âge tardif face au problème de la libération de l'homme." In Christian Wenin, ed., *L'homme et son univers au Moyen Age*, 1:3–15. Philosophes médiévaux, 26–27. Louvain: Institut Supérieur de Philosophie, 1986.

Tester, S. J. *A History of Western Astrology.* Woodbridge, Suffolk: Boydell Press, 1987.

Thomas, Keith. *Religion and the Decline of Magic.* New York: Scribner's, 1971.

Thorndike, Lynn. "Four British Manuscripts of Scientific Works by Pierre d'Ailly." *Imago mundi* 16 (1962): 157–60.

———. "A Highly Specialized Medieval Library." *Scriptorium* 7 (1953): 81–88.

———. *A History of Magic and Experimental Science.* 8 vols. New York: Columbia University Press, 1923–58.

———. *The Sphere of Sacrobosco and Its Commentators.* Chicago: University of Chicago Press, 1949.

———. "The True Place of Astrology in the History of Science." *Isis* 46 (1955): 273–78.

———. *University Records and Life in the Middle Ages.* New York: Columbia University Press, 1944.

Thorndike, Lynn, and Pearl Kibre, eds. *A Catalogue of Incipits of Medieval Scientific Writings in Latin.* Rev. ed. Cambridge, Mass.: Medieval Academy of America, 1963.

Tobin, Matthew. "Une collection de textes prophétiques du XVe siècle: Le manuscrit 520 de la Bibliothèque Municipale de Tours." *Mélanges de l'Ecole Française de Rome. Moyen Age* 102 (1990): 417–23.

Trinkaus, Charles. "Coluccio Salutati's Critique of Astrology in the Context of His Natural Philosophy." *Speculum* 64 (1989): 46–68.

Tschackert, Paul. *Peter von Ailli: Zur Geschichte des grossen abendländischen Schisma und der Reformconcilien von Pisa und Constanz.* Gotha: Friedrich Andreas Perthes, 1877. Reprint. Amsterdam: Rodopi, 1968.

Tuckerman, Bryant. *Planetary, Lunar, and Solar Positions A.D. 2 to A.D. 1649 at Five-day and Ten-day Intervals.* Memoirs of the American Philosophical Society, 59. Philadelphia: American Philosophical Society, 1964.

Unterreitmeier, Hans. "Deutsche Astronomie/Astrologie im Spätmittelalter." *Archiv für Kulturgeschichte* 65 (1983): 21–41.

Valois, Noel. "Un ouvrage inédit de Pierre d'Ailly, le *De persecutionibus ecclesiae.*" *Bibliothèque de l'Ecole des Chartes* 65 (1904): 557–74.

Vauchez, André. "Les théologiens face aux prophéties à l'époque des papes d'Avignon et du Grand Schisme." *Mélanges de l'Ecole Française de Rome. Moyen Age* 102 (1990): 577–88

Vickers, Brian. "Analogy Versus Identity: The Rejection of Occult Symbolism, 1580–1680." In Brian Vickers, ed., *Occult and Scientific Mentalities in the Renaissance,* pp. 95–164. Cambridge: Cambridge University Press, 1984.

Vescovini, Graziella Federici. *Astrologia e Scienza: La crisi dell'aristotelismo sul cadere del Trecento e Biagio Pelacani da Parma.* Florence: Enrico Vallecchi, 1979.

Volli, Ugo. *La Retorica delle Stelle. Semiotica dell'Astrologia.* [Rome]: Espresso Strumenti, 1979.

Wadstein, Ernst. *Die eschatologische Ideengruppe: Antichrist—Weltsabbat—Weltende und Weltgericht, in den Hauptmomenten ihrer christlich-mittelalterlichen Gesamtentwickelung.* Leipzig: O. R. Reisland, 1896.

Watts, Pauline Moffit. "Prophecy and Discovery: On the Spiritual Origins of Christopher Columbus's 'Enterprise of the Indies.'" *American Historical Review* 90 (1985): 73–102.

Wedel, Theodore Otto. *The Medieval Attitude towards Astrology Particularly in England.* New Haven: Yale University Press, 1920.

Wenin, Christian, ed. *L'homme et son univers au Moyen Age.* Philosophes médiévaux, 26–27. Louvain: Institut Supérieur de Philosophie, 1986.

White, Lynn, Jr. "Medical Astrologers and Late Medieval Technology." *Viator* 6 (1975): 295–308.

Wilcox, Donald J. *The Measure of Times Past: Pre-Newtonian Chronologies and the Rhetoric of Relative Time.* Chicago: University of Chicago Press, 1987.

Williams, Ann, ed. *Prophecy and Millenarianism: Essays in Honor of Marjorie Reeves.* Essex: Longman, 1980.

Woody, Kennerly. "Dante and the Doctrine of the Great Conjunctions." *Dante Studies* 95 (1977): 119–34.

Zambelli, Paola. "Albert le Grand et l'astrologie." *Recherches de théologie ancienne et médiévale* 49 (1982): 141–58.

————, ed. *"Astrologi hallucinati": Stars and the End of the World in Luther's Time.* Berlin and New York: Walter de Gruyter, 1986.

Zinner, Ernst. *Verzeichnis der astronomischen Handschriften des deutschen Kulturgebietes.* Munich: C. H. Beck, 1925.

Index

Abel, 68

Abraham, 68

Abu Ma'shar (Albumasar), 5, 29, 31, 45, 50, 51, 53, 54, 55, 57, 62, 65–71, 73, 77, 78, 79, 106, 114–15

Adam, 72, 109

Adelard of Bath, 29, 30

Advent, sermons for: (d'Ailly, undated), 48, 51, 55; (d'Ailly, 1385), 48, 49–50, 51, 55–56, 57, 58, 98–100, 104, 110; (d'Ailly, 1414), 83, 110, 119

Aegidius Cantoris, 116

Ailly, Pierre d'. See d'Ailly, Pierre

Albertus Magnus, 30, 55

Alchabitius. See Qabisi, al- (Alchabitius)

Alexander V (pope), 8

Alexander the Great, 68, 107

Alfonsine tables, 11, 65, 70, 73–74, 76

All Saints' Day: sermon for (d'Ailly), 6, 52, 75

Almagest (Ptolemy), 13, 16, 29

Antichrist: and age of the world, 72, 109; arrival of, 4, 22, 41, 52, 53, 54, 59, 60, 69, 74, 84, 85, 90, 92, 94–101, 102–7, 111–14, 115, 119–20, 122; reign of, 87; religious sect of, 38, 59, 62, 81

anticlericalism, 116, 117

Antipope, 92

Apocalypse: postponement of, 102–21, 122; predictions of, 3, 4, 22, 36, 43, 50–52, 57–60, 77, 85–101

apocalyptic literature, 4, 54, 92. See also names of individual works

apogees, 14–15

Apologetice defensiones astronomice veritatis (Apologetic defenses of astrological truth) (d'Ailly), 38, 39–40, 59, 73

Apologia concilii Pisani (d'Ailly), 59

Aquinas, Thomas. See Thomas Aquinas, Saint

Aristotelian physics, 25, 32, 34, 41

Aristotle, 25, 29, 30, 34, 44, 45, 46, 58, 60

Arnold of Villanova, 50, 90–91, 98, 101, 164n.47

aspects, 19–20

astral causality, 25–26, 49, 63, 124

astral determinism, 31, 37, 44, 124

astral influences: on history, 61–74; on the human body, 30–31, 33, 37, 44, 47, 48, 54, 63, 124; on religion, 31, 37–38, 46, 52, 53, 56, 59, 61, 62, 103

astrologia, 27, 45

astrological charts, 17–19; of the Creation, 65–66, 69, 73, 82, 174n.28, 177n.83; and genethlialogy, 15–16, 17, 28, 45

astrological medicine, 22, 31, 33, 40, 58, 79

astrology: Christian criticism of, 25–29; d'Ailly's defense of, 36–42; elections in, 17, 27, 30; general predictions in, 17, 20, 49, 50, 79; history as the proof of, 74–77, 104; interrogations in, 17; nativities in, 17, 28, 30, 33, 39, 45; Parisian criticisms of, 32–36, 44; purpose of, 15; revival of interest in, 29–32

astronomia, 27, 45

Augustine of Hippo, Saint: on the Apocalypse, 88, 96, 97, 99, 101; on astrology, 25–27, 29, 36, 38, 41, 49, 51, 56, 63–64; on history, 80, 82, 88

Augustus, 68

Auriol, Peter, 111–12

autumnal equinox, 12

Avicenna. See Ibn Sina

Babylonian astrology, 15

Babylonian Captivity, 8, 58, 68

Bacon, Roger, 30–31, 37, 38, 41, 43, 53–57, 59, 62, 79, 80, 98–99, 100, 103–4, 105

Baldwin, 112

Bate of Malines, Henry, 53, 57, 74, 177n.73

Battani, al-, 14

Becket, Thomas, 69

Bede, Saint, 63, 64, 76, 87, 88, 173n.17, 184n.7

Benedict XIII (pope), 8, 58, 59, 94, 187n.43

Bernard, Saint, 101; sermon on (d'Ailly), 97–98, 104, 107

Bernard Silvestris, 29

Bezold, Friedrich von, 77–78, 79

Bible. Acts of the Apostles, 87–88, 90, 96, 97, 99, 104; and age of the world, 63; apocalytic information in, 86–88; Daniel, 77, 86, 87, 90, 94, 186n.36; Ezekiel, 90; Luke, 99, 110, 172n.5; Matthew, 87, 99; Revelation, 53, 54, 60, 86–87, 96–97, 111–12, 119; 2 Thessalonians, 4, 87, 92, 94, 96, 97, 100, 103, 105, 106, 107, 110, 115, 127, 128

Biel, Gabriel, 125

Blanchard, Jean, 7, 102
body: astral influences on, 30–31, 33, 37, 44, 47, 48, 54, 63, 124
body parts: astrological locations associated with, 17, 22
Boethius, 46–47, 51, 153n.61
Bonaventure, 80
Boniface VIII (pope), 90
Brincken, A. D. von den, 82

Cain, 68
Campanus of Novara, 11
Canons (Eusebius), 63
Carey, Hilary, 6
Carneades, 147–48n.1
Carthage: founding of, 68
Cato, 33
Cecco d'Ascoli, 78–79
Cedar of Lebanon prophecy, 184n.16
Centiloquium (pseudo-Ptolemaic), 29, 48, 64
Chaldean religious sect, 38, 50, 55, 62
Charlemagne, 111, 184n.8
Charles V (king of France), 8, 32, 33, 58
Charles VI (king of France), 8, 58
Chaucer, Geoffrey, 5
Christianity: astrological descriptions of, 31, 38, 59, 62
chronology: and age of the world, 62–74, 75, 88, 104, 108, 109–10
chronosophies, 80–81
Cicero, Marcus Tullius, 33, 47
City of God (Augustine), 25, 26–27, 38, 49, 64, 80, 88, 101
Clement VI (pope), 195–96n 30
Clement VII (pope), 7, 8, 58, 96
Cohn, Norman, 115
Columbus, Christopher, 3, 23–24
Commentary on the Dream of Scipio (Macrobius), 65
Conciliarism, 9, 118, 120
Conciliator (Peter of Abano), 78
Concordantia astronomie cum hystorica narratione (Concordance of astrology with historical narration) (d'Ailly), 38, 53, 54, 57, 59, 60, 66–69, 71, 72, 76, 82, 104–5, 106, 108, 112, 113
Concordantia astronomie cum theologia (d'Ailly). See Vigintiloquium, or Concordantia astronomie cum theologia (Concordance of astrology with theology) (d'Ailly)
configurations, 34, 35
conjunctions: great, defined, 21–22; mean vs.

true, 70–71, 72, 74; predictive value and timing of, 20–22, 31, 35, 38, 41, 50–57, 61–74, 78, 79, 81–83, 98, 103, 104–5, 108, 109, 113
Consolation of Philosophy (Boethius), 46–47, 51
Contra curiositatem studentium (Gerson), 161n.24
Coopland, G. W., 5
cosmology, 11
Council of Constance, 4, 6, 8, 43, 52, 53, 59, 60, 83, 105, 106, 108–11, 115, 116, 119, 122. See also Great Schism
Council of Graga, 28
Council of Pisa, 8, 52, 56, 59, 105
Council of Toledo, 28
Creation: horoscope of, 65–66, 69, 73, 82, 174n.28, 177n.83, time of, and the Apocalypse, 88
Crucifixion, eclipse at, 46
Crusades, 69, 112
Cyril, 99, 112, 118

d'Ailly, Pierre: anti-apocalyptic message of, 115–19, 120–21; apocalypticism of, 95–101, 102; availability of works by, 43, 131–35; birth of, 7; chronology of works by, 43, 136–37; death of, 8, 60; defense of astrology by, 36–42; development of his thoughts on astrology, 43–54, 61–84; education of, 7, 10, 44, 58; honors bestowed on, 7, 58, 102, 123; impact of, on Columbus, 3; his interest in astrology, 3–4, 6, 22, 23, 44, 57–60; named as cardinal, 8; political thought of, 9; role of, in Great Schism, 8; sources for his thoughts on astrology, 54–57; theology of, 9, 22, 122–30; tomb of, 100
De anima (d'Ailly), 46
De antichristo (Anon.), 109–10, 177n.78, 196n.38
De antichristo (Arnold of Villanova), 101
De antichristo (John of Paris), 113
De commensurabilitate vel incommensurabilitate motuum coeli (On the commensurability or incommensurability of the heavenly motions) (Oresme), 33
De concordia Astronomicae veritatis et narrationis historicae (d'Ailly), 108
De concordia discordantium astronomorum (On the concordance of discordant astronomers) (d'Ailly), 74, 76
Decretum (Gratian), 29
De falsis prophetis I (d'Ailly), 100–101, 115
De falsis prophetis II (d'Ailly), 37, 38, 40, 48, 50, 51, 56, 58, 100

De figura inceptionis mundi et coniunctionibus mediis sequentibus (On the horoscope of creation and the following mean conjunctions) (d'Ailly), 57, 59, 73–74, 76, 83

De generatione et corruptione (Aristotle), 19

De legibus et sectis contra superstitiosos astronomos (On the laws and the sects, against the superstitious astrologers) (d'Ailly), 6, 28, 37–39, 52–57, 59, 75, 103, 125, 126

De magnis coniunctionibus (Abu Ma'shar), 57, 62, 77

De motu octave spere (Thabit ibn Qurra), 15

Demonstracions contre Sortileges (Deschamps), 35

De persecutionibus ecclesie (On the persecutions of the church) (d'Ailly), 57, 60, 111–15, 129

De radiis (al-Kindi), 158–59n.131

De reformatione ecclesiae (On the reformation of the church) (d'Ailly), 102–3, 105, 115

Deschamps, Eustache, 35, 92

De semine scripturarum (pseudo-Joachim), 118

De sphera (On the sphere) (Sacrobosco), 10–11, 45, 47, 58, 60

De tempore adventus Antichristi (On the time of the advent of Antichrist) (Arnold of Villanova), 90

De vetula (pseudo-Ovid), 51, 55, 171n.3

Devota meditatio seu expositio super psalmum: In te Domine speravi (d'Ailly), 170n.105

Devota meditatio super Ave Maria (d'Ailly), 170n.105

Devota meditatio super psalmum: Judica me Deus (d'Ailly), 170n.105

divination, 28, 31, 35, 36, 49

Dominic, Saint: sermon on (d'Ailly), 96, 97, 100, 104

ecliptic, 12, 15, 70

Egyptian astrology, 15, 167n.82

Egyptian religious sect, 38, 62

Elucidarium (Elucidation) (d'Ailly), 53, 57, 59, 69, 72–73, 75, 76, 78, 81, 82, 108, 109–10, 113

End of the World. See Apocalypse

Enoch, 40

epicycles, 13, 70

Epilogus mappe mundi (Epilogue to the map of the world) (d'Ailly), 11

Epistola ad novos Hebraeos (Letter to the new Hebrews) (d'Ailly), 55, 56

equant point, 13–14

equatory (equatorium), 15

equinoxes: precession of, 12, 13–15, 45, 50. See also motion of the eighth sphere

Erythraean Sibyl, 50

Etymologiae (Isidore of Seville), 27–28, 156n.103

Eusebius, 63, 64, 88, 172n.5, 184n.7

evil, human, 26

exaltation, 19, 65

Fargani, al- (Alfraganus), 11

fatalism vs. free will, 25–26, 30, 32, 33, 37, 41, 42, 44, 46–47, 49, 63, 79, 80, 124, 127

Ferdinand II (king of Aragon), 3

Ferrer, Boniface, 59

Festinger, Leon, 42

fetal development, 40

Firmicus Maternus, Julius, 28, 65–66

First Punic War, 68

fixed stars, 11, 12, 13, 15

flagellants, 116–17, 120

Flint, Valerie I. J., 6, 28–29

Flood, 40, 64–69, 72, 73, 75, 76, 82, 173n.21

four elements, 16, 34, 35, 46

Fourth Lateran Council, 69

four triplicities, 16, 21, 62, 72, 74

Francis of Assisi, Saint: sermon on (d'Ailly), 96–97, 197n.48

Frederick I (Holy Roman emperor), 69

Frederick II (Holy Roman emperor), 30

Frederick of Brunswick (heretical preacher), 115

free will: vs. fatalism, 25–26, 30, 32, 33, 37, 41, 42, 44, 46–47, 49, 63, 79, 80, 124, 127

genethlialogy, 15–16, 17, 28, 45

geographical areas: astrological locations associated with, 17, 40

Gerard of Borgo San Donnino, 90

Gerard of Cremona, 11

Gerson, Jean, 35, 37, 52, 59, 60, 93, 116–17, 120, 191n.91

Gervais, Maître Chrétien, 32

glossa ordinaria, 87, 98

Godfrey de Bouillon, 112

Gog and Magog, 89, 92

Grange, Jean de la, 7

Gratian (Franciscus Gratianus), 29

great conjunctions, doctrine of. See conjunctions

Great Schism (1378–1414): changing interpretations of, 4, 7–8, 22, 43, 51, 57–60, 76, 77, 84–86, 92–95, 102–3, 105, 107–9, 111, 112, 116–19, 122, 127–28, 129

Great Year (magnus annus), 33, 50, 164n.48

Greek astrology, 5, 15, 16, 167n.82
Greek astronomy, 12
Gregory I (the Great), Saint, 97, 101
Gregory XI (pope), 8, 107
Gregory, Tullio, 80–81
Guenée, Bernard, 6, 82, 123
Gui, Bernard, 82

habitus, 46
Hemmerlin, Felix, 88
Henry IV (emperor), 111, 112
Henry of Harclay, 91
Henry of Kirkestede, 92
Henry of Langenstein (or Hesse), 25, 32, 35–36, 38, 39–40, 41, 52, 93, 118–19
Hermann of Carinthia, 81
Hermes Trismegistus, 50
Hildegard of Bingen, 89, 96–97, 99, 100, 101, 103, 105, 108, 110, 112, 118, 119
Hindu astronomy and astrology, 65, 71, 81, 173n.22
Hipparchus, 14
Hippocrates, 26
Hippolytus, 184n.7
Hirsch-Reich, Beatrice, 118
history: astral influences on, 61–74; as the proof of astrology, 74–77, 104
Holkot, Robert, 22
homines intelligentiae (men of intelligence) sect, 116, 120
horoscope. See astrological charts
horoscopus (ascendant), 17, 18
Hugh of St. Victor, 156n.103
Hus, John, 116

Ibn Ezra, Abraham (Avenesra), 53, 72, 177n.73
Ibn Sina (Avicenna), 45
Ibn Yusuf, Ahmed, 162n.30
idol worshipers, 62
Imago mundi (d'Ailly), 54, 56, 57, 59, 75
incommensurability, 34, 41
Indian astrology, 5, 15, 16
Innocent III (pope), 69
Introductorium maius (Abu Ma'shar), 29, 30
irrational numbers, 34
Isabella I (queen of Castile), 3
Isidore of Seville, 27–28, 29, 31, 41, 63, 64, 76, 101, 156n.103

Jean de Bruges, 66, 193–94n.14
Jean de France (duc de Berry), 8
Jean sans Peur, 8

Jerome, Saint, 63, 87, 98
Jesus Christ: astrological predictions concerning, 28, 31, 38, 48–55, 62, 64, 68, 72, 78, 81–82, 98, 109
Joachim of Fiore, 86, 89–90, 92–94, 97, 99–100, 103, 105, 108, 110, 112, 118, 119
Johannes de Muris (Jehan des Murs), 53, 195–96n.30
Johannes Trithemius, 79
John XXIII (pope), 8, 107–8, 111, 115, 116, 117
John of Ashenden, 53, 57, 66, 69, 176n.64
John of Paris (Joannes Quidort Parisiensis), 9, 91, 113, 114, 190n.80
John of Salisbury, 47, 49, 161n.25
John of Saxony, 11
John of Seville, 11
Joseph, 68
Judaism, 38, 59, 62
Julian the Apostate, 111
Julius Africanus, 63
Jupiter: characteristics of, 16; conjunctions of, 20–22, 31, 38, 50, 52–57, 61–74, 78, 81, 82, 98, 103, 104–5, 108, 109, 113

Kaliyuga, 65
Kennedy, Edward S., 5
Kennedy, Leonard, 126
Khwarizmi, al-, 29, 65
Kindi, al-, 158–59n.131

Laistner, M. L. W., 28
Landes, Richard, 88, 104
Le Goff, Jacques, 80
Leopold of Austria, 53, 57, 71
Lerner, Robert, 88
liars, sect of, 62
Lieberman, Max, 55, 59
Livre de divinacions (Oresme), 33, 34, 56
Lloyd, Geoffrey, 42
Louis (duc d'Orléans), 8, 59
Luhrmann, T. M., 42
Luna, Pedro de la. See Benedict XIII (pope)

Macrobius, 65
Magi, 28, 45, 51
magic, 23, 28, 37, 42
magnus orbis, 71–72, 77
Magog. See Gog and Magog
Mani, 68
Martin V (pope), 8, 59
Mary (mother of God), 38, 54, 55

Masha'allah (Messahalla), 5, 53, 66, 78, 174n.33
Mathias of Janov, 96
Maurice (emperor), 111
medius motus, 70
Messahalla. *See* Masha'allah (Messahalla)
Meteorologica (Aristotle), 29, 45, 46, 58, 60
Methodius, 88–89
Methuselah, 68
Mézières, Philippe de, 7, 35, 56
mobility. astrological locations associated with, 16–17
Moeller, Bernd, 10
Monzon, Juan de, 7, 102
Mosaic law. *See* Judaism
Moses, 68, 72, 109
motion of the eighth sphere, 11, 13–15, 90, 91, 106, 113, 193–94n.14, 194n.17
Muhammad, 68, 73, 74, 76, 112, 114
Murray, Alexander, 83
Muslim religious sect, 38

nativities, 17, 28, 30, 33, 39, 45, 51; and genethlialogy, 15–16, 17, 28, 45
Nero (Roman emperor), 68
Neugebauer, Otto, 4–5
Nicholas of Lyra, 91, 111, 113, 165n 53
Nicolas of Clamanges, 92, 170n.102
nominalism, 143n.23, 205n.14
Normore, Calvin, 127
North, John D., 5

Oakley, Francis, 125
occult, 23, 32
omens, 15
Opus maius (Bacon), 30, 54–57, 59, 62, 98–99, 103
Oraculum Cyrilli, 92, 101
Oratio dominica anogogice exposita (d'Ailly), 170n.105
Oration on Matthew (d'Ailly), 101
Oresme, Nicole, 25, 32–36, 38–41, 44, 49, 51, 52, 56
Orosius, 64
Ovid, 50

pagan religion, 62
Pascoe, Louis, 85
periodizations, 77, 82
Peter Lombard, 7, 8, 124, 125–26, 127, 128
Peter of Abano, 78
Peter of Auvergne, 91
Petrus Comestor, 79

Petrus de Alliaco. *See* d'Ailly, Pierre
Philip II (duke of Burgundy), 8, 59
Pico della Mirandola, 35
Pierre de Limoges, 109
Pingree, David, 5
plagiarism, 9–10, 157n.118
plague, 32, 76, 92
planets: characteristics of, 16, 45; movement of, 12–13
Pomian, Kryzstof, 80–81
positivism, 77
potentia absoluta (God's absolute power), 9, 41, 124–26, 127, 128
potentia ordinata (God's ordained power), 9, 41, 124–26, 127, 128
Préaud, Maxime, 6
precession of the equinoxes, 11, 13–15, 45, 50. *See also* motion of the eighth sphere
predictions, general, 17, 20, 49, 50, 79
Principium in cursum Biblie (d'Ailly), 44–45, 51
Priscillianists, 28
Pro declaratione decem dictarum figurarum (To explain the ten said horoscopes) (d'Ailly), 57, 73
prophecy, 48, 51, 54, 58, 64, 99, 122–23, 127–29
prophetic dreams (*somnia*), 48
prophets: true and false, 93, 95, 99, 100–101
Pruckner, Herbert, 5
Pseudo-Haly, 48
Pseudo-Joachim, 99–100, 101, 102, 118
Pseudo-Methodius, 88–89, 106–7, 112, 114
pseudoprophets, 87
Pseudo-Ptolemy, 29, 48, 64
Ptolemy (Claudius Ptolemaeus), 11, 13, 16, 29, 48, 64, 91

Qabisi, al- (Alchabitius), 11, 15, 17, 19, 45, 53, 57, 65
Quaestio contra divinatores horoscopis (Question argued against those divining by horoscopes) (Oresme), 33, 34
Quaestio de cometa (Henry of Langenstein), 35
Questiones on Sacrobosco (d'Ailly), 45
Quodlibeta (Quodlibetal questions) (Oresme), 33, 34

Rede, William, 176n.64
Reeves, Marjorie, 118
Regiomontanus (Johann Müller), 53
religion: astral influences on, 31, 37–38, 46, 52, 53, 56, 59, 61, 62, 103

retrograde motion, 13, 105
Revelationes (pseudo-Methodius), 89, 106–7
Rhabanus Maurus, 160n.8
Richard of Wallingford, 5
Robert of Geneva See Clement VII (pope)
Robine, Marie, 93
Rome: founding of, 68, 82
Roquetaillade, Jean de (John of Rupescissa), 92, 93, 95

Sacrobosco, Johannes de (John of Holywood), 10–11, 14, 45–46, 47, 58, 60
Salembier, Louis, 6
Salimbene of Parma, 79
Salutati, Coluccio, 41
salvation, 77
Saracen religion, 62
Sarton, George, 4
Sassanian astrology, 5, 16, 65, 71
Saturn: characteristics of, 16; conjunctions of, 20–22, 50, 52–57, 61–74, 78, 81, 82, 98, 104–5, 108, 109, 113
Saul, 68
Schmidt, Conrad, 117
Schumaker, Wayne, 23
Scot, Michael, 79
Sentences (Peter Lombard), 7, 9, 124, 125–26, 127, 128
seven ages, 77, 90
sex: astrological locations associated with, 16, 40
Sibyls, 86, 118
Sigismund (king of Hungary), 59
Solomon, 40
Somnium Viridarii, 35
Songe de vieil pelerin (Mézières), 35, 56
soul. See body: astral influences on; free will
Southern, Sir Richard, 89
Speculum astronomiae (Albertus Magnus), 30, 55
Speculum historiale (Vincent of Beauvais), 63, 79
Speier, Jacob von, 53
Statius, Publius Pappinius, 33
Strzempino, Thomas von, 151n.45
Summa astrologiae (John of Ashenden), 69
Summa theologiae (Thomas Aquinas), 32
summer solstice, 12
sun: apogee of, 15, 91; movement of, 12, 13
Super Hieremiam (pseudo-Joachim), 101
Symon de Phares, 5

Telesphorus of Cosenza, 93, 118, 193n.11
Tempier, Bsp. Etienne, 32, 33, 41, 44, 48
Terrena, Guido, 91

Tester, S. J., 5–6
Tetrabiblos (Quadripartitum) (Ptolemy), 16, 29
Thabit ibn Qurra, 15, 91, 190n.80
thema mundi, 65–66
Theorica planetarum (Campanus of Novara), 11
Theorica planetarum (attrib. to Gerard of Cremona), 11
Thomas, Keith, 23, 83
Thomas Aquinas, Saint, 25, 30–32, 34, 36, 37, 41, 44, 49, 51, 54, 56, 60, 80, 124
Thorndike, Lynn, 4
three status (Joachim), 90, 92
Tiburtine Sibyl, 88
time: astrological vs. God's, 80–83
Toledo Letter, 184n.16
Tractatus contra astrologos (Oresme), 38, 40, 56
Tractatus contra coniunctionistas (Treatise against the conjunctionists) (Henry of Langenstein), 35
Tractatus contra iudiciarios astronomos (Treatise against judicial astronomers) (Oresme), 33, 34–35, 49
Tractatus de antichristo (John of Paris), 91
Tractatus de imagine mundi et varia ejusdem auctoris et Joannis Gersonis opuscula (Treatise on the image of the world, and various works by the same author and by Jean Gerson) (d'Ailly), 3, 23–24
translationes imperii, 82
trepidation, 15
trigons, 16
Troy: fall of, 68, 74
Tschackert, Paul, 6
Turrel, Pierre, 193–94n.14
twelve hours, 77
twelve houses (domus), 18, 19
twins: fate of, 26–27, 34

Urban V (pope), 107
Urban VI (pope), 8

Vade mecum in tribulatione (Handbook for times of tribulation) (Roquetaillade), 92
Varro, 83
Vauchez, André, 93
vernal equinox, 12, 13, 65
Vespasian, 68
Vigintiloquium, or Concordantia astronomie cum theologia (Concordance of astrology with theology) (d'Ailly), 36–37, 39, 54, 57, 59, 62–66, 69, 71, 72, 75, 76, 81, 82, 83, 104, 108
Villani, Giovanni, 78

Vincent Ferrer, Saint, 94–95, 116–17, 120
Vincent of Beauvais, 63, 64, 76, 79, 173n.17

Walter of Odington, 66
Watts, Pauline Moffit, 24
Wickersheimer, Ernest, 5
Wilhelmus de Hildernissem, 116

William (a French monk), 93
William of Auvergne, 96, 97, 103
William of Ockham, 9, 125, 128
winter solstice, 12

zodiac: signs of, 12, 15, 16, 22, 45, 53, 109
Zoroaster, 68

9 780691 600512